Pedagogic Research in Geography Higher Education

There are many books about teaching in Geography, but this is the first volume to deal specifically with pedagogic research and its methods and practices. Pedagogic research concerns the processes of learning and the development of learners. It is a learner-centred activity that aims to evaluate and improve the ways that students both learn and learn to manage, control and comprehend their own learning processes, first as Geographers in Higher Education but, equally, as future educated citizens.

This book explores: original research and critical perspectives on how Geographers learn; critical evaluations of both new and traditional frameworks and methods used for pedagogic research in Geography; and case studies on the promotion of self-authorship and learner autonomy in key Geography Higher Education contexts, such as fieldwork and undergraduate project work.

This book is a compilation of articles from various issues of the *Journal of Geography in Higher Education.*

Martin Haigh, Debby Cotton and **Tim Hall** are all active members of the Editorial Board of the *Journal of Geography in Higher Education* (*JGHE*), and active researchers in their own right. In recent years they have been leading the development of the *JGHE*'s Pedagogic Research strand, and this volume represents the best of the results of this work.

Pedagogic Research in Geography Higher Education

Edited by
Martin Haigh, Debby Cotton and Tim Hall

Routledge
Taylor & Francis Group
LONDON AND NEW YORK

First published 2017
by Routledge

2 Park Square, Milton Park, Abingdon, Oxfordshire OX14 4RN
711 Third Avenue, New York, NY 10017

Routledge is an imprint of the Taylor & Francis Group, an informa business

First issued in paperback 2018

British Library Cataloguing in Publication Data
A catalogue record for this book is available from the British Library

ISBN: 978-1-138-96208-8 (hbk)
ISBN: 978-0-367-03012-4 (pbk)

Typeset in Times New Roman
by RefineCatch Limited, Bungay, Suffolk

Publisher's Note
The publisher accepts responsibility for any inconsistencies that may have arisen during the conversion of this book from journal articles to book chapters, namely the possible inclusion of journal terminology.

Contents

Part III: Case Studies

Part IV: Aspects of Pedagogic Research-informed Teaching

Citation Information

The chapters in this book were originally published in the *Journal of Geography in Higher Education*. When citing this material, please use the original volume, issue number and page numbering for each article, as follows:

Chapter 1
Editorial: Pedagogic research in geography higher education
Martin Haigh, Debby Cotton and Tim Hall
Journal of Geography in Higher Education, volume 39, issue 3 (2015), pp. 299–312

Chapter 2
Researching the hidden curriculum: intentional and unintended messages
Debby Cotton, Jennie Winter and Ian Bailey
Journal of Geography in Higher Education, volume 37, issue 2 (2013), pp. 192–203

Chapter 3
Invitational Education: Theory, Research and Practice
Martin Haigh
Journal of Geography in Higher Education, volume 35, issue 2 (2011), pp. 299–309

Chapter 4
AQAL Integral: a holistic framework for pedagogic research
Martin Haigh
Journal of Geography in Higher Education, volume 37, issue 2 (2013), pp. 174–191

Chapter 5
Is Q for you?: using Q methodology within geographical and pedagogical research
Paul N. Wright
Journal of Geography in Higher Education, volume 37, issue 2 (2013), pp. 152–163

Chapter 6
Using Observational Methods to Research the Student Experience
Debby R.E. Cotton, Alison Stokes & Peter A. Cotton
Journal of Geography in Higher Education, volume 34, issue 3 (2010), pp. 463–473

Chapter 7
Visual methodology as a pedagogical research tool in geography education
Bryan Wee, Amy DePierre, Peter Anthamatten and Jon Barbour
Journal of Geography in Higher Education, volume 37, issue 2 (2013), pp. 164–173

Chapter 8
Using student interviews for becoming a reflective geographer
Hanne Kirstine Adriansen and Lene Møller Madsen
Journal of Geography in Higher Education, volume 38, issue 4 (2014), pp. 595–605

Chapter 9
Using Focus Group Research to Support Teaching and Learning
Heather Winlow, David Simm, Alan Marvell and Rebecca Schaaf
Journal of Geography in Higher Education, volume 37, issue 2 (2011), pp. 292–303

Chapter 10
The Power of Debate: Reflections on the Potential of Debates for Engaging Students in Critical Thinking about Controversial Geographical Topics
Ruth L. Healey
Journal of Geography in Higher Education, volume 36, issue 2 (2012), pp. 239–257

Chapter 11
Construction of Student Groups Using Belbin: Supporting Group Work in Environmental Management
Mark Smith, Giles Polglase & Carolyn Parry
Journal of Geography in Higher Education, volume 36, issue 4 (2012), pp. 585–601

Chapter 12
Reflective Journals as a Tool for Auto-Ethnographic Learning: A Case Study of Student Experiences with Individualized Sustainability
Jon Anderson
Journal of Geography in Higher Education, volume 36, issue 4 (2012), pp. 613–623

Chapter 13
Mapping the Journey Toward Self-Authorship in Geography
Niamh Moore, Eric J. Fournier, Susan W. Hardwick, Mick Healey, John MacLachlan & Jörn Seemann
Journal of Geography in Higher Education, volume 35, issue 3 (2011), pp. 351–364

Chapter 14
Gaia: "thinking like a planet" as transformative learning
Martin Haigh
Journal of Geography in Higher Education, volume 38, issue 1 (2014), pp. 49–68

Chapter 15
Going mobile: perspectives on aligning learning and teaching in geography
Claire H. Jarvis, Jennifer Dickie and Gavin Brown
Journal of Geography in Higher Education, volume 37, issue 1 (2013), pp. 76–91

Chapter 16
Considerations of How to Study Learning Processes when Students use GIS as an Instrument for Developing Spatial Thinking Skills
Lene Møller Madsen & Camilla Rump
Journal of Geography in Higher Education, volume 36, issue 1 (2012), pp. 97–116

Chapter 17
Students individual engagement in GIS
Lene Møller Madsen, Frederik Christiansen and Camilla Rump
Journal of Geography in Higher Education, volume 38, issue 2 (2014), pp. 251–265

Chapter 18
Writing in Geography: Student Attitudes and Assessment
Vanessa Slinger-Friedman & Lynn M. Patterson
Journal of Geography in Higher Education, volume 36, issue 2 (2012), pp. 179–195

Chapter 19
Capstone portfolios and geography student learning outcomes
Joann Mossa
Journal of Geography in Higher Education, volume 38, issue 4 (2014), pp. 571–581

Chapter 20
Re-framing the Geography Dissertation: A Consideration of Alternative, Innovative and Creative Approaches
Jennifer Hill, Pauline Kneale, Dawn Nicholson, Shelagh Waddington & Waverly Ray
Journal of Geography in Higher Education, volume 35, issue 3 (2011), pp. 331–349

Chapter 21
Embedding Research-Based Learning Early in the Undergraduate Geography Curriculum
Helen Walkington, Amy L. Griffin, Lisa Keys-Mathews, Sandra K. Metoyer, Wendy E. Miller, Richard Baker & Derek France
Journal of Geography in Higher Education, volume 35, issue 3 (2011), pp. 315–330

Chapter 22
Reviewing the 'Research Placement' as a Means of Enhancing Student Learning and Stimulating Research Activity
Fiona Tweed & Rob Boast
Journal of Geography in Higher Education, volume 35, issue 4 (2011), pp. 599–615

Editorial Note

The papers collected in this volume are arrayed in a sequence that runs from overview through pedagogic methodology to research methods, case studies, and, finally, structural and curricular matters. Although all of these papers come from the *Journal of Geography in Higher Education*, and all are heavily peer reviewed (as detailed in Haigh, 2013), many come from different sections of the journal. These different sections are flagged at the start of each paper, for example, Chapter 1, which introduces and overviews this collection, was published as an 'Editorial'. At the core of this volume are two thematic collections that are labelled 'Symposium'. The first of these arises from a special session on "Communicated Geographies" at the Royal Geographical Society's International Annual Conference, which was created to explore how pedagogic research methods could enhance research into learning and teaching in Geography in Higher Education (Haigh et al., 2013). This was billed as a 'Resources' Symposium because it aimed to support the 'Resources' section of the journal, which features papers that focus on research techniques, methods and occasionally methodologies. Four papers in this collection originate from this section of the journal and are flagged with the word 'Resources'. Deeper in this collection are three further papers labelled 'Symposium' that deal, mainly, with institutional frameworks and curricular matters. These are products from a workshop of the International Network on Learning and Teaching that focused on "Changing Geography Practice: Towards a Scholarship of Academic Practice" (Solem, 2011). Finally, nine further chapters have no special label because they are drawn from the body of the journal. Most of these are case studies of Geography research that are selected because they employ novel pedagogies and show how these may be applied.

References

Haigh, M. (2013). Writing successfully for the *Journal of Geography in Higher Education. Journal of Geography in Higher Education, 37*(1), 117–135. [Which is available open access from: http://www.tandfonline.com/doi/abs/10.1080/03098265.2012.692158#.UjV86JpwYY0].

Haigh, M., Hall, T., & Cotton, D.R.E. (2013). Researching 'Communicated Geographies': introduction. *Journal of Geography in Higher Education, 37*(2), 147–151.

Solem, M. (2011). Introduction: towards a scholarship of academic practice. *Journal of Geography in Higher Education, 35*(3), 311–314.

INTRODUCTION

Pedagogic research in geography higher education

Introduction

Pedagogic research, a term whose meaning is much debated, is often conceived as concerning the investigation of student learning and how this is affected by teaching and assessment practices. This editorial explores some recent contributions in the *Journal of Geography in Higher Education* (JGHE) to pedagogic research and wonders if there may exist, or may emerge, a distinctive school of pedagogic research in Geography Higher Education? It reviews recent developments in Pedagogic Research in Geography and suggests some priorities for future research.

For some, "Pedagogical research simply means finding out more about how learning takes place so that we, as teachers, can direct our energies into approaches which are more likely to be successful" (Reid, 2003, p. 3). Hence, pedagogic research can be recognized because it concerns the mode of learning more than it concerns the subject matter being learnt. So, if the main purpose of a paper is Geography, by this definition, it is probably not pedagogic research. By contrast, if a paper is primarily an exploration of the processes of learning, then it is pedagogic research even if the vehicle for that learning is Geography. The distinction, a difference in emphasis, is sometimes far from clear cut. For example, in 2014, Haigh published a paper entitled: "Gaia: 'thinking like a planet' as transformative learning", which explores issue of transformative learning, especially its potentially negative impacts on learners who do not "transform" effectively. This is pedagogic research. However, the paper also deals with Gaia Theory and its potential for education in Physical Geography and Earth Systems Science. Arguably, in this case, the content and pedagogy are inter-related – and a holistic, transformative approach to learning is intrinsically bound to the topic of study. This is indicative of much research into Geographical education, not least where this concerns that signature pedagogy of Geography – fieldwork.

One of the foundational arguments for *JGHE* was to establish Geography Higher Education as research field of equivalent quality and relevance to that in all other areas of Geography. It was argued that, since almost all University Geography researchers also teach Geography, using these research skills to enhance the quality of teaching and better the production of new Geographers was a very important activity. Close links between staff research, staff teaching and student learning are typical of Higher Education. These could be reconstructed in ways that are more productive for Geography's learners by using Geographical research skills in pedagogic contexts (Jenkins, 2003), so exploiting the huge overlap between the research methods employed by geographers and those employed by education professionals for the benefit of the discipline.

Subsequently, the linking of disciplinary research and teaching has become a major theme within Geography (as other disciplines) both in the UK and in the USA, where the idea was one of the ten changes in the undergraduate curriculum promoted by the USA's Boyer Commission of 1998 (Healey, 2005; Jenkins & Healey, 2009). However, research into teaching and learning, practising pedagogic research, has become a minor theme within a movement that now focusses more on enabling learners to produce new geographical research through a kind of apprenticeship (Hill, Kneale, Nicholson, Waddington, & Ray, 2011; Walkington & Jenkins, 2008). Of course, this process contains many pitfalls, some shared with other forms of problem-based and inquiry-based learning (Spronken-Smith & Walker, 2010). These include overconcentration on the local and minutiae, and less savoury problems – such as the possible exploitation of learners to advance their instructor's Geographical research, a risk of focussing resources on "top performing" students at the expense of others, and the production of unquestioning "Mini-me" mimics of current practice. Nevertheless, Tweed and Boast, who explore a course that involved learners directly as research assistants found the approach encourages "student ownership of learning and develops an active stakeholder culture….[as well as] a more vibrant research environment" (Tweed & Boast, 2011, p. 613). It is argued that this approach to research, discovery-based learning, does more to include students in the research culture of universities, helps learners read research papers more critically, helps them hone creative, reflective, analytical and communication skills, including graphicacy, and build emotional intelligence (Walkington et al., 2011). Walkington et al. (2011)'s paper also provides a nicely worked example of how to conduct research with teachers in the context of a short workshop using a "liquid" "World Café" format (cf. Steyart & Lisoir, 2006).

Since its inception, the *JGHE* has been focussed on the development of a better, more student-centred Geographical education. Moore et al. (2011) envisage a kind of make-over for Geography curricula that emphasizes the development of student's capabilities as "independent, critical thinkers and responsible active citizens" (Moore et al., 2011, p. 362). They argue that developing the learner's capabilities for "self-authorship" should be recognized as "a guiding principle for curriculum reform and revision" (Moore et al., 2011, p. 352). Others, worrying about the accountability of Geographical Education to its stakeholders, both graduates and society at large, have the narrower vision of reshaping Geography curricula and learning practices in ways that enhance the employability of its products (Arrowsmith, Bagoly-Simó, Finchum, Oda, & Pawson, 2011; Jenkins & Healey, 1995).

Back in 2003, when Glynis Cousin and Mick Healey (2003) introduced the Pedagogic Research Methods theme of the *Journal of Geography in Higher Education*'s Resources Section, they detected resistance to the use of pedagogic research methods, especially qualitative methods, which they relate to a disciplinary perception of their reliability and validity. Later volumes demonstrate that such concerns have been dispelled. However, there remains a resistance to the unfamiliar, which was built into the rubric for this Section. Despite the need to connect the microcosm of Pedagogic Research within Geography with that of the other arenas of pedagogic research, importing a Pedagogic Research Method that does not already have a Geography pedigree is discouraged. It remains a requirement of the *JGHE*'s Pedagogic Research Methods papers that they demonstrate prior use within Geography. Effectively, this restriction encourages a derived character for Pedagogic Research in Geography Higher Education. Of course, this problem is larger than Geography; for example Cullen et al.'s (2002) review of pedagogic research worries that "communities of research and discourse across the sectors of post-secondary education are often quite closed, self-referencing, with poor transfer and transition across boundaries either of research or of knowledge and its application." The key argument for the development of a discipline specific pedagogic research is the "interplay between the generic and the contextual" and the need to give "generic … ways of thinking meaning" in Geographical learning contexts (Prosser, 2008, p 3.)

In recent years, the broad emphasis of Geography Higher Education has shifted from the better production of Geographical understanding, especially through active learning and fieldwork (Healey, Pawson, & Solem, 2010), to the better development of learners in more general terms. In sum, the shift of attention has moved from what they learn to how they learn, and from how they learn Geography to how they develop as people. This development has been coeval with larger changes within Geography. For many decades, Geography was dominated by what Ken Wilber calls "flatland" thinking that considers only the objective exterior point of view (Haigh, 2013a). However, since Geography's "Cultural Turn" of the 1980s and its slighter "Affective Turn" in the 1990s, Geographers have become more comfortable in dealing with subjectivity and both the collective and individual interior. These changes have shifted Geographical Pedagogic Research further from its scientific and social-scientific roots, not towards the more rigorous statistical approaches of psychology, but toward more qualitative and humanities-oriented styles. Much of this seems to echo the values and action-reflection-related practitioner-production of "Living Theory", even "Living Global Citizenship", in education (Potts, 2014; Whitehead, 1989).

What is the geographical contribution to pedagogic research?

Previously, the *JGHE* published compilations of papers on particular themes in book form (Chalkley, Haigh, & Higgitt, 2009; Healey et al., 2010). This editorial asks what might a similar volume on "Pedagogic Research in Geography Higher Education" include. It explores some of the more interesting Pedagogic Research Method-focussed papers published in the journal since 2010, including some from the Pedagogic Research Methods theme of the journal's Resources section and all of the papers in the Pedagogic Research Method-focussed Symposium "Communicated Geographies", which addressed ways of enhancing the development, evaluation and impact of Geographical communication and its interaction with the ways that Geographers learn (Haigh, Hall, & Cotton, 2013).

Most broadly, these papers highlight the degree to which Geographers import methodologies from education and other disciplines such as Psychology and Anthropology. For example, "Q" Theory emerges from Psychology, although it has been recommended to Human geographers where it is said to "dovetail with the interpretive and cultural turn in geographic theory and is a powerful tool for anti-essentialist approaches to subjectivity and for constructivist inquiries into categories of nature, urban form, and scientific activity" (Eden, Donaldson, & Walker, 2005, Robbins & Krueger, 2000, p. 638). However, Wright (2013) uses the approach to map the self-assessments of a group of learners on a subject in terms of degrees of agreement, and analyses of the outcome with Factor Analysis, thus quantifying the array of viewpoints.

In a similar vein, Smith, Polglase, and Parry (2012) evaluate the introduction of Belbin Team management theory's ideas of role theory on enhancing group-work, which is as ubiquitous as it is problematic for learning and teaching in Geography. Belbin was used to help learners identify their own group-work problems after an initial experience had surfaced those problems: poor communication, uneven commitment and participation, poor or absent leadership and interpersonal problems. Belbin Team Roles are used in business to create productive working relationships and high-performing teams, to build personal self-awareness and self-efficacy and build mutual trust and understanding. Nine roles are identified, each with its strengths and weaknesses, these are the Plant or problem solver, Resource Investigator or networker, Monitor Evaluator – who tries to provide objective guidance, Co-ordinator – who wants to lead and delegate, Implementer – who aims to get things done, Completer Finisher – who engages with the fine print and detail, Team-worker – who strives for cohesion, Shaper – who thrives on pressure and overcoming obstacles and the Specialist who has particular skills (Belbin, 2010). Smith et al. (2012) show that knowledge of team role theory helps learners depersonalize their team problems and recognize the significance of this activity for their future employment. The authors use Belbin's labels to classify the resulting styles of their work groups. Their work has been deployed by later workers who define six TREO (team role experience and orientation) dimensions linked to the five domains of personality (Mathieu, Tannenbaum, Kukenberger, Donsbach, & Alliger, 2014).

Much Geographical Pedagogic Research is devoted to "methods", often engaging the potential of new technologies that facilitate better pedagogic understanding as a secondary consequence. These include the use of social media and audio-visual technologies as ways of understanding the thought processes of learners and of engaging those learners in deeper critical introspection. For example, reflective learning journals can involve the new opportunities of social media and blogs. Jarvis, Dickie, and Brown (2013, p. 78) use a student developed group "mediascape", which is "a multimedia display on a GPS-enabled smartphone device that is automatically triggered as the holder passes through an area" to facilitate fieldwork in Dublin. They note how this avoids the linear reduction of a field experience in subsequent write-ups and admits a more complete expression of the learner's experience. Using a constructively aligned assessment process, they show how this strategy helps learners to "visualize and synthesize intellectually challenging theoretical concepts about the relationality of urban space" (Jarvis et al., 2013, p. 89).

Similarly, the making of video documentaries can help learners enhance their media literacy because the process helps them understand how films are constructed (Dando & Chadwick, 2014). Mavroudi and Jöns (2011, pp. 588–589) use video documentary making and reflective journal writing to promote deeper learning and ethical reflection in learners, although their paper emphasizes the technical issues of student engagement and assessment. France and Wakefield (2011) report on practical issues with using digital stories in teaching and learning, noting the potential for such an approach to harness skills that many students already possess, but also highlighting the need for ethical considerations in the use of video recording. Notwithstanding this concern, Wee, DePierre, Anthamatten, and Barbour (2013) have demonstrated the value of visual methods as a pedagogic research method to help understand learners' "sense of place" and the meanings that learners attach to geographical concepts. This, in turn, helps instructors design curricula and support learning.

Of course, an abiding worry in geographical Education is why some learners ask so few questions of their subject matter. This is an especial problem in technical areas like GIS, which may be why the *JGHE* recently devoted a special issue to ethics in GIS (Carr, 2014). Madsen and Nielsen (2013) studied the questions asked by GIS learners, work inspired by research on the value of student-generated questions within science education, and found that most student-generated questions requested basic information and were independent of the students' self-image of their learning approach. Reflective "Wonderment" questions that showed learners seeking to extend their Geographical knowledge were rare. However, elsewhere, exploring ways of studying the development of spatial thinking through GIS, Madsen and Rump (2012) find that GIS shapes the learners' perception of the world and worry about the role of GIS as a lens. Their study shows that, while the majority either prioritize the instrumentality of the task or treat the GIS as a game, more than a third of learners focussed on their geographical understanding. They wonder about how much the others were distracted by the provision of a cook-book-style course manual. Madsen and

Rump (2012) conclude that those who used the course manual most closely thought least about the Geography of what they were doing. However, those who reflected on the subject of the task and those who treated the work as a game did move towards becoming real spatial thinkers (Madsen, Christiansen & Rump, 2014).

Slinger-Friedman and Patterson (2012) suggest that many Geography learners are more comfortable with maps and visual images that with writing. They differentiate between writing for communication and writing for learning, which involves critical reflection, active engagement and the attempt to clarify concepts – as in the learning journal approach (Cook, 2000; McGuinness, 2009). By contrast, writing for communication is "transactional" and closer to surface than deep-learning. Slinger-Friedman and Patterson (2012) use a model controlled, quantitative, research design to explore the effects of "writing to learn" interventions on two classes, finding that between half and two-thirds agreed that their writing to learn activities improved their comprehension and engagement, which resulted in significantly greater appreciation of the courses. The intervention, however, had no significant impact on examination results based on multiple choice tests, which probably do not test higher order thought processes.

In many systems, especially in the UK, the major writing exercise is the dissertation produced in the final year. Hill et al. (2011, p. 344) evaluate the potential of alternative models for the conventional dissertation, in ways that "challenge students to demonstrate their higher order cognitive, affective and inter-personal skills" including models based on group-work, work-based learning, and that involve mapping and visualization as much as written skills.

Mossa (2014) evaluates the advantages of using capstone portfolios to assess learning outcomes. Here, learners are invited to select 4–6 examples of their best work in Geography and to justify that selection in terms of their development of the Department's learning outcomes on critical thinking and communication skills. The range of materials allowed extends far beyond the conventional assessed paper to include maps, graphics, examples of data analysis, Powerpoint presentations, web-pages, videos, professional blogs and so on. These portfolios were mutually advantageous for both learners, who used the portfolio in negotiations with future employers and teachers, who used insights from their assessing the portfolios to tune up their delivery and curriculum (Mossa, 2014).

Healey (2012, p. 242) explores the value of debates, not only as an active-learning technique to helping learners weigh up the arguments on two sides of a controversial subject, but also as a device for deepening critical thinking in learners. Engagement with a debate helps learners by "encouraging them to find their own path through a topic". The weakness of the approach is that it tends to polarize thinking into the two sides, where alternative role-play approaches such as stakeholder panels or the "Public Inquiry" format allow for less clarity but greater diversity (Livingstone, 1999). However, especially at introductory level, the approach proves a valuable way of engaging learners and Healey (2012) charts an array of ways the technique can be employed to enhance Geographical education and for helping learners realize the value of critical thinking skills.

Exploring the learner experience

More fundamental than understanding what learners learn is exploring the way that they learn and the ways that they situate themselves within their learning. This involves discovering what drives learning and identifying the levers that affect the development of reflective practices. This can be problematic within a discipline that tends to emphasize objectivity, especially within Physical Geography/GIS, and that tends to operate at the landscape or social scales more than the subjective interior. AQAL or AQ mapping is one vehicle that may be used to help learners parse out the subjective and objective and to express both their interior aspirations and perceptions of how they are perceived at both the individual and collective levels (Haigh, 2013a). AQ maps deconstruct every situation into four component quadrants to divide the felt interior from the observed performative exterior, first for the individual and then for the social group. In language terms, these four describe each situation from, first' the interior, the insider "I" and "We" perspectives, and then from the exterior, the "S/He/It" or "They/many/Its" perspectives of the external observer. The approach allows the learners and researcher to distinguish what is wished and felt from what actually happens.

Of course, reflective learning journals, written in the first person, are an effective and traditional way of helping learners put themselves in the picture, which is equally valuable for pedagogic researchers both as data and as a way of assessing both their students and, indeed, their own research processes (McGuinness, 2009; McGuinness & Brien, 2007). Anderson (2012) evaluates the value of reflective journals in promoting reflection and critical self-awareness in learners. The context was fostering reflection on the problems and potentials for achieving sustainability at the personal level using auto-ethnography to connect an environmental problem to the personal, cultural, social and political. Of course, auto-ethnographic learning is valuable because it shifts its focus from the interior to the exterior quadrants, between looking outward and looking inwards and engages the author is creating their own narrative. Dyson (2007, p. 38) notes: "the style was ... the only way to present, in a meaningful and mindful way, the cultural phenomenon that I was living and researching. In stepping back, I also realized that I was changing as an individual and a researcher ..." before self-consciously adopting the research journey as a metaphor and discussing a personal landscape of transformation (cf. Haigh, 2014). Similarly, Anderson (2012) notes that such journals often emerge as a space of transformative self-education, emancipation and empowerment for the learner, while also granting powerful insights for the pedagogic researcher.

Of course, another good way of exploring the learner's feelings and perceptions of performance at every level is through interviews and, once again, the autoethnographic approach may yield positive results. Adriansen and Madsen (2014) argue that interviewing students is an effective way of fostering reflexive practice in teachers. They analyse the ethics of such events considering the problem of the power relations involved and recommending as a means of creating co-ownership of the experience, the visual co-construction methodology of Kvale (1996). This timeline-based strategy was previously used by Adriansen for life-history research (Adriansen, 2012).

Cotton, Stokes, and Cotton (2010) advocate applying more of the observational methods used by Anthropology to examine the behaviour of learners and their interactions in class or field-settings. They argue that these methods avoid some of the common pitfalls of using post-experience questionnaires, which are subject to learner selectivity, memory limitations, post-hoc revision and stereotyping. They also note the pitfalls of the approach, which include the production of a mass of unstructured data and the fact that observation can change the behaviour of those being observed. They conclude that, while they may be complex and time-consuming, methods such as participant observation produce rich data and merit greater use in Geographical pedagogic research at least as an alternative to the ubiquitous questionnaire.

Focus groups are a pedagogic research tool that aims to make the task of student interviewing more efficient and that aims to promote reflective participatory discussion between learners (Steyart & Lisoir, 2006). Winlow, Simm, Marvell, and Schaaf (2013) build on the work of Breen (2006), whose paper is among the most downloaded in JGHE history. Their case study of Foundation Year students in the UK revisits the design and management of focus groups sessions, beginning with its ethics, before concluding that the approach can provide insights into the curriculum that questionnaires and individual interviews cannot.

Healey et al. (2010, p. 1) emphasize the desirability of student engagement and of reshaping the "learning spaces" of the curriculum around the learner to encourage and develop both their personal creativity and problem-solving skills. In recent years, Biggs and Tang (2011) ideas that there should be a constructive alignment between teaching and assessment, as well as other aspects of the educational environment, have had great influence in Geography. A similar, perhaps more holistic approach is suggested by Invitational Theory, an idea emerging from Humanistic psychology and the Safe-Schools movement in the USA (Haigh, 2011; Purkey & Stanley, 1991). Invitational Education seeks the conscious production of positively inviting educational environments through the systematic alignment of people, place, programmes, policies and procedures. Pedagogic research linked to the theory explores the degree to which different learning environments invite positive participation by learners (cf. Haigh, 2008).

Other curricula

However, deeper exploration reveals that Pedagogic Research has a still more important function, which is to explore the socio-cultural construction of the curriculum and its subliminal messages of social acceptability and control. In such terms, Pedagogic Research concerns "the organizational, discursive and transmission processes in all pedagogic agencies" and the ways these interact with the learner's "selective acquisition" of understanding, skills and attitudes as well as their symbolic agency in a process of cultural reproduction–production (Bernstein, 1996, p. 3). Hence, a deeper function of Pedagogic Research in Geographical Education is to explore the production of the Geography curriculum.

Here, Geography is unusual in having deep roots in a curriculum that lies outside conventional campus settings. Fieldwork is a signature pedagogy of Geography that embeds the discipline in real-world experiences (Spronken-Smith, 2013). It is generally considered far more important by Geography academics than either by learners or institutions (e.g. Griffin, 2013). This may be one reason why a huge amount of Geographical Education research concerns the construction of field courses; some of the JGHE's contribution has recently appeared as a special on-line compendium. Meanwhile, Morrissey, Clavin, and Reilly (2013) explore the interactions between GIS and fieldwork in the form of a participation in the reconstruction of Sarajevo, with its multiple layers of development discourse. The paper is Geography focussed but mentions participatory knowledge and Field-Based learning as supplementary curricula that are distinct from conventional classwork.

MacPherson (2011) is among those that recognize that the true character of a Curriculum is quite complicated. She portrays the Curriculum as an iceberg, where only a small fraction is visible at the surface and much more is to be found in the, typically unexplored, depths beneath (Figure 1). Hitherto, most conventional pedagogic research in geography tackles only the formally planned or explicit curriculum and the differences between this and the curriculum received by learners. It is quite unusual to find papers that address the learning that is received from peers and very rare, outside of international education, for the non-formal curriculum of learning in the family and community to be

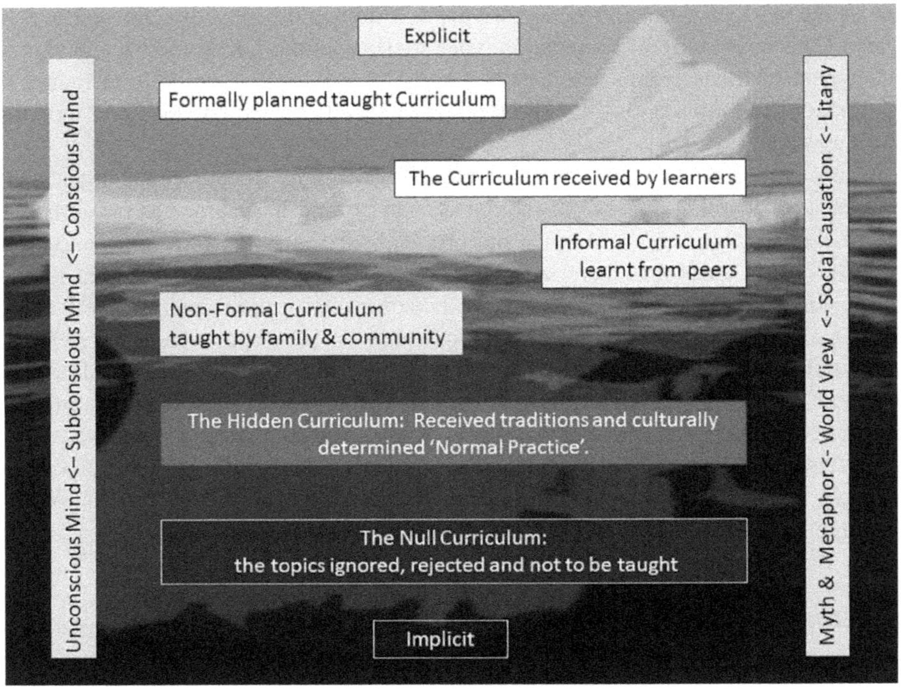

Figure 1. Curriculum iceberg (adapted from MacPherson, 2011).

addressed, at least in Higher Education. Influences of gap year projects, which are undertaken in the year before college, are discussed (Ansell, 2008; Blackburn, Clark, & Pilgrim, 2005). There are some nods to work-based learning (Jenkins & Healey, 1995; McEwen et al., 2005; Shepherd, 1998). There is a larger literature on service learning, volunteer working with NGOs and other community organizations, especially from the USA (Dorsey, 2001; Eflin & Sheaffer, 2006; Haigh, 2006; Oldfield, 2008). However, the total contribution is neither large nor hugely cited.

Deeper layers are still less well explored. The culture of a Higher Education community contains many traditions that are unquestioned and inherited. These subconscious elements of the curriculum define what is taught, how it should be taught, practised and assessed, and also how teachers and learners are supposed to interact. Collectively, this creates a Hidden Curriculum of practice and assessment that is never taught (Joughin, 2010). In Geography, Cotton, Winter and Bailey (2013) provide guidance on how to research the role of this "hidden curriculum" and so explore the implicit messages a university's operational environment and values may send out on subjects such as sustainability, as exemplified by Winter and Cotton (2012).

The structure of the hidden curriculum differs radically between universities, more so between different national systems, and even more radically when taken across larger cultural boundaries. Indeed, the existence of the hidden curriculum becomes most obtrusive in the context of international education, when learners and teachers find themselves having completely different expectations of their interactions. Some authors have attempted to mine these differences for the importation of concepts and structures that may be used to enhance teaching and learning in their own educational systems, for example MacPherson's (2011) work with the Tibet Buddhist "Refuge Tree" curriculum model or Haigh's (2009) work with Samkhya's Three Modes of Nature.

Less visible is the hidden curriculum's role in the self-preservation of power elites and its consequences, which include the preservation of gender disparities, social class structures and patterns of cultural dominance such as Eurocentrism (cf. Gough, 1998). Through such means "White cultures are able to maintain their invisibility and also control of economic, social and cultural relations" (Borg, Mayo, & Sultana, 1998, p. 376). In Geography, Brown, Collard, and Hoogeveen (2014) apply critical pedagogic theory to their analysis of the problems of negotiating the "teaching statements" required by appointments and promotions panels with a teaching philosophy that does not conform to the dominant neoliberal expectations of Western universities. Curiously, they count "mentoring", which is another approach that is overtly and covertly employed to create conformity, as a "feminist praxis" and their paper closes in a confusion of "hopeful contradiction".

Deepest and least explored is the Null Curriculum, which includes all of those subjects that simply are ignored, avoided or rejected as being beyond the pale (Flinders, Noddings, & Thornton 1986; MacPherson, 2011). Eisner (1994) argues that what is not in the curriculum may be as significant as anything actually included and may lead to impoverishment of the taught programme. For example, although Geographers are active players in the world, influencing regional, social and environmental decision-making, the idea that they should be taught ethics, so enabling them to become reflective practitioners, is a relatively new concern for the taught curriculum in Higher Education (Boyd et al., 2008; c.f. Buttimer, 1974; Smith, 1995).

In sum, there is much work that Pedagogic Research in Geography Higher Education could and should do to explore and contest the character of the Null Curriculum, the embedded cultural traditions and assumptions that provide the Hidden Curriculum, and full role and significance of the Non-formal Curriculum in the production of Geographical learning and the deep engagement of learners. Of course, many Geographers have recognized the role of learning in Non-formal contexts, especially where it affects the success of active learning and fieldwork strategies. However, much more could be done to evaluate the full effect on the development of learners and their capabilities.

Conclusion

Geography may have a flag-carrying role among disciplines engaged with pedagogic research and, more generally, the scholarship of teaching and learning. It contains several and distinctive signature pedagogies, such as fieldwork (Spronken-Smith, 2013) and is deeply engaged with mapping and GIS. It embodies a commitment to active learning (Healey et al., 2010) and to innovation in learning and teaching. In theory, these processes might have initiated the emergence of uniquely Geographical methods and frameworks of pedagogic research in Geography. However, the existence of a Geographical pedagogic research style remains questionable.

Citation analysis, of course, suggests a different story. Tight (2008, p. 598) notes that: "Citations ... are in part about identifying the community or tribe you belong to, or, perhaps, that you would like to join". Geographers tend to cite other Geographers amidst a fairly limited array of outside influences. In citation terms, the result may be mapped as a discrete Geographical school, or maybe several regional schools, of pedagogic research in Higher Education. Of course, Higher Education research is, anyway, prone to compartmentalization, Tight (2014) records journal self-citation rates, that is where articles in one journal cite the same journal, in six leading journals of research into Higher Education, ranging from 28% up to almost 58%. However, in Geography's case, this apparent independence is largely illusory. The "fuzzy set" of Geographical Pedagogic Research, like Geography as a whole, shares major overlapping interests and qualities with similar work going on in Geography's sister disciplines in the Environmental Sciences, Social Sciences and Humanities and business fields like Tourism. It also shares the same "derived" character of its methods and methodologies, which are not Geography creations but drawn selectively from Education, Psychology and so on. Meanwhile, there is a huge area of critical inquiry that remains to be addressed by Geographical Pedagogic Research, which may be especially important to a discipline with a global scope. This concerns the impacts of the "Other Curricula": Non-formal, Null and Hidden, on learners and learning in Geography.

However, even if it is by no means unique, Geography has a style of pedagogic research that is quite distinctive. This mainstream Geography approach to Pedagogic Research and Geographical Education is spelled out in the guidance offered to those intending to write for the JGHE, a document reverse engineered from the JGHE's "better" empirical papers and from the aspirations of several generations of editors (Haigh, 2013b). It describes a Geographical Pedagogic Research that is practitioner-based, directly engaged with student-active learning, and focussed on the practical, case-study-based exploration and evaluation of innovative approaches to learning in the classroom in the field and increasingly on-line.

Martin Haigh
Department of Social Sciences, Oxford Brookes University, Oxford, UK

Debby Cotton
Educational Development, Plymouth University, Plymouth, UK

Tim Hall
Department of Applied Social Studies, University of Winchester, Winchester, UK

References

Adriansen, H. K. (2012). Timeline interviews: A tool for conducting life history research. *Qualitative Studies*, *3*, 40–55.

Adriansen, H. K., & Madsen, L. M. (2014). Using student interviews for becoming a reflective geographer. *Journal of Geography in Higher Education*, *38*, 595–605. doi:10.1080/03098265.2014.936310

Eflin, J., & Sheaffer, A. L. (2006). Service-learning in watershed-based initiatives: Keys to education for sustainability in geography? *Journal of Geography*, *105*, 33–44. doi:10.1080/00221340608978656

Anderson, J. (2012). Reflective journals as a tool for auto-ethnographic learning: A case study of student experiences with individualized sustainability. *Journal of Geography in Higher Education*, *36*, 613–623. doi:10.1080/03098265.2012.692157

Ansell, N. (2008). Third World gap year projects: Youth transitions and the mediation of risk. *Environment and Planning. D, Society and Space*, *26*, 218–240. doi:10.1068/d83j

Arrowsmith, C., Bagoly-Simó, P., Finchum, A., Oda, K., & Pawson, E. (2011). Student employability and its implications for geography curricula and learning practices. *Journal of Geography in Higher Education*, *35*, 365–377. doi:10.1080/03098265.2011.563379

Belbin, R. M. (2010). *Team roles at work* (2nd ed). Oxford: Butterworth Heinnemann.

Bernstein, B. B. (1996). *Pedagogy, symbolic control, and identity: Theory, research, critique*. Lanham (MD): Rowman & Littlefield.

Biggs, J., & Tang, C. (2011). *Teaching for quality learning at university* (4th ed.). Maidenhead: SRHE – Open University Press.

Blackburn, G. A., Clark, G., & Pilgrim, D. (2005). The gap year for geographers. *Geography, 90*, 32–41.

Borg, C., Mayo, P., & Sultana, R. (1998). Revolution and reality: An interview with Peter McLaren. In W. Pinar (Ed.), *Curriculum: Toward new identities* (pp. 355–377). Abingdon: Routledge (2013).

Boyd, W., Healey, R., Hardwick, S., Haigh, M., Klein, P., Doran, P., … Bradbeer, J. (2008). None of us sets out to hurt people: The ethical geographer and geography curricula in higher education. *Journal of Geography in Higher Education, 32*, 37–50. doi:10.1080/03098260701731462

Breen, R. L. (2006). A practical guide to focus-group research. *Journal of Geography in Higher Education, 30*, 463–475. doi:10.1080/03098260600927575

Brown, S., Collard, R. C., & Hoogeveen, D. (2014). Pedagogical declarations: Feminist engagements with the teaching statement. *Journal of Geography in Higher Education, 38*, 148–154. doi:10.1080/03098265.2013.845741

Buttimer, A. (1974). *Values in geography*. Washington, DC: Association of American Geographers.

Carr, J. (2014). Introduction: Towards a methodology for teaching the ethics of geocoded data privacy. *Journal of Geography in Higher Education, 38*, 441–443. doi:10.1080/03098265.2014.971402

Chalkley, B., Haigh, M., & Higgitt, D. (Eds.). (2009). *Education for sustainable development: Papers in honour of the United Nations Decade of Education for Sustainable Development (2005–2014)*. Abingdon: Routledge.

Cook, I. (2000). Nothing can ever be the case of 'us' and 'them' again: Exploring the politics of difference through border pedagogy and student journal writing. *Journal of Geography in Higher Education, 24*, 13–27. doi:10.1080/03098260085108

Cotton, D. R., Stokes, A., & Cotton, P. A. (2010). Using observational methods to research the student experience. *Journal of Geography in Higher Education, 34*, 463–473. doi:10.1080/03098265.2010.501541

Cotton, D., Winter, J., & Bailey, I. (2013). Researching the hidden curriculum: Intentional and unintended messages. *Journal of Geography in Higher Education, 37*, 192–203. doi:10.1080/03098265.2012.733684

Cousin, G., & Healey, M. (2003). Pedagogic research methods in geography higher education. *Journal of Geography in Higher Education, 27*, 341–347. doi:10.1080/0309826032000145098

Cullen, J., Hadjivassiliou, K., Hamilton, E., Kelleher, J., Sommerlad, E., & Stern, E. (2002). *Review of current pedagogic research and practice in the fields of post-compulsory education and lifelong learning: Final report, revised*. London: Tavistock Institute of Human Relations.

Dando, C. E., & Chadwick, J. J. (2014). Enhancing geographic learning and literacy through filmmaking. *Journal of Geography, 113*, 78–84. doi:10.1080/00221341.2013.846394

Dorsey, B. (2001). Linking theories of service-learning and undergraduate geography education. *Journal of Geography, 100*, 124–132. doi:10.1080/00221340108978428

Dyson, M. (2007). My story in a profession of stories: Auto ethnography – an empowering methodology for educators. *Australian Journal of Teacher Education, 32*, art.3. doi:10.14221/ajte.2007v32n1.3

Eden, S., Donaldson, A., & Walker, G. (2005). Structuring subjectivities? Using Q methodology in human geography. *Area, 37*, 413–422. doi:10.1111/j.1475-4762.2005.00641.x

Eisner, E. W. (1994). *Educational Imagination: On design and evaluation of school programs* (3rd ed.). Upper Saddle River, NJ: Merrill Prentice Hall. Retrieved from http://people.cehd.tamu.edu/~pslattery/documents/EducationalImagination.pdf

Flinders, D. J., Noddings, N., & Thornton, S. J. (1986). The null curriculum: Its theoretical basis and practical implication. *Curriculum Inquiry, 16*, 33–42. doi:10.2307/1179551

France, D., & Wakefield, K. (2011). How to produce a digital Story. *Journal of Geography in Higher Education, 35*, 617–623. doi:10.1080/03098265.2011.560658

Gough, A. (1998). Beyond Eurocentrism in science education: Promises and problematics from a Feminist Poststructuralist perspective. In W. Pinar (Ed.), *Curriculum: Toward new identities* (pp. 185–210). Abingdon: Routledge.

Griffin, K. (2013). *Determining "best practice" methodologies in the pedagogy of tourism related fieldwork*. Institute of Technology, Learning, Teaching & Technology Centre. Retrieved from http://arrow.dit.ie/cgi/viewcontent.cgi?article=1002&context=ltcoth

Haigh, M. (2006). Promoting environmental education for sustainable development: The value of links between higher education & non-governmental organizations (NGOs). *Journal of Geography in Higher Education*, *30*, 327–349. doi:10.1080/03098260600717422

Haigh, M. (2008). Coloring in the emotional language of place. *Journal of Invitational Theory and Practice*, *14*, 25–40. Retrieved from http://eric.ed.gov/?q=Coloring+in&id=EJ839524.

Haigh, M. (2009). Fostering cross-cultural empathy with non-Western curricular structures. *Journal of Studies in International Education*, *13*, 271–284. doi:10.1177/1028315308329791

Haigh, M. (2011). Invitational education: Theory, research and practice. *Journal of Geography in Higher Education*, *35*, 299–309. doi:10.1080/03098265.2011.554115

Haigh, M. (2013a). AQAL Integral: A holistic framework for pedagogic research. *Journal of Geography in Higher Education*, *37*, 174–191. doi:10.1080/03098265.2012.755615

Haigh, M. (2013b). Writing successfully for the Journal of Geography in Higher Education. *Journal of Geography in Higher Education*, *37*, 117–135. doi:10.1080/03098265.2012.692158

Haigh, M. (2014). Gaia: "thinking like a planet" as transformative learning. *Journal of Geography in Higher Education*, *38*, 49–68. doi:10.1080/03098265.2012.763161

Haigh, M., Hall, T., & Cotton, D. R. E. (2013). Researching communicated geographies: Introduction. *Journal of Geography in Higher Education*, *37*, 147–151. doi:10.1080/03098265.2013.793521

Healey, M. (2005). Linking research and teaching to benefit student learning. *Journal of Geography in Higher Education*, *29*, 183–201. doi:10.1080/03098260500130387

Healey, M., Pawson, E., & Solem, M. (2010). Introduction. In M. Healey, E. Pawson, & M. Solem (Eds.), *Active learning and student engagement: International perspectives and practices in geography in higher education* (pp. 1–8). Abingdon: Routledge.

Healey, R. L. (2012). The power of debate: Reflections on the potential of debates for engaging students in critical thinking about controversial geographical topics. *Journal of Geography in Higher Education*, *36*, 239–257. doi:10.1080/03098265.2011.619522

Hill, J., Kneale, P., Nicholson, D., Waddington, S., & Ray, W. (2011). Re-framing the geography dissertation: A consideration of alternative, innovative and creative approaches. *Journal of Geography in Higher Education*, *35*, 331–349. doi:10.1080/03098265.2011.563381

Jarvis, C. H., Dickie, J., & Brown, G. (2013). Going mobile: Perspectives on aligning learning and teaching in geography. *Journal of Geography in Higher Education*, *37*, 76–91. doi:10.1080/03098265.2012.694069

Jenkins, A. (2003). Designing a curriculum that values a research-based approach to student learning in geography, earth and environmental sciences (GEES). In S. Gaskin (Ed.), *'Linking teaching and research and undertaking pedagogic research in geography, earth and environmental sciences (GEES)': Planet* (special edition 5, 1). Retrieved from http://www.gees.ac.uk/planet and www.gees.ac.uk/planet/#PSE5.

Jenkins, A., & Healey, M. (Eds.) (1995). JGHE symposium: Linking the geography curriculum to the worlds of industry. *Journal of Geography in Higher Education*, *19*, 177–181. doi:10.1080/03098269508709301

Jenkins, A., & Healey, M. (2009). *Developing the student as a researcher through the curriculum. Innovations in practice* (Vol. 3). Liverpool: John Moores University.

Joughin, G. (2010). The hidden curriculum revisited: A critical review of research into the influence of summative assessment on learning. *Assessment & Evaluation in Higher Education*, *35*, 335–345. doi:10.1080/02602930903221493

Kvale, S. (1996). *InterViews: An introduction to qualitative research writing*. Thousand Oaks, CA: Sage.

Livingstone, I. (1999). Role-playing planning public inquiries. *Journal of Geography in Higher Education*, *23*, 63–76. doi:10.1080/03098269985605

MacPherson, S. (2011). *Education and sustainability: learning across the diaspora, indigenous, and minority divide*. Florence, KY: Routledge.

Madsen, L. M., Christiansen, F., & Rump, C. (2014). Students individual engagement in GIS. *Journal of Geography in Higher Education*, *38*, 251–265. doi:10.1080/03098265.2014.910758

Madsen, L. M., & Nielsen, T. T. (2013). Learning to do geography? University students posing questions in GIS laboratory exercises. *Norsk Geografisk Tidsskrift-Norwegian Journal of Geography*, *67*, 157–161. doi:10.1080/00291951.2013.803259

Madsen, L. M., & Rump, C. (2012). Considerations of how to study learning processes when students use GIS as an instrument for developing spatial thinking skills. *Journal of Geography in Higher Education*, *36*, 97–116. doi:10.1080/03098265.2011.576336

Mathieu, J. E., Tannenbaum, S. I., Kukenberger, M. R., Donsbach, J. S., & Alliger, G. M. (2014). Team role experience and orientation: A measure and tests of construct validity. *Group & Organization Management*, *40*, 6–34. doi:1059601114562000

Mavroudi, E., & Jöns, H. (2011). Video documentaries in the assessment of human geography field courses. *Journal of Geography in Higher Education*, *35*, 579–598. doi:10.1080/03098265.2011.559578

McEwen, L., Duck, R., Haigh, M., Smith, S., Wolfenden, L., & Kelly, K. (2005). Evaluating the 'postgraduateness' of vocational taught Masters environmental courses: Student perspectives. *Planet*, *14*, 8–12. doi:10.11120/plan.2005.00140008

McGuinness, C., & Brien, M. (2007). Using reflective journals to assess the research process. *Reference Services Review*, *35*, 21–40. Retrieved from http://www.emarketing.net.cn/upload/file/2008/04/16/001208276597786.pdf. doi:10.1108/00907320710729346

McGuinness, M. (2009). Putting themselves in the picture: Using reflective diaries in the teaching of feminist geography. *Journal of Geography in Higher Education*, *33*, 339–349. doi:10.1080/03098260902742425

Moore, N., Fournier, E. J., Hardwick, S. W., Healey, M., MacLachlan, J., & Seemann, J. (2011). Mapping the journey toward self-authorship in geography. *Journal of Geography in Higher Education*, *35*, 351–364. doi:10.1080/03098265.2011.563378

Morrissey, J., Clavin, A., & Reilly, K. (2013). Field-based learning: The challenge of practising participatory knowledge. *Journal of Geography in Higher Education*, *37*, 619–627. doi:10.1080/03098265.2013.794333

Mossa, J. (2014). Capstone portfolios and geography student learning outcomes. *Journal of Geography in Higher Education*, *38*, 571–581. doi:10.1080/03098265.2014.958659

Oldfield, S. (2008). Who's serving whom? Partners, process, and products in service-learning projects in South African urban geography. *Journal of Geography in Higher Education*, *32*, 269–285. doi:10.1080/03098260701514215

Potts, M. (2014). Living global citizenship: A new pedagogy for citizenship education. *Educational Journal of Living Theories*, *7*, 102–118. Retrieved from http://ejolts.net/current

Prosser, M. (2008). The scholarship of teaching and learning: What is it? A personal view. *International Journal for the Scholarship of Teaching and Learning*, *2*(Article 2), 1–4. Retrieved from http://digitalcommons.georgiasouthern.edu/cgi/viewcontent.cgi?article=1098&context=ij-sotl&sei-redir=1&referer=http%3A%2F%2Fscholar.google.co.uk%2Fscholar%3Fq%3Dpedagogic%2Bresearch%2BAustralia%26btnG%3D%26hl%3Den%26as_sdt%3D1%252C5#search=%22pedagogic%20research%20Australia%22

Purkey, W. W., & Stanley, P. H. (1991). *Invitational teaching, learning, and living*. Washington, DC: National Education Association, Analysis & Action Series. Retrieved from http://files.eric.ed.gov/fulltext/ED340689.pdf

Reid, N. (2003). *Getting started in pedagogical research in the physical sciences*. York: Higher Education Academy. Retrieved from http://www-new2.heacademy.ac.uk/assets/ps/documents/practice_guides/practice_guides/ps0076_getting_sarted_in_pedagogic_research_in_the_physical_sciences_aug_2004.pdf

Robbins, P., & Krueger, R. (2000). Beyond bias? The promise and limits of Q method in human geography. *The Professional Geographer*, *52*, 636–648. doi:10.1111/0033-0124.00252

Shepherd, I. D. (1998). Work Experience: who needs it? *Journal of Geography in Higher Education*, *22*, 135–145. doi:10.1080/03098269886155

Slinger-Friedman, V., & Patterson, L. M. (2012). Writing in geography: Student attitudes and assessment. *Journal of Geography in Higher Education*, *36*, 179–195. doi:10.1080/03098265.2011.599369

Smith, D. M. (1995). Moral teaching in geography. *Journal of Geography in Higher Education*, *19*, 271–283. doi:10.1080/03098269508709316

Smith, M., Polglase, G., & Parry, C. (2012). Construction of student groups using Belbin: Supporting group work in environmental management. *Journal of Geography in Higher Education, 36*, 585–601. doi:10.1080/03098265.2012.692156

Spronken-Smith, R. (2013). Toward securing a future for geography graduates. *Journal of Geography in Higher Education, 37*, 315–326. doi:10.1080/03098265.2013.794334

Spronken-Smith, R., & Walker, R. (2010). Can inquiry-based learning strengthen the links between teaching and disciplinary research? *Studies in Higher Education, 35*, 723–740. doi:10.1080/03075070903315502

Steyart, S., & Lisoir, H. (Eds.). (2006). *Participatory methods toolkit: A practitioner's manual* (2nd ed.). Brussels: King Baudouin Foundation/Vlaams Instituut voor Wetenschappelijk en Technologisch Aspectenonderzoek (viWTA). Retrieved from http://www.kbs-frb.be/uploadedFiles/KBS-FRB/Files/EN/PUB_1540_Participatoty_toolkit_New_edition.pdf

Tight, M. (2008). Higher education research as tribe, territory and/or community: A co-citation analysis. *Higher Education, 55*, 593–605. doi:10.1007/s10734-007-9077-1

Tight, M. (2014). Working in separate silos? What citation patterns reveal about higher education research internationally. *Higher Education, 68*, 379–395. doi:10.1007/s10734-014-9718-0

Tweed, F., & Boast, R. (2011). Reviewing the 'research placement' as a means of enhancing student learning and stimulating research activity. *Journal of Geography in Higher Education, 35*, 599–615. doi:10.1080/03098265.2011.559579

Walkington, H., Griffin, A. L., Keys-Mathews, L., Metoyer, S. K., Miller, W. E., Baker, R., & France, D. (2011). Embedding research-based learning early in the undergraduate geography curriculum. *Journal of Geography in Higher Education, 35*, 315–330. doi:10.1080/03098265.2011.563377

Walkington, H., & Jenkins, A. (2008). Embedding undergraduate research publication in the student learning experience: Ten suggested strategies. *Brookes E-journal of Learning and Teaching, 2* (3), 1. Retrieved from http://bejlt.brookes.ac.uk/issue/volume-two-issue-three/.

Wee, B., DePierre, A., Anthamatten, P., & Barbour, J. (2013). Visual methodology as a pedagogical research tool in geography education. *Journal of Geography in Higher Education, 37*, 164–173. doi:10.1080/03098265.2013.775568

Whitehead, J. (1989). Creating A Living Educational Theory from questions of the kind, 'how do I improve my practice?'. *Cambridge Journal of Education, 19*, 41–52. doi:10.1080/0305764890190106

Winlow, H., Simm, D., Marvell, A., & Schaaf, R. (2013). Using focus group research to support teaching and learning. *Journal of Geography in Higher Education, 37*, 292–303. doi:10.1080/03098265.2012.696595

Winter, J., & Cotton, D. (2012). Making the hidden curriculum visible: Sustainability literacy in higher education. *Environmental Education Research, 18*, 783–796. doi:10.1080/13504622.2012.670207

Wright, P. N. (2013). Is Q for you? Using Q methodology within geographical and pedagogical research. *Journal of Geography in Higher Education, 37*, 152–163. doi:10.1080/03098265.2012.729814

SYMPOSIUM

Researching the hidden curriculum: intentional and unintended messages

Debby Cotton[a], Jennie Winter[a] and Ian Bailey[b]

[a]PEDRIO, Teaching and Learning Directorate, Plymouth University, Plymouth, UK; [b]School of Geography, Earth and Environmental Sciences, Plymouth University, Plymouth, UK

In this paper, we examine the concept of the hidden curriculum, its importance to pedagogic research in geography and research methods which might be used to investigate it. We review three case studies of research projects which have explored the hidden curriculum in geographical contexts, and use these to illustrate the strengths and weaknesses of different methods. We conclude by suggesting that there is an increasing need for research exploring new elements of the hidden curriculum being created by current changes in the political and economic context affecting geography in higher education in the UK and other countries.

Introduction

What is the hidden curriculum?

The hidden curriculum is a complex and ambiguous term which has been used in a range of different ways since it was coined by Philip Jackson in 1968. In "Life in Classrooms," Jackson explored what he saw as the disconnect between what is overtly taught in educational institutions and what pupils actually learn—the "unpublicised features of school life" (Jackson, 1968, p. 17). In higher education, this may be made up of the societal, institutional or lecturers' values that are transmitted unconsciously to students. Much hidden curriculum research has been concerned with its undesirable aspects, such as the tendency to reproduce the inequalities of wider society. A classic example is the ethnographic study by Willis (1978), "Learning to Labour" which—as well as looking at working class subcultures—described the ways in which the organization of secondary schooling contributed to the preparation of working class pupils for a lifetime of work in factories. However, the hidden curriculum is not inevitably negative: "the notion of conspiracy is not always implied" (Portelli, 1993, p. 344). Apple and King (1977), for instance, differentiate between "weak" and "strong" conceptions of the hidden curriculum, using "weak" to refer to features inherent in educational processes, such as student socialization and professionalization (including punctuality, neatness and obedience), that may provide benefits for wider society and community cohesion, and "strong" to describe processes of social and cultural reproduction that serve to ensure the "preservation of existing social privilege, interests, and knowledge of one element of the population at the expense of less powerful groups" (Apple & King, 1977, p. 34).

However construed, hidden curriculum research has been used to inform a broad-ranging critique of covert social and political influences on the curriculum. Work exploring weak aspects of the hidden curriculum has largely focused on the student experience, investigating student responses to institutional rhetoric (Snyder, 1971); how students develop as members of a disciplinary community (Margolis & Romero, 1998) and the ways students are inculcated into maintaining the social order through conformity as a means of achieving academic success (Jackson, 1968; Margolis & Romero, 1998). Research into the stronger dimension has explored issues such as concern about educational institutions tasked with promoting equality that may in fact reinforce gender stereotypes (Stanworth, 1981); the predominance of white institutions, values and belief systems (Brandt, 1986; Mullard, 1982); how education reproduces class dominance, exploitation and inequality (Giroux, 1983; Hargreaves, 1967; Lacey, 1970; Willis, 1978) and structural correspondences between everyday educational experiences and capitalist social relations (Bowles & Gintis, 1976). To date, the majority of this work has focused on secondary education, although there is increasing interest in researching higher education, where less densely codified curricula provide a landscape of potential for multiple hidden curricula to exist (Ahola, 2000; Cotton & Winter, 2010; Margolis & Romero, 1998).

Why does the hidden curriculum matter?

Skelton (1997) observed that the "initial wave of critical research into the hidden curriculum... achieved its primary purpose of revealing covert political interference" (p. 177). Indeed, as early as 1978, Hargreaves (1978) posed the question: "from whom, one wonders, is the hidden curriculum still hidden?" (p. 97). However, arguably, many of the ways that education reinforces and reproduces inequalities remain relevant (Giroux, 1983; Kentli, 2009), and the ways in which the hidden curriculum is produced and maintained are poorly understood. Lack of awareness leads to what Willis (1978) describes as a "reductionist tautology," whereby actors within the system believe that "nothing can be done until the basic structures are changed but the structures prevent us making any changes" (p. 186). Furthermore, external forces bring in new ideas and practices. In particular, higher education has undergone a radical transformation as a result of the state-induced massification of the sector that has challenged traditional liberal perspectives by vocationalizing disciplines and placing increasing emphasis on assessment as the pivotal yardstick of academic achievement. As Apple (1980) observed, "the reproductive function [of education] posits a mirror relationship between the norms and values taught ... and those 'required' in the economy" (p. 47). In education this is partly manifested through equality and opportunity being replaced by efficiency and value for money (Ball, 1990). The formal curriculum in this new milieu is increasingly driven not only by the needs of employers but also by the wider socio-economic context of the "super-complexity" of the 21st century (Barnett, 2000). Exploring these tensions through the concept of the hidden curriculum may offer insights into how competing factors (such as the importance of the discipline, tradition and academic autonomy versus newer political and economic drivers) inform the student experience.

Within this context, what stimulates, motivates and maintains the hidden dimensions of higher education curricula continues to be pertinent, particularly considering the evolving student identity as both consumer and product of higher education. As Skelton (1997) asks: "Do current conceptions of curricula prepare students for the millennium? Do they reflect and anticipate the challenge of living in a world which is rapidly changing and becoming ever more complex, fragmented and uncertain?" (p. 177). To what extent are students able to make sense of and respond to exposure to contradictory information, values, beliefs and practices? Are they cognizant and critical or are they overdetermined, passive recipients of hidden curriculum messages? (Skelton, 1997). The importance of such questions is reinforced by the fact that the lessons the hidden curriculum teaches are experienced daily and embedded over the years students attend education (Bloom, 1972). Documenting features of the hidden curriculum thus remains an essential task in helping to understand the learner's experience of schooling and assists educators in challenging aspects of the hidden curriculum that may subvert the formal goals of higher education.

Why research the hidden curriculum in geography?

Recent years have seen renewed interest among geographers in the geographies of education and educating in geography. A specific aspect of pedagogy which might merit further investigation in geographical contexts is conflict in the curriculum. Many geographical topics are contentious or controversial in some way, and the way in which these are handled sends important messages to students about means of participation within unequal societies (Apple, 1990). The examples selected in this paper are of research in sustainability education—which is itself a contested topic—but the hidden curriculum is also integral to other aspects of geography such as development or population and migration studies to name but two. Anderson (2001, p. 30) notes three approaches to researching the hidden curriculum:

- Investigating forms of discrimination, indoctrination and cultural inculcation within the formal curriculum of education systems;
- Explaining the ways and extent to which the environment of educational institutions structures learning conditions and processes;
- Exemplifying "the unstated rules necessary for successful completion of formal education."

The first two aspects are particularly important for geographers and might include research on the processes of selecting topics, sources and perspectives for degree programmes; the teaching and learning contexts of higher education (lectures, seminars, fieldwork) and their impact on student experiences; and lecturers' beliefs and attitudes and their impact on teaching and learning.

One example of the way in which the hidden curriculum influences the selection or emphasis of topics in geography is the discipline's preoccupation with disciplinary assertion and with demonstrating the importance of geography to the understanding of sustainability issues (Purvis & Grainger, 2004). This may encourage a focus on facets of sustainability that most strongly illustrate the geographical dimension, either as a stated agenda or as an implicit theme. Thus, issues of scale, place, connection, difference, globalization, networks and uneven development may feature more prominently compared with economic, engineering and architectural studies of sustainability (Bailey, 2007, 2008; Castree, 2006). Although this may be partly in response to the expectations of students taking a geography degree, many students do not hold the same allegiance to their discipline as academics, and are more interested in issues *per se* than their geographical properties. This may create conditions for a "battle of ideologies" within the educational setting, or simply mutual misunderstanding whereby staffs are frustrated by students' reluctance to connect with their discipline. Part of the "weak" conception of the hidden curriculum involves lecturers' expectations about students' developing as geographers, an aspiration which may not be shared by the students themselves.

Although not exclusive to geography, the discipline also offers a rich arena for non-formal learning experiences. The contribution of fieldwork is one such context, of which the discipline is rightly proud, which acts as a catalyst for experiential learning (Fuller, Edmondson, France, Higgitt, & Ratinen, 2006). Although fieldwork may explicitly aim to elucidate specific processes or develop skills, and its experiential component might be intended (as part of a more deliberate curriculum unstated in the formal module outline) to increase students' passion for the topic, even seasoned field workers are often surprised at unexpected lessons gained (Dunphy & Spellman, 2009; Healey & Jenkins, 2000; Hope, 2009). These may range from the snippet of knowledge to the sensory experiences of visiting exotic places, for instance, discovering aspects of culture while conducting unguided questionnaire surveys in an unfamiliar locality. It is likely that many such connections remain unarticulated by students who nevertheless draw impromptu lessons which, rightly or wrongly, become lifelong beliefs.

Finally, there are issues related to the attitudes and values of academic staff teaching the subject who may be keen to instil passion for geography or specific sustainability issues. Even for studious observers of positional neutrality in education (and more directly, opponents of sustainability indoctrination), research findings suggest that the important social, ecological and economic issues which sustainability addresses cannot be treated with ambivalence (Cotton, 2006a). Given that the lecturer's perspective may substantially shape the worldviews of individual students, this is a crucial area for research. The aim of researching the hidden curriculum is to make the undisclosed agendas of formal education more detectable. Only by making the hidden curriculum visible can pedagogic researchers and educators better understand the structures which enable some students to succeed and others to be less successful, or the influence of teaching context on what students learn:

> once revealed, the hidden curriculum becomes negotiable and visible to all participants ... allowing for remediation, change, defence, improvement, and informed dialogue.... (Anderson, 2002, p. 117)

What methods might be used to research the hidden curriculum?

The hidden curriculum raises a number of issues for researchers. It is by definition tacit and not immediately accessible. A range of methods have been used in hidden curriculum research and it is difficult to generalize about 'most appropriate' methods. There is, nonetheless, some evidence in the literature that qualitative approaches may be more suited to researching the hidden curriculum than quantitative approaches, since the researcher is more open to unknowns and subtle distinctions (Vallance, 1980, p. 138). Margolis and Romero (1998), for instance, used wide-ranging interviews to allow respondents to "define the issues they faced" (p. 5) in graduate school. Although questionnaires can help to establish perception and behaviour trends (see Ahola, 2000;

Cotton, Warren, Maiboroda, & Bailey, 2007), they are less likely to yield the depth of response needed really to appreciate why and how aspects of the hidden curriculum take shape, diffuse and are perceived (Anderson, 2002). The fact that characteristics of the hidden curriculum are also highly dependent on context also militates against numerical generalization. Each educational institution is characterized by both shared and particularistic components (Lynch, 1989) and, moreover, impacts are experienced and interpreted differently depending on the drivers, actors and resources operating at particular points in space and time (Latour, 1987).

Among the qualitative methods used, observation is widely regarded as an important tool for revealing the nuances of the hidden curriculum. Lynch (1989) notes that through observational fieldwork, she was able to "get an insight into the school ethos which would not have been possible with postal questionnaires" (p. 36). It is notable also that Willis' (1978) work in schools was achieved by using an ethnographic approach with participant observation over significant time periods spent in the research context. Edwards and Westgate (1994) argue that observing tutor and students' talk is a crucial source of evidence in revealing both manifest and hidden curricula:

> As we listen and as we talk, we learn what it is necessary to know, do and say in that area of social life or that setting, and can display the competence necessary to be accepted as a member. (p. 15)

They nevertheless note that "formidable difficulties" exist in analysing natural conversations because of the sheer mass of data generated and the unstructured nature of observed talk (Edwards & Westgate, 1994, p. 16). Despite such practical problems, observation may still be more useful than interviews or questionnaires alone since hidden curriculum research entails the search for meanings and contexts which may not be immediately visible to actors in that context. The "most valuable research approaches will focus on in-depth interviews combined with observations", according to Anderson (2002, p. 125).

When combining interviews and observational methods, another approach which may prove useful is the use of stimulated recall. This technique, developed by Bloom (1953), aims to aid recall of events and thoughts using audio or video stimuli. Underpinning the method of stimulated recall is the premise that a respondent may be able to "relive" an event with vividness and accuracy if presented with clues or stimuli from the original situation (Bloom, 1953). In pedagogic research, this usually involves the researcher observing a lesson, lecture, seminar or online interaction, and then using an audio or video extract or a transcript of an online discussion to prompt discussion in interview. Although there is some controversy over whether using a think-aloud approach is more likely to access authentic thoughts or prompt post-hoc rationalization (Yinger, 1986), there are occasions when it can be beneficial in encouraging tutors to reflect upon specific teaching events or incidents.

The combination of observation and interviews is one example of the use of methodological triangulation to access the hidden curriculum. Methodological triangulation involves the combining of two or more data collection methods, and allows the researcher to check understandings of events with participants and note differences between data collected using different methods (Cousin, 2009). For example, research relying on interviews may risk accessing only "espoused theory" rather than "theory in use" (Argyris & Schön, 1974) and overestimate the occurrence of socially acceptable behaviours amongst the research population. Malone and Tranter (2005) note the importance of triangulation for enhancing reliability and validity, and also the potential for methodological triangulation to reduce the impact of researcher bias. Among the methods used by Malone and Tranter to explore the hidden curriculum in informal learning and play in the school grounds were drawings, behaviour mapping, interviews, photography, observation and surveys. Importantly, they also examined both pupils and teachers' perspectives, using participant triangulation to test their interpretations. One difficulty with accessing ideological patterns within an educational context is that the perspective of participants will generally mean that only certain aspects of situations are accessible to them. In teaching contexts, the perspectives of the students and the tutor may differ appreciably, but both (in isolation and conjunction) are useful in understanding manifestations of the hidden curriculum (as perceived by different individuals) and the multiple ways in which it can be transmitted.

In summary, a wide range of methods may—and arguably should—be used in conjunction to research the hidden curriculum. In the following examples we illustrate from our own research the ways in which different methodological approaches might be used in this context. The three examples focus respectively on:

- *Curriculum:* how choices of what to include and exclude from teaching form part of the hidden curriculum of higher education;
- *Pedagogy:* the ways in which teachers' views may form part of the hidden curriculum through pedagogic choices made;
- *Institution:* research examining the impact of institutional and campus environments on what students learn through the hidden curriculum.

In each example, we draw out the strengths and weaknesses of methods used for researching this area.

Example 1: Lecturers' views of sustainability in higher education (Curriculum)

Research methods used: questionnaires and interviews

In the first example (Cotton *et al.*, 2007; Cotton, Bailey, Warren, & Bissell, 2009) we aimed to explore lecturers' views on sustainability in the higher education curriculum. The research was undertaken as a single institution case study using mixed quantitative and qualitative methods. An online questionnaire was used for the first stage to gain a wide range of views from lecturers teaching sustainability issues across multiple disciplines. The major advantage of this method was that it yielded a large number of responses (328 in total) from which to undertake statistical analysis, but the disadvantage was that it provided relatively little information from each respondent and responses were somewhat constrained (although this was mitigated by using open-ended response items). The research also relied on self-reporting of beliefs and behaviours by respondents rather than first-hand observation of behaviours and practices as one would in an observational study. The quantitative approach nevertheless revealed some unexpected findings. For example, it was anticipated that lecturers in some disciplines would be inclined to see sustainability as more relevant to their discipline than would others (e.g. geographers and environmental scientists compared with historians or nurses). However, statistical analysis did not reveal statistically significant differences between disciplines in respondents' perceptions of the relevance of sustainability to their disciplines. The only correlation found was between a belief that sustainability was important in general and a belief that it was relevant to respondents' disciplines. The implication of this finding in terms of the hidden curriculum is that it suggests that the choice of what to include in the curriculum (or what it is relevant to include in the curriculum) is heavily mediated by lecturers' wider beliefs and attitudes.

So, rather than seeing a discipline as a set body of knowledge which needs to be imparted to students through the higher education curriculum, it should be viewed as a more flexible entity shaped by the interests, expertise and values of the individual tutors designing and delivering their curriculum. Whilst this aspect was not explored further in the study, there is a scope for further exploration about not only what is included in the curriculum of different disciplines in different institutions but also what is *not* included (the null curriculum).

The second part of the project involved interviews with a selection of lecturers who responded to the survey (Cotton *et al.*, 2009). Respondents were chosen to represent a wide range of disciplines, experience, gender and views on sustainability. The interviews provided further insights into the ways the hidden curriculum might function in higher education. One frequently encountered theme concerned the covert inclusion of sustainability in the curriculum. Staff talked about sustainability "creeping in the curriculum." When pressed to explain, they outlined a series of ways in which occurred, including: the use of examples to illustrate a more general issue; using small amounts of time during a lecture session to expound personal views about sustainability and the choice of texts or cultural artefacts that offered opportunities for commentary on sustainability issues. These respondents were teaching programmes which were not overtly about sustainability from reading documentation on the formal curriculum, such as definitive module records and module handbooks; yet tutors were finding ways of making links to sustainability in what they taught. The fact that words like "creeping," "surreptitious" and "undercover" were used indicates that lecturers were aware of this being a somewhat unauthorized aspect of their curriculum. During reflection on their teaching practices during interviews, many lecturers explained in more depth the ways they felt that their views might influence the curriculum.

Example 2: teaching controversial environmental issues in geography (Pedagogy)

Research methods used: observation and interviews

This research was conducted in secondary schools and focused on the teaching of controversial environmental issues in A level geography (Cotton, 2006a, 2006b). It used a qualitative approach that combined classroom observations (recorded using a radio-microphone) with interviews with pupils and teachers to enable triangulation of both methods and participants. It also utilized stimulated recall, using clips from audio-recordings or transcripts to prompt teachers and pupils to talk in more detail about specific classroom events. The advantages of combining observations with interviews were that the researcher had a record of shared experiences about specific events and could check the accuracy of respondents' recollections. This proved to be crucial since the pupil and teacher descriptions of certain situations were very different, even in terms of what was said and by whom, and even more so in the interpretation of events. In this context, it was useful to have a transcript of discussions alongside the interview data. It was also clear that the use of stimulated recall enabled teachers and students to provide their perspective on classroom events and interactions. Although these interviews undoubtedly involved an element of "meta-analysis and reflection that was most likely absent in the original event" (Yinger, 1986, p. 271), they were critical in identifying those elements of the hidden curriculum which were not clear to the teacher or researcher during fast-paced interchanges in the classroom.

One key finding from the study was that whilst teachers believed that they should maintain a neutral or balanced approach during classroom discussions of controversial environmental issues, this was problematic in practice and their own views were revealed as an element of the hidden curriculum. One potential explanation is that teaching controversial issues introduces a conflict within the normal rules of classroom interaction, which themselves form part of the hidden curriculum of all lessons. These rules, broadly speaking, are not only that the teacher controls what knowledge is deemed acceptable and relevant but also that the teacher should not impose his/her views on the students where "facts" are contested or uncertain or where values are involved (Edwards & Mercer, 1987). The result was that teachers' opinions were expressed in indirect ways, through questioning or through control of participants' turns in the discussion. For example, the use of rhetorical questions to challenge students' opinions enabled the teacher to maintain an air of neutrality (and open questioning) whilst actually expressing a personal viewpoint. On occasions, the teachers also strategically managed students' turns in discussion to ensure that a range of viewpoints (especially those they agreed with) were aired. They would ask specific pupils for their opinions on a topic, knowing that the selected pupil was likely to express an argument they wanted to be emphasized. Whilst the teachers used a variety of strategies to avoid expressing their own opinions, line-by-line analysis of observation transcripts, combined with interview data about the teachers' own views, revealed the impossible nature of this task. In contrast to their espoused beliefs about not expressing personal viewpoints, the data suggest that the influence of the teachers' attitudes was greater than they either intended or, in all probability, realised.

Example 3: Students' views of sustainability on campus (Institution)

Research methods used: video diaries and interviews

The third example is from recent research examining the ways and extent to which the university campus impacts on students' understanding of sustainability through informal learning experiences. This project was developed in response to increasing interest in the relationship between different curriculum dimensions and education for sustainable development (Jones, Selby, & Sterling, 2010) and illustrates the ways in which, particularly with sustainability, there is potential for what happens outside the classroom challenge what students are formally taught. Institutions often conduct aspects of their business and estates management in an unsustainable manner, for instance through poor energy management in buildings, lack of recycling facilities or the operation of procurement policies that prioritise cost effectiveness over issues such as fair trade or local and organic food products. Several authors argue that such contradictions result in students receiving discordant messages about sustainability (e.g. Djordjevic & Cotton, 2011; Hopkinson, Hughes, & Layer, 2008; Sterling, 2001). This tension between campus and curriculum has been used to advocate the whole institutional approaches to embedding education for sustainable development which have been pioneered in different formats at the Universities of Bradford (Hopkinson *et al.*, 2008), Gloucester (University of Gloucester, 2009) and Plymouth (Jones *et al.*, 2010). Nonetheless, there is still much work to be done at these institutions to ensure that sustainability permeates all aspects of university life.

The research in question employed an instrumental case study approach (Stake, 1995) focusing on a leading sustainability university. Using a broadly ethnographic approach and drawing on students as participant observers in their everyday context, the researchers obtained video diaries of sustainability as manifested in the campus environment and utilized semi-structured interviews using stimulated recall. The student researchers were asked to walk around campus and record their experiences on video with a commentary.

They were also asked to interview other students on video to elicit their views about sustainability on campus. The benefit of using students as researchers was that it provided a first-hand account of their views on the university campus, in contrast to the more widely reported views of academics or sustainability managers. Students responded well to the medium and the approach of combining observations with interviews provided a relatively undiluted, if not necessarily natural, focus on the student perspective. The student researchers were subsequently interviewed, using stimulated recall through clips from the video-diaries, about their experiences in order to elicit a detailed account of their perspectives and reflections on the research. Undertaking the videos enabled participants to identify evidence about sustainable and unsustainable practices in the university campus. They were also encouraged to reflect on their experiences and make connections between sustainability discourses presented in the classroom and the implementation of sustainability related initiatives within a large organization. This process generated critical dialogue about rhetoric and practice and, in some cases, was transformative for participants who reviewed their own perspectives and behavioural responses to sustainability issues. The research methods in this case proved to be a useful pedagogic vehicle for enhancing students' capacity to critique sustainability within an organization in transition (Mackie & Jones, 2010) and highlighted the tensions that exist between the formal curriculum and what students learn informally through the institutional environment.

Conclusions

The purpose of this paper has been to explore some of the benefits and pitfalls of different approaches to researching the hidden curriculum in higher education. Unsurprisingly, given the indistinct and subjective nature of the hidden curriculum, none of the research methods examined completely evade the hazards of navigating researcher preconceptions or the incomplete recollections and different interpretations held by "providers" and "recipients" of the hidden curriculum. However, experiences of research in this area generally lead towards the use of qualitative techniques and triangulation (particularly methodological and participant) to capture the meanings ascribed to events and situations by research participants. In particular, stimulated recall through the use of video diaries and observations proved useful in encouraging participants to recall and discuss specific elements of the hidden curriculum and their impact on both formal teaching practices and broader educational experiences.

The three examples given illustrate a range of ways in which pedagogic research can help to render the hidden curriculum more visible and comprehensible. Research in this area is unquestionably complex, not least because the hidden curriculum is not a singular concept but, rather, encapsulates a diversity of premeditated, inadvertent, transmitted and received "lessons" that intersect and co-exist with the formal aims and learning outcomes of degree courses across a multiplicity of disciplines and institutions. As such, researchers are inevitably forced to focus on particular facets of the hidden curriculum (in our case, sustainability) and particular contexts while at the same time being conscious that the foci, methods and research locations selected all involve elements of selectivity and, arguably, conscious or unconscious agendas that may influence the ways in which the hidden curriculum is construed and comprehended.

That said, the wider institutional and political changes currently shaping higher education make a clearer understanding of the "contemporary" and evolving hidden curriculum that are highly valuable and important part of understanding and managing educational processes. The impacts of these changes on geography as a discipline are difficult to predict, but it is worth reflecting on geography's variable funding status under the classification of STEM subjects (science, technology, engineering and mathematics), which, among other things, impact on teaching and staff–student ratios and hints at broader political (and institutional) influences on the curriculum through the expansion, contraction and refocusing of funding. Another topical area concerns the employability agenda, which has both placed pressure on curriculum time and shaped the orientation of degrees. New university fee structures and the trend towards more "enterprising" and partnership-oriented higher education institutions are equally constitutive of new rules for higher education affecting geography—and other less explicitly vocational disciplines. Further research into this area might profitably focus on the impact of changes to the funding climate and students' perspectives resulting from the increasingly consumer model being applied to higher education. One might plausibly suggest that the current changes may impact even more deeply on the hidden curriculum as certain disciplines and particular key skills are viewed as being more important than others. Given these pressures on higher education, a clearer understanding of the nature and impact of the hidden curriculum is crucial to how disciplines like geography shape their future.

References

Ahola, S. (2000). Hidden curriculum in higher education: Something to fear or to comply to? Paper presented at the innovations in higher education conference, September 2, Helsinki.

Anderson, T. (2001). The hidden curriculum in distance education: An updated view. *Change*, *33*, 28–35.

Anderson, T. (2002). Revealing the hidden curriculum of E-learning. In C. Vrasidas & G. V. Glass (Eds.), *Distance education and distributed learning* (pp. 115–133). Greenwich, CT: Information Age Publishing.

Apple, M. W. (1980). The other side of the hidden curriculum: correspondence theories and the labour process. *Journal of Education*, *162*, 47–66.

Apple, M. W. (1990). *Ideology and curriculum*. London: Routledge.

Apple, M. W., & King, N. P. (1977). What do schools teach? In R. H. Weller (Ed.), *Humanistic education* (pp. 29–63). Berkley, CA: McCutchan.

Argyris, C., & Schön, D. A. (1974). *Theories in practice*. San Francisco, CA: Jossey Bass.

Bailey, I. (2007). Climate policy implementation: Geographical perspectives. *Area*, *39*, 415–417.

Bailey, I. (2008). Geographical work at the boundaries of climate policy: A commentary and complement to Mike Hulme. *Transactions of the Institute of British Geographers*, *33*, 420–423.

Ball, S. (1990). Markets, inequality and urban schooling. *The Urban Review*, *22*, 85–100.

Barnett, R. (2000). Supercomplexity and the curriculum. *Studies in Higher Education*, *25*, 255–265.

Bloom, B. S. (1953). Thought processes in lectures and discussions. *Journal of General Education*, *7*, 160–169.

Bloom, B. S. (1972). Innocence in education. *School Review*, *80*, 333–352.

Bowles, S., & Gintis, H. (1976). *Schooling in capitalist America*. New York: Guildford Press.

Brandt, G. L. (1986). *The realization of anti-racist teaching*. London: Falmer Press.

Castree, N. (2006). Commentary: From neoliberalism to neoliberalisation: Consolations, confusions, and necessary illusions. *Environment and Planning A*, *38*, 1–6.

Cotton, D. R. E. (2006a). Teaching controversial environmental issues: Neutrality and balance in the reality of the classroom. *Educational Research*, *48*, 223–241.

Cotton, D. R. E. (2006b). Implementing curriculum guidance on environmental education: The importance of teachers' beliefs. *Journal of Curriculum Studies*, *38*, 67–83.

Cotton, D., Bailey, I., Warren, M., & Bissell, S. (2009). Revolutions and second-best solutions: Education for sustainable development in higher education. *Studies in Higher Education*, *34*, 719–733.

Cotton, D. R. E., Warren, M. F., Maiboroda, O., & Bailey, I. (2007). Sustainable development, higher education and pedagogy: A study of lecturers' beliefs and attitudes. *Environmental Education Research*, *13*, 579–597.

Cotton, D. R. E., & Winter, J. (2010). 'It's not just bits of paper and light bulbs': A review of sustainability pedagogies and their potential for use in higher education. In P. Jones, D. Selby &

S. Sterling (Eds.), *Sustainability education: Perspectives and practice across higher education* (pp. 39–54). London: Earthscan.
Cousin, G. (2009). *Researching learning in higher education: An introduction to contemporary methods and approaches*. London: Routledge.
Djordjevic, A., & Cotton, D. R. E. (2011). Communicating the sustainability message in higher education institutions. *International Journal of Sustainability in Higher Education, 12*, 381–394.
Dunphy, A., & Spellman, G. (2009). Geography fieldwork, fieldwork values and learning styles. *International Research in Geographical and Environmental Education, 18*, 19–28.
Edwards, D., & Mercer, N. (1987). *Common knowledge: The development of understanding in the classroom*. London: Routledge.
Edwards, A. D., & Westgate, D. P. G. (1994). *Investigating classroom talk*. London: The Falmer Press.
Fuller, I., Edmondson, S., France, D., Higgitt, D., & Ratinen, I. (2006). International perspectives on the effectiveness of geography fieldwork for learning. *Journal of Geography in Higher Education, 30*, 89–101.
Giroux, H. A. (1983). Theories of reproduction and resistance in the new sociology of education: A critical analysis. *Harvard Educational Review, 53*, 257–293.
Hargreaves, D. (1967). *Social relations in the secondary school*. London: Routledge and Kegan Paul.
Hargreaves, D. (1978). Power and the paracurriculum. In C. Richards (Ed.), *Power and the curriculum: Issues in curriculum studies* (pp. 97–108). Driffield: Nafferton.
Healey, M., & Jenkins, A. (2000). Kolb's experiential learning theory and its application in geography in higher education. *Journal of Geography, 99*, 185–195.
Hope, M. (2009). The importance of direct experience: A philosophical defence of fieldwork in human geography. *Journal of Geography in Higher Education, 33*, 169–182.
Hopkinson, P., Hughes, P., & Layer, G. (2008). Sustainable graduates: Linking formal, informal and campus curricula to embed education for sustainable development in the student learning experience. *Environmental Education Research, 14*, 435–454.
Jackson, P. W. (1968). *Life in classrooms*. New York: Holt, Reinhart and Winston.
Jones, P., Selby, D., & Sterling, S. (2010). *Sustainability education: Perspectives and practice across higher education*. London: Earthscan.
Kentli, F. D. (2009). Comparison of hidden curriculum theories. *European Journal of Education Studies, 1*, 83–88.
Lacey, C. (1970). *Hightown grammar*. Manchester: University of Manchester Press.
Latour, B. (1987). *Science in action: How to follow scientists and engineers through society*. Milton Keynes: Open University Press.
Lynch, K. (1989). *The hidden curriculum: Reproduction in education, a reappraisal*. London: The Falmer Press.
Mackie, M., & Jones, P. (2010). *The campus as a learning resource: A briefing paper*. Plymouth: University of Plymouth.
Malone, K., & Tranter, P. (2005). 'Hanging out in the schoolground': A reflective look at researching children's environmental learning. *Canadian Journal of Environmental Education, 10*, 212–224.
Margolis, E., & Romero, M. (1998). The department is very male, very white, very old and very conservative: The functioning of the hidden curriculum in graduate sociology departments. *Harvard Educational Review, 68*, 1–32.
Mullard, C. (1982). Multi-racial education in Britain: From assimilation to cultural pluralism. In J. Tierney (Ed.), *Race, immigration and schooling* (pp. 120–133). London: Holt International.
Portelli, J. P. (1993). Exposing the hidden curriculum. *Journal of Curriculum Studies, 25*, 343–358.
Purvis, S. & Grainger, A. (Eds.). (2004). *Exploring sustainable development: Geographical perspectives*. London: Earthscan.
Skelton, A. (1997). Studying hidden curricula: Developing a perspective in the light of post-modern insights. *Pedagogy, Climate and Society, 5*, 177–193.
Snyder, B. R. (1971). *The hidden curriculum*. New York: Knopf.
Stake, R. E. (1995). *The art of case study research*. London: Sage.
Stanworth, M. (1981). *Gender and schooling*. London: Hutchinson.
Sterling, S. (2001). *Sustainable education: Re-visioning learning and change*. Totnes: Green Books.

University of Gloucester (2009). Education for sustainability framework for the University of Gloucester., Retrieved from http://www.glos.ac.uk/about/sustainability/education/Pages/efs.aspx

Vallance, E. (1980). The hidden curriculum and qualitative inquiry as states of mind. *Journal of Education*, *162*, 138–151.

Willis, P. (1978). *Learning to labour*. New York: Colombia University Press.

Yinger, R. J. (1986). Examining thought in action: A theoretical and methodological critique of research on interactive teaching. *Teaching and Teacher Education*, 2, 263–282.

RESOURCES

Invitational Education: Theory, Research and Practice

MARTIN HAIGH
Department of Anthropology and Geography, Oxford Brookes University, UK

ABSTRACT *Invitational Theory argues that learning is enhanced when learners are positively encouraged or 'invited' into the educational experience. Arising from perceptual and self-concept theory, Invitational Pedagogy is constructed on four principles: respect for people, trust, optimism and intentionality, and upon five pillars: people, places, policies, programmes and processes. Its main contributions address the psychogeography and the constructive alignment of learning environments and the training of teachers and counsellors. Key research methods include the invitational/disinvitational index and the 5-P relay.*

Introduction: Invitational Thinking

Swami Vivekananda wrote, "Education is the manifestation of the perfection already in Humans . . . the only duty of the teacher . . . is to remove all obstructions from the way . . . That is our duty, to clear the way" (Vivekananda, 1894, in Vivekananda, 1989, Vol. 4, p. 358). In higher educational practice, all many learners need is an opportunity; they are self-motivated, driven by internal fire and self-belief in their personal goals. Others need help. They must feel persuaded that the time is right, the place is right, the people are supportive and that learning is good. If not, they may remain inactive, frozen and embarrassed until the mental obstacles that inhibit them are removed. Removing obstacles to engagement with education, this is the purpose of a learning invitation.

An invitation is a signal that requests companionship or consideration. Invitational Theory is an applied field that seeks, intentionally, to modify the sum of the (often non-verbal) signals and signs that affect human self-belief and to create from these a system of educational practice. Invitational Theory has roots in John Dewey's 'democratic ethos' (Dewey & Ratner, 1939), but it is constructed upon foundations in perceptual and self-concept theory (Purkey & Stanley, 1991). Its goal is positive intentionality; the constructive alignment of the messages transmitted by, and contained within, any given learning environment (Biggs, 2003). It aims to expunge the negatives that inhibit learners.

This paper introduces Invitational Theory as a framework that provides the ways and means of researching the geographies of learning places and especially the 'affectiveness' of learning environments, which is the way a learning environment makes you feel. It provides a way of recognizing why certain learning processes, such as fieldwork, successfully engage learners (Haigh & Gold, 1993; Meijles & Van Hoven, 2010). It provides a methodology for the intentional construction of learning invitations and illustrations of how the impact of these invitations may be evaluated. It introduces Invitational Theory as a new lens for research into Geographical Higher Education and indicates where it already intersects with the discipline. Finally, it contrasts Invitational Pedagogy with Appreciative Inquiry (AI), an apparently similar but mainly unrelated approach from the Positive Psychology stable and from the Action Research toolkit.

Invitational Research and Practice

Invitational Theory constructs itself on four principles: respect for people and their differences, trust expressed through cooperation and a sense of community, optimism about the untapped potential contained within each learner and intentionality, which means taking those steps needed to create each learning invitation and to address it to each learner (Purkey, 1991a; Purkey & Novak, 2008). In Missouri, Burns (2007) found that better performing schools display more invitational qualities and, qualitatively, that respect and trust are the qualities teachers feel most essential for creating an effective organization, while their leaders emphasized only 'trust'. In South Africa, Steyn (2003) used semi-structured interviews and focus groups to study the effect of introducing Invitational Practice into an award-winning Afrikaans medium school. She found that the keywords for the success of this cultural change were respect, trust, optimism and intentionality, with establishing trust through leadership as its central component (Covey, 2006).

Intentionality is the keyword of Invitational Theory. Many invitational learning environments are created through good practice unconsciously and many disappear the same way. The aim of Invitational Education is to understand what it takes to create and sustain invitational environments. This analysis is guided by Purkey's (1999; Table 1) five 'Powerful Ps' of invitational practice: people, places, policies, programmes and processes, whose invitational qualities must be woven into the culture of a learning situation or institution. At the institutional level, Fink (1992) admitted a sixth P–political will power. Collectively, these ally closely with three of Fink's (2008) familiar 'four fundamental tasks of teaching': 'designing effective learning experiences', 'interacting with learners' and 'supportive course management' but not so much with 'course content'.

Purkey and Novak (1996) expressed the 'P's as five principles: (1) people are able, valuable, responsible and should be treated as such; (2) educating should be a cooperative activity; (3) a Gandhian ideal, that there is no separation between the means and the end product; (4) that the role of the teacher is to help learners realize their untapped potential and (5) this can best be realized by places, policies, programmes and processes that are designed to invite development and people who intentionally invite engagement. They argued that when the 'P's evoke positive feelings in a learner, that person is said to be 'invited'. Equally, when they evoke negative feelings, that learner is 'disinvited'.

Table 1. Five powerful 'P's of Invitational Education (after Purkey, 1999)

P1. People	Instructors and learners work together as a family, with courtesy and respect, to build long-term relationships and to manage stress and conflict
P2. Places	Places exude a positive, cared for and caring atmosphere, clean, sustainability conscious, welcoming and including, so that those who use them feel some sense of ownership and of belonging
P3. Policies	Policy and policy-making respect people and their needs, provide an environment of mutual support and a learning community ethos
P4. Programmes	All programmes emphasize community engagement and service, wellness, well-being and the enrichment of all involved especially the learner and their self-development
P5. Processes	All management and operational processes function in inclusive, democratic ways that support the self-developing ethos of the learning community

Invitational educational research concerns the construction of the positive attributes of learning situations. For example, in a traditional Geography subject context, Haigh (2008) engaged in research that tries to reconnect learners with their sense of place, one of those five 'P's, by inviting them to analyse the way in which the different places interact with their emotional states. The context of this work is an upper-level Geography course, 'The Ethical Geographer', which explores ethics, emotional intelligence and empathy, environment and future employment (Boyd *et al.*, 2008; Mayer & Salovey, 1993). Haigh used fieldwork as a vehicle for exploring the feelings induced by different places and invited learners to reflect, creatively, on the way that different places affect the way they feel (Tuan, 2004). Building on precedents in Situationism and Psychogeography (Debord, 1969; Coverley, 2006), student teams go out to map parts of their local habitat in terms of their emotional impact. Responding to an intercultural understanding agenda, Haigh's method for this assessment is borrowed from India's Sāmkhya Philosophy (Harzer, 2005). Sāmkhya conceives the universe as the product of consciousness and among its core concepts are the three Gunas. In Sāmkhya, these Gunas, or 'Strands of Nature', are said to control and colour everything—much as dots of just three primary colours make up each colour photograph. The three Gunas are Sattva, which is light, pure, reflective and serene, Rajas which is active, creative/destructive and dynamic, and Tamas, which is inert, veiled and obstructing (Jacobsen, 1999). Learners, in teams and through discussion, create a map of their local environment using the Gunas as their palette. Later, individually, they are asked to consider what their group considered made one particular place inviting and another disinviting and why? They are then asked to consider how their personal habitat, often the campus or their college room, could be made into a more inviting place for either (Sattvic) reflection or (Rajasic) creativity and interaction. Research into the effectiveness of this intervention involved the analysis of end-of-semester questionnaires, the contents of team project report posters and individual reflective learning diaries. Participants reported enjoying the novelty of this exercise and, in some cases, that it initiated new creative thinking about the places they inhabit and the way they organize their lives (Haigh, 2008). Of course, Geographers may be allowed a special interest in 'Place' but all five 'Powerful Ps' and their interactions are legitimate targets for Geographical research.

Table 2. Four invitational levels (after Purkey & Schmidt 1990, pp. 21–30)

Intentionally disinviting	Demotivation is a powerful tool of control and it is widely used in academic contexts. Its 'toxic assets' include activities designed to discourage, demean, defeat, demoralize, disempower and discriminate against. Its manifestations include the constructive use of overloading, stress and assessment to promote insecurity, a focus on shortcomings and sanctions and all activities intended to undermine the self-concept and confidence of learners and teachers alike
Unintentionally disinviting	Inadvertent demotivation is a common outcome of target-driven systems that are often well-meant but obsessed with policies and procedures. Negative and discouraging messages are transmitted through control systems that display a lack of confidence in those involved, through systems that emphasize 'the stick', that create stress through overloading and that display disregard, disrespect, distrust, insincerity, pessimism and a lack of concern for those participating in the learning process
Unintentionally inviting	A motivational learning environment has appeared by chance. It is positive, supportive and collegial. It demonstrates respect, trust, optimism, self-belief, etc. in its human interactions. The only problem is that this outcome is a side effect, it is 'just the way it is', and there is no conscious effort or design in either its creation or preservation. It is thus vulnerable to inconsistencies and retrogression through inappropriate decision-making by those who, quite simply, do not comprehend what they manage and hence are incapable of either its maintenance or development
Intentionally inviting	A motivational learning environment is created by conscious attention to the four principles and five 'powerful Ps' of Invitational Education. Its ambience is optimistic, respectful and trustworthy. It is empowering, engaging and energizing. Its participants are enthusiastic, genuine, open and imbued with self-belief. Its managers are supportive, nurturing, democratic, ethical encouraging and self-aware. Its places are caring and cared for. Its true intention is to encourage every participant to make the most of their own potentials and this is understood by all

Invitational research concerns the construction of motivational learning situations and the detection and elimination of demotivational learning situations, both at the personal and the collective levels (Foote, 2010). These situations fall into four categories, which are summarized on Table 2, and which provide a framework. Together, these four categories provide a framework for research and analysis. Where used to explore the outcomes of learning interventions and to develop strategies for improving learning environments, this structure is often applied in the form of the inviting/disinviting index (IDI), a quantitative tool used to evaluate behavioural impulses in educational contexts. The IDI questionnaire is based on four subscales: inviting to self (self-confidence), inviting to others (e.g. congratulating others on success), disinviting to self (self-inhibiting) and disinviting to others (negativity, blame games, threats) (Schmidt *et al.* 1998). Table 3 displays these four levels expanded into questionnaire items. The example comes from a study that explores the relationship between invitational behaviour on the part of learners and various motivational and achievement indexes (Valiante & Pajare, 1999).

Table 3. IDI questionnaire subscales (after Valiante & Pajare, 1999) (with actual numbering of the question on the questionnaire)

Subscale: Inviting self
8. I am quick to recognize my own value
11. I plan time for enjoyable activities that I can do on my own
13. I congratulate myself on my successes
18. I forgive myself for my misbehaviour and mistakes
20. I am impressed with my own abilities

Subscale: Inviting to others
1. I like to include other people in enjoyable activities
4. I congratulate others on their successes
6. I forgive others for their misbehaviour and mistakes
9. I am impressed with the abilities of other people
15. I am quick to recognize the value of other people

Subscale: Disinviting self
2. I am hard on myself when I think I've done something wrong
5. I neglect my own needs
7. It affects me for a long time when I think I've done something stupid
10. I am overly critical of myself
14. I don't pay attention to my own needs

Subscale: Disinviting to others
3. I criticize others when I think it is needed
12. I neglect the needs of other people
16. I tell other people when I think they have done something stupid
17. I do not pay much attention to other people's needs
19. I blame others when I think they did something wrong

Here, statistical analysis found high positive invitational IDI scores to have an indirect correlation with achievement through correlations with motivation and self-regulatory practice (Valiante & Pajare, 1999). The study also found a direct negative relationship between self-invitational behaviour and age; younger learners scoring higher than older learners, an observation echoed in many other studies (Valiante & Pajare, 1999). However, in contrast to earlier research, this study found significant gender differences between male and female (K6–K8) learners, with the latter being more invitational and less disinvitational to others. This moved Valiante and Pajare (1999) to link their findings to the engendered 'ethic of caring theories' of Gilligan (1982). Elsewhere, an interview study of the factors that affected the success of female academics in Mathematics and Technology found that while self-belief was nurtured by a range of family, academic and work-related influences, the pivot was often "an invitation received from an influential person, often during a critical time, who helped them develop their beliefs about their capabilities" (Pajare & Zeldin, 1999, p. 60). Burns (2007) found no significant difference between the invitational qualities of male and female educational leaders.

Intersects Between Invitational Education and Geography

To date, Invitational Theory has had small impact on research in Geographical Higher Education, impinging mainly through teacher training work designed to create safer and more learner-supportive micro-geographies in secondary schools (Stanley *et al.,* 2004). However, it has greater influence on professional fields such as healthcare; hence,

MacDonnell and Andrews (2006) focused on invitational theory's exploration of the means of creating safe environments in sexual healthcare contexts using the perspectives of Feminist Geography.

Invitational Theory is also promoted as one of the ways of helping teachers achieve empathy and deal more effectively with human relationships, not least in cross-cultural contexts (Alberts *et al.*, 2010; Patterson & Purkey, 1993; Ivers *et al.,* 2008). Alverno College's (1995) innovative teacher training curriculum, which has been influential in Geography, used Purkey and Novak's (1996) methods to develop affective teaching that focuses on the positive development of each student's self-concept through the development of learning practices that invite active learner participation (cf. Haigh, 1996; Jenkins, 1996). Invitational Theory stresses three qualities in teacher development: empathy, non-possessive warmth and genuineness (Patterson & Purkey, 1993). This is why it has also been employed to deal with the ethical educational challenges created by the commercialization of educational institutes (Purkey & Novak, 1998) and as a best practice approach to the peer observation of teaching (Osborne & Purkey, 1995; Blackmore, 2005).

Using Invitational Theory in Pedagogic Research

From a Geographer's point of view, some of the most interesting applications of Invitational Theory relate to the wider problems of constructive alignment between the classroom message and the management ethos of institutions (cf. Biggs, 2003). Here, Invitational Theory has a major role to play (Bluestein, 2001). Typically, Invitational Theory's five 'P's and four levels (Tables 1 and 2) provide the framework for the generation of research hypotheses, which are explored by methods typical of qualitative and quantitative social science research (e.g. Steyn, 2006, phenomenology; Steyn, 2003—interviews; Asbill & Gonzalez, 2000—questionnaires and statistical analysis). However, in this institutional context, the most distinctive Invitational research tool is the 60-item *Revised Invitational Schools Survey*, which is also applicable to HE levels (Smith & Barnard, 2004). This standard questionnaire evaluates the influence of the five 'P's and the results from its application identify 'place', followed by 'people', as the leading factors governing the perception that an institution 'invites learning'. Smith and Barnard (2004) also suggested that the strength of such perceptions increases with respondent's age and possibly maturity.

Another distinctive and useful Invitational research methodology is the P5-Relay, which is a 'World Café' style exercise used to brainstorm the qualities and effectiveness of any particular educational situation. Five desks are set aside, one for each of the 'Powerful Ps' and a recorder set at the table. Next, groups of those who are stakeholders in the target environment move from table to table identifying the degree to that their current circumstances are inviting or disinviting. Once this is accomplished, the groups move on toward identifying solutions (Stanley, 2006; for method see Purkey, 1991b).

AI vs. Invitational Education: Comparison and Critique

Presently, when Invitational Education is introduced, mention is often made of 'Appreciative Inquiry'. AI is another member of the Positive Psychology stable but there seems to be little direct connection between it and Invitational Education or its literature

(Seel, 2008). AI is an approach to the management of organizational change while Invitational Educational Theory emerges from perceptual psychology and educational practice. However, it may be worth exploring the intersections of these two approaches.

Invitational practice is about creating a courteous and supportive educational habitat in which everyone flourishes. It aims to enable stakeholders to achieve their full potential through participation in a shared vision that is guided by optimism and finds diversity normal (Burns, 2007). It is often applied to the reform of educational organizations. By contrast, AI is all about the reform of organizations. It begins with the assumption that all organizations are alive with creative, constructive possibilities and it aims to apply a paradigm of conscious good that stimulates the coevolution of people and their organizations (Cooperrider, 1990). AI is a form of Social Constructionism, which locates meaning, value and action in social communications in interaction. Like Invitational Education, it is also a branch of Action Research. However, the AI key is the art of asking constructive, unconditionally positive, questions that help an organization understand and anticipate its potential. Where Invitational Education employs vehicles like the 5-P relay to help guide the reform of Educational systems, AI offers its more spiritual 4-D cycle, where the Ds are Discovery, Dream, Design and Destiny and inquiry derives from 'affirmative topic choice', where that topic is something that brings life to the organization. While Invitational Education emphasizes quiet affirmation, the slightly more wild-eyed AI stresses enthusiasm (cf. Bunch, 1982). In the language of Sāmkhya, Invitational Education is more Sattvic and AI more Rajassic (Haigh, 2008). Its practice, however, is grounded in the recognition of an organization's 'positive core'. It builds from positive examples in an organization's past, where 'history' is largely absent from Invitational Education thought. Its application involves the realization of a collective dream through new structures based on a new way of knowing, a new 'art of the possible' that AI discourses should generate. Its vehicle is the creation of a 'Positive Change Network' of transformative cooperation, and its success evolves from making organizational change that seems inspired rather than mechanical (Cooperrider *et al.*, 2003; Piderit *et al.*, 2007). AI is beginning to seep into pedagogic research (and hopefully in due course—practice), especially those parts that relate to administration, leadership and school improvement, but there is not yet a huge literature (Yballe & O'Connor, 2000, Willoughby & Tosey, 2007, Calabresse *et al.*, 2008).

Meanwhile, Invitational Theory is criticized for being apolitical, saccharine and romanticized (cf. Giroux, 1983). Although it is popular with classroom practitioners, critics argue that it inherits the individualist ethic of humanistic psychology and prefers social etiquette and self-concept to the discourses of ideology and power. It needs to be more than an apology for the *status quo* (Maclaren, 1986). By contrast, Shuayb *et al.* (2009) advised that AI only works if it is supported by deep commitment from all participants, especially leaders, and that it is unsuited for research into areas where participants have little experience or where there are problematic social issues—since it requires a new way of being that is confounded by negative attitudes (Elliot, 1999). This need for the collective positive is not shared by Invitational Education. Certainly, Invitational Education may be used to frame research across whole organisations but it is equally applicable to individual learning situations. More importantly, Invitational Education was designed specifically for work in negative environments and among those who are disempowered, alienated and disengaged.

Conclusion

Invitational Education is an approach to affective education that is constructed on the belief that encouragement and self-belief are critical factors in promoting better learning and that learning thrives more effectively in an environment that is supportive and caring. Its core concept is the 'Learning Invitation'—a courteous request to engage with education. However, the approach is a pedagogic contribution to a larger array of strategies that emphasize the social construction of self-esteem and enthusiasm (Ellis, 2001), including Positive Psychology (Boniwell, 2006) and AI (Cooperrider & Whitney, 2005), all of which involve positive thinking, participative and transformative action research and which tackle self-destructive tendencies in organizational culture.

For many years, Geography departments have, more or less subconsciously, deployed fieldwork programmes as their own particular learning invitation as their prominence at recruitment events bears witness. However, many Geography courses that employ opt-in inquiry-based or problem-based learning strategies have similar aspects (Pawson *et al.* 2006). Invitational Theory's main benefit for Geography is that it provides a framework that allows these learning invitations to be made intentionally—by design rather than by chance. In addition, it offers a range of research methods that may be adapted for the holistic evaluation of learning environments and the assessment of the effectiveness of any particular learning invitation. In science, a good hypothesis is said to be that which encourages the creation of new and useful questions. In this sense, Invitational Theory is good because it encourages researchers to address new questions and provides researchers with questions that need to be answered, such as what affects the engagement and disengagement of learners with different aspects of their learning?

Currently, while its heyday may be past, the front line in Invitational Education research involves attempts to substantiate its effectiveness claims. It is one thing to boost 'self-belief' or create intentionally invitational environments but far harder to measure its impact on learning outcomes beyond an intuition that "the harder you try the more likely you are to succeed" (Riner, 2003, p. 42). However, Invitational Education's holistic approach to the whole educational environment chimes in very closely with the concepts of constructive alignment (Biggs, 2003). It contains some simple, easily applied, research tools including the IDI, 5-P relay and the Revised Invitational Schools Survey questionnaire, which are useful for the diagnosis and transformation of learning environments. Its theoretical base is easily grasped and easily accessed, with much information available freely on the Web. Finally, the fact that it emphasizes 'place' and the micro-geographies of learning and teaching, makes it deserving of attention from those who would meld geography and educational practice.

References

Alberts, H. C., Hazen, H. D. & Theobald, R. B. (2010) Classroom incivilities: the challenge of interactions between students and instructors in the US, *Journal of Geography in Higher Education*, 34(3), pp. 439–462.

Alverno College (1995) *Handbook for Education Students Part II Conceptual Frameworks* (Milwaukee, WI: Alverno College). Available at http://www.alverno.edu/current_students/ed_handbook_2.pdf (accessed November 2008).

Asbill, K. & Gonzalez, M.-L. (2000) Inviting leadership: Teacher perceptions of inviting principal practices, *Journal of Invitational Theory and Practice*, 7(1), pp. 16–29. Available at http://www.invitationaleducation.net/journal/v7n1.pdf (accessed July 2009).

Biggs, J. (2003) *Teaching for Quality Learning at University*, 2nd edn (Buckingham: Society for Research into Higher Education & Open University Press).

Blackmore, J. A. (2005) A critical evaluation of peer review via teaching observation within higher education, *International Journal of Educational Management*, 19, pp. 218–232.

Bluestein, J. (2001) *Creating Emotionally Safe Schools: A Guide for Educators and Parents* (Deerfield, FL: Health Communications, Inc.).

Boniwell, I. (2006) *Positive Psychology in a Nutshell* (London: Personal Wellbeing Centre).

Boyd, W., Healey, R., Hardwick, S., Haigh, M., Klein, P., Doran, P., Trafford, J. & Bradbeer, J. (2008) 'None of Us Sets Out To Hurt People': The ethical geographer and geography curricula in higher education, *Journal of Geography in Higher Education*, 32, pp. 37–50.

Bunch, R. (1982) *Two Ears of Corn: A Guide to People-centered Agricultural Improvement* (Oklahoma City, OK: World Neighbors).

Burns, G. J. (2007) *Invitational Leadership in Public Schools* (Columbia, MO: University of Missouri Columbia, Faculty of the Graduate School, Doctor of Education Dissertation). Available at http://edt.missouri.edu/Winter2007/Dissertation/BurnsG-033007-D6624/research.pdf (accessed October 2009).

Calabrese, R., Roberts, B., McCleod, S., Niles, R. & Berry, M. (2008) Emerging technologies in global communication: Using appreciative inquiry to improve the preparation of school administrators, *International Journal of Educational Management*, 22, pp. 696–709.

Cooperrider, D. L. (1990) Positive image, positive action, in: S. Srivastva & D. L. Cooperrider (Eds) *Appreciative Management and Leadership*, pp. 91–125 (San Francisco, CA: Jossey-Bass). See also: D. L. Cooperrider, F. Peter, P. F. Sorensen, Jr., F. Therese, T. F. Yaeger & D. Whitney (Eds) *Appreciative Inquiry: An Emerging Direction for Organization Development* (Champaign, IL: Stipes). Available at http://www.stipes.com/aichap2.htm (accessed August 2009).

Cooperrider, D. L. & Whitney, D. K. (2005) *Appreciative Inquiry: A Positive Revolution in Change* (San Francisco, CA: Berrett Koehler).

Cooperrider, D. L., Whitney, D. K. & Stavros, J. M. (2003) *Appreciative Inquiry Handbook: The First in a Series of AI Workbooks for Leaders of Change* (Bedford Heights, OH: Lakeshore Communications).

Coverley, M. (2006) *Psychogeography* (Harpenden, Herts: Pocket Essentials).

Covey, S. M. R. (2006) *The Speed of Trust: The One Thing that Changes Everything* (New York: Simon & Schuster).

Debord, G. (1969) *The Society of the Spectacle* (London: Rebel Press).

Dewey, J. & Ratner, J. (1939) *Intelligence in the Modern World: John Dewey's Philosophy* (New York: Modern Library).

Elliot, C. (1999) *Locating the Energy for Change: An Introduction to Appreciative Inquiry* (Winnipeg, Canada: International Institute for Sustainable Development (IISD)). Available at http://www.iisd.org/pdf/appreciativeinquiry.pdf (accessed October 2009).

Ellis, A. (2001) *Research on Educational Innovations*, 3rd edn (Larchmont, NY: Eye on Education).

Fink, D. (1992) The sixth 'P'-politics, *Journal of Invitational Theory and Practice*, 1(1), pp. 21–27. Available at http://www.invitationaleducation.net/journal/v11p21.htm (accessed June 2009).

Fink, L. D. (2008) Evaluating teaching: A new approach to an old problem, *To Improve the Academy: Resources for Faculty, Instructional, and Organizational Development*, 26, pp. 3–21 (San Francisco, CA: Jossey-Bass).

Foote, K. (2010) Creating a community of support for graduate students and early acreer academics, *Journal of Geography in Higher Education*, 34(1), pp. 7–20.

Gilligan, C. (1982) *In a Different Voice: Psychological Theory and Women's Development* (Cambridge, MA: Harvard University Press).

Giroux, H. (1983) *Theory and Resistance in Education: A Pedagogy for the Opposition* (South Hadley, MA: Bergin & Garvey).

Haigh, M. (1996) Empowerment, ethics, environmental action: A practical exercise, *Journal of Geography in Higher Education*, 20, pp. 399–411.

Haigh, M. (2008) Coloring in the emotional language of place, *Journal of Invitational Theory and Practice*, 14, pp. 25–40. Available at http://www.invitationaleducation.net/journal/JITP%20V%2014%202008.pdf (accessed June 2009).

Haigh, M. & Gold, J. R. (1993) The problems with fieldwork: A group-based approach towards integrating fieldwork into the undergraduate geography curriculum, *Journal of Geography in Higher Education*, 17(1), pp. 21–32.

Harzer, E. (2005) Sāmkhya, pp. 8089-8093, in: L. Jones (Ed.) *Encyclopedia of Religion*, Vol. 12, 2nd edn (Detroit, MI: Macmillan Reference USA).

Ivers, J. J. Sr, Ivers, J. J. Jr & Ivers, N. N. (2008) Cross-cultural instruction, consciousness raising, and inviting heightened self-esteem, *Journal of Invitational Theory and Practice*, 14, pp. 11–24. Available at http://www.invitationaleducation.net/journal/JITP%20V%2014%202008.pdf (accessed June 2009).

Jacobsen, K. A. (1999) *Prakrti in Sāmkhya-yoga: Material Principle, Religious Experience, Ethical Implications* (New York: Peter Lang Verlagsgruppe & Delhi, Motilal Banarsidass).

Jenkins, A. (1996) Discipline-based educational development, *International Journal for Academic Development*, 1, pp. 50–62.

MacDonnell, J. A. & Andrews, G. J. (2006) Placing sexuality in health policies: Feminist geographies and public health nursing, *GeoJournal*, 65, pp. 349–364.

Maclaren, P. (1986) Interrogating the conceptual roots of Invitational Education—A review of Purkey and Novak's 'Inviting School Success', *Interchange*, 17(4), pp. 19–25.

Mayer, J. D. & Salovey, P. (1993) The intelligence of emotional intelligence, *Intelligence*, 17, pp. 433–442.

Meijles, E. & Van Hoven, B. (2010) Using the *rural atelier* as an educational method in landscape studies, *Journal of Geography in Higher Education*, 34(4), pp. 541–560.

Osborne, W. L. & Purkey, W. W. (1995) A model faculty peer review process for counselor education programme, *Journal of Counselling & Development*, 73, pp. 654–658.

Pajare, F. & Zeldin, A. L. (1999) Inviting self-efficacy revisited: The role of invitations in the lives of women with mathematics-related careers, *Journal of Invitational Theory and Practice*, 6(1), pp. 48–68.

Patterson, C. H. & Purkey, W. W. (1993) Preparation of humanistic teachers for next century schools, *Journal of Humanistic Education and Development*, 30, pp. 147–155.

Pawson, E., Fournier, E., Haigh, M., Muniz, O., Trafford, J. & Vajoczki, S. (2006) Problem-based learning in geography: Towards a critical assessment of its purposes, benefits and risks, *Journal of Geography in Higher Education*, 30(1), pp. 103–116.

Piderit, S. K., Fry, R. E. & Cooperrider, D. L. (2007) *Handbook of Transformative Cooperation: New Designs and Dynamics* (Stanford, CA: Stanford University Press).

Purkey, W. W. (1991a) *An Introduction to Invitational Theory* (Radford, NC: International Alliance for Invitational Education). Available at http://www.invitationaleducation.net/ie/index.htm (accessed June 2007).

Purkey, W. W. (1991b) 5-P Relay. Available at http://www.invitationaleducation.net/ie/PDFs/5-P%20Relay.pdf5 (accessed July 2009).

Purkey, W. W. (1999) *Creating Safe Schools through Invitational Education* (Greensboro, NC: ERIC Clearinghouse on Counselling and Student Services), ERIC Digest ED435946. Available at http://www.ericdigests.org/2000-3/safe.htm (accessed August 2007).

Purkey, W. W. & Novak, J. M. (1996) *Inviting School Success: A Self-Concept Approach to Teaching, Learning, and Democratic Practice*, 3rd edn (Florence, KY: Wadsworth, Inc).

Purkey, W. W. & Novak, J. M. (1998) An invitational approach to ethical practice in teaching, *The Educational Forum*, 63, pp. 37–43.

Purkey, W. W. & Novak, J. M. (2008) *Fundamentals of Invitational Education* (Kennesaw, GA: International Alliance for Invitational Education).

Purkey, W. W. & Schmidt, J. J. (1990) *Invitational Learning for Counselling and Development* (Ann Arbor, MI: University of Michigan, ERIC Counselling & Personnel Services Clearing House).

Purkey, W. W. & Stanley, P. A. (1991) *Invitational Teaching, Learning, and Living* (Washington, DC: National Education Association Analysis & Action Series). Available as ERIC Document ED340689 at http://www.eric.ed.gov/ERICDocs/data/ericdocs2sql/content_storage_01/0000019b/80/23/80/71.pdf (accessed August 2007).

Riner, P. S. (2003) The intimate correlation of Invitational Education and effective classroom management, *Journal of Invitational Theory and Practice*, 9, pp. 41–55. Available at http://www.invitationaleducation.net/journal/V9n0.pdf (accessed June 2009).

Schmidt, J., Shields, C. & Ciechalski, J. (1998) The inviting-disinviting index: A study of validity and reliability, *Journal of Invitational Theory and Practice*, 5, pp. 31–42. Available at http://www.invitationaleducation.net/journal/v5n1.pdf (accessed October 2009).

Seel, R. (2008) *Introduction to Appreciative Inquiry* (Chedgrave, Norfolk: New Paradigm Consulting), Available at http://www.new-paradigm.co.uk/ (accessed August 2009).

Shuayb, M., Sharp, C., Judkins, M. & Hetherington, M. (2009) *Using Appreciative Inquiry in Educational Research: Possibilities and Limitations* (Slough, UK: National Foundation for Educational Research Report). Available at http://www.nfer.ac.uk/nfer/publications/AEN01/AEN01.pdf (accessed October 2009).

Smith, K. H. & Barnard, J. (2004) The psychometric properties of the Invitational School Survey (ISS): An Australian Study, *Journal of Invitational Theory & Practice*, 10, pp. 7–26. Available at http://www.invitationaleducation.net/journal/v10n0.pdf (accessed June 2009).

Stanley, P. H. (2006) Using the 5 P Relay in task groups, *The Journal for Specialists in Group Work*, 31(1), pp. 25–35.

Stanley, P. H., Juhnke, G. A. & Purkey, W. W. (2004) Using an Invitational Theory of Practice to create safe and successful schools, *Journal of Counseling and Development*, 82(3), pp. 302–309.

Steyn, T. (2003) Adhering to the assumptions of Invitational Education: A case study, *South African Journal of Education*, 27(2), pp. 265–281, Available at http://ajol.info/index.php/saje/article/viewFile/44144/27659 (accessed July 2009).

Steyn, G. M. (2006) A qualitative study of the aspects influencing the implementation of Invitational Education in schools in the United States of America, *Journal of Invitational Theory and Practice*, 12, pp. 17–36. Available at http://www.eric.ed.gov/ERICDocs/data/ericdocs2sql/content_storage_01/0000019b/80/29/9d/1c.pdf (accessed July 2009).

Tuan, Yi Fu (2004) *Place, Art and the Self* (Santa Fe, NM: Center for American Places).

Valiante, G. & Pajare, F. (1999) The inviting/disinviting index: Instrument validation and relation to motivation and achievement, *Journal of Invitational Theory and Practice*, 6(1), pp. 28–47.

Vivekananda, S. (1894) What We believe In, in Vivekananda, Swami (1989), *The Complete Works of Swami Vivekananda*, 4, pp. 358 (Kolkata: Advaita Ashrama).

Willoughby, G. & Tosey, P. (2007) Imagine Meadfield: Appreciative Inquiry as a process for leading school improvement, *Educational Management Administration and Leadership*, 35, pp. 499–520.

Yballe, L. & O'Connor, D. (2000) Appreciative pedagogy: Constructing positive models for learning, *Journal of Management Education*, 24, pp. 474–483.

SYMPOSIUM

AQAL Integral: a holistic framework for pedagogic research

Martin Haigh

Department of Social Sciences, Oxford Brookes University, Oxford, UK

AQAL is a methodology for the holistic mapping of multiple perspectives and worldviews. Developed by Ken Wilber and colleagues, AQAL Integral analysis is founded in AQ quadrant mapping, which assesses four viewpoints for every situation. These are the Interior Subjective intentional (I) perspective, the Interior Collective cultural (We) perspective, the Exterior Individual (It, she/he) behavioural perspective and the Exterior Collective (They, many, its) social perspective. Each AQ map may be further elaborated as part of a many-layered system. AQAL Integral has a small but growing footprint in Education and, independently, in Geography – especially Urban Studies and GIS.

Introduction

Geography at large (and, equally, Geography pedagogy) is often concerned with evaluating different "ways of seeing" and different ways of interpreting the world (Cosgrove, 2008; Gold, Haigh, & Jenkins, 1993; Haigh, Revill, & Gold, 1995; Thrift, 2009). Such studies aim to help learners understand that different people perceive the world in very different ways, because of fundamental differences in their belief systems, cultural traditions, role, gender, social class and many more (cf. Pepper, 1942). By tradition, Geographers use maps to enable their understanding of the world, and this paper introduces a kind of mapping technique that grants a new perspective on the ways that different viewpoints and worldviews may be comprehended (Wilber, 2005). This approach is useful for pedagogic research because it allows an overview of the different perspectives that are involved in learning. However, the method applies equally well to other kinds of social interactions and, in the process, provides a holistic research method of considerable and largely unexplored potential (see Sima, 2010, on Cosgrove, 2008).

AQAL emerges from the ideas of the Integral philosopher Ken Wilber (2001). According to Reynolds (2006, p. 4) "Wilber's all encompassing integral philosophy has become a clarifying beacon for people throughout the world... [and]... 'Where Wilber's at'... is what's now known as the AQAL approach". AQAL is a means of mapping human experience that helps learners engage with pluralism and the interrelationships between different viewpoints and worldviews (Esbjörn-Hargens & Zimmerman, 2009; Wilber,

2000). The method offers researchers a route through the complexity of human interactions that permits integration, synthesis and finding "the One in the Many" (Filipsone, 2009, p. 105). The acronym "AQAL" means "All Quadrants All Levels, Lines, States and Types". At its heart is the four-quadrant AQ map.

The quadrants: AQ mapping

"AQ" mapping is an "elegant, infinitely adaptable, and panoramic stand-alone method for defining a problem. It is independently powerful and useful as a means to explore and describe the human territory of experience" (Cook-Greuter, 2005, p. 1).

Simply, an AQ map is a quadrant that separates four key viewpoints on human experience along two axes: individual or collective, interior or exterior (Table 1). The Individual Interior Upper Left (UL) quadrant concerns consciousness and the personal "what I experience" viewpoint while the Collective or Social Interior Lower Left quadrant (LL) is about community, society, social tradition (social dharma) and culture; it is about "what we should do". The Individual Exterior Upper Right (UR) quadrant concerns behaviour, namely "what he/she or it does", while the Collective or Social Exterior Lower Right (LR) quadrant contains the behavioural attributes and systems of the collective; it is "what they do" usually expressed in terms of empirical observations and statistics. Davis (2008) separates the subjective interior from objective exterior perspectives by suggesting that the first is more about seeking meaning and self-justification, often in a cultural context, while the exterior is more about analysing what is happening and why? Supporters of the AQAL "Integral" approach claim that a complete understanding of any human situation is possible only when all four of these viewpoints are considered. Table 1 explores these quadrants and the worldview that each contains.

The key to achieving a successful AQ mapping is contained by the tense employed (Wilber, 2005). If "I" is the subject then the UL quadrant is most appropriate and if "we" then the LL quadrant. If the subject is another individual who might be described in the

Table 1. Introducing the four quadrants of Integral Theory (after Wilber & Marrero, 2007).

	INTERIOR	EXTERIOR
INDIVIDUAL	*Intentional, the Upper Left (UL) or "I think" quadrant:* this contains the Individual Interior – the thoughts and beliefs, feelings, emotions and values of the individual, which are manifested as intentionality and ambition. This is the subjective, inward view, of the self.	*Behavioural the Upper right (UR) or "It–That, S/he does" quadrant:* this is the Individual Exterior. It is the individual seen from the outside, in the Third person and concerns all subjects that could be labelled as an objective individual 'It'.
SOCIAL	*Communal – Cultural, the Lower Right (LL) or "We think" quadrant:* this is the Social Interior, where the beliefs, values and culture of the collective manifest as the intentions and expectations of the class or academy as a whole. This is the realm of the cultural presumptions that define so-called hidden curricula. Research explores this through critical theory, ethnography or studies of cultural change.	*Systems, the Lower Left (LR) or "Many Its – They do" quadrant:* this is the Social Exterior – the external structures and inter-objective systems of the collective.

Table 2. Some geographies mapped in AQAL's four quadrants.

	INTERIOR	EXTERIOR
INDIVIDUAL	*Intentional, the Upper Left (UL) or "I think" quadrant:* Cultural Geographies that emphasise the individual "eye" or that is written in the first person; some Ethical and Moral Geographies; in Physical Geography: some experiential Deep Ecology.	*Behavioural the Upper right (UR) or "It–That, S/he does" quadrant:* Behavioural Geographies that operate at the level of the individual, biographical studies of Geographers or other key individuals; in Physical Geography: case studies of individual objects such as a glacier or river or events such as a tornado.
SOCIAL	*Social Cultural, the Lower Right (LR) or "We think" quadrant:* Social, Cultural and some Political Geographies, much of Development Studies, Area Studies and Critical Social Theory, Anthropological and Ethnographic studies, most Humanities oriented studies including studies based in Western or Eastern philosophies, in Physical Geography: some work on environmental perception and hazard management.	*Systems, the Lower Left (LL) or "They and Many It-s do" quadrant:* Economic and Transport Geographies, much Political and all Planning-oriented Geographies, most traditional Regional Geography, most of the "Quantitative", map-based or distribution-oriented Geographies, Geographical Information Sciences, and almost the whole of Physical Geography (pure and applied).

third person, "he, she or it", then the AQ quadrant is UR, or if there are many individuals and the appropriate pronoun is plural, "they" or "those", then LR. As Cook-Greuter (2005, p. 7) comments: "AQ represents seemingly universally used perspectives to describe human experience. There is no claim that these are the only possible perspectives, only that they are the most commonly observable . . . "

The process of AQ mapping may be illustrated by considering some of the subdivisions of Geography (Table 2). The bulk of traditional Geography, Human, Physical (see Esbjörn-Hargens & Zimmerman, 2009) and Geography Education, fits into the (social) scientific, collective exterior quadrant (LR). Appropriately, most is written in the detached third person (Haigh, in press). By contrast, aspects of modern Social and Cultural Geography tend to serve the (LL) interior collective quadrant that views social groups from an engaged or insider perspective, while the new wave of Affective Geographical studies that deal with feelings and emotions, sometimes in educational contexts, commonly address the feelings of individuals (UL). In these contexts, the subjective first person tense is often employed.

Case study of using AQ mapping in fieldwork

The educational value of AQ mapping is most easily demonstrated by means of a worked example. Table 3 summarises the comments of a class of Geography undergraduates who were set the task of reflecting upon the performance of a fieldwork exercise. This exercise involved a tree planting event, presently undertaken as part of a final year, Honours level, course called "The Ethical Geographer" (Boyd et al., 2010). The exercise is part of the "Education for Sustainability" component of this course and links with a wider discourse, namely Oxford Brookes University's recognition of "Global Citizenship" as one of the five "Graduate Attributes" it will inculcate in all of its learners (Haigh & Clifford, 2011). The exercise involves learners in planting trees, in composing and attaching a statement of their personal hopes for the future to some of

Table 3. Student AQ mapping of a fieldwork exercise: tree planting (Haigh, 2004).

	INTERIOR	EXTERIOR
INDIVIDUAL	I became more self-conscious of my actions and their effects (EL12). I gained a sense of achievement knowing I had done something for the environment (KA12). Viewing my performance, I believed that I was doing the task efficiently but not really thinking about the meanings behind the exercise (ME12). It meant a lot to me that I was outside giving something back to the environment (AL12). Self-satisfaction – giving back from that which I have taken (JH12). I was focused on the bigger picture of the task rather than results (JH212).	She worked with her team enthusiastically and engaged with her peers (KA12). He focussed on the task more than the group he was part of (ME12). That individual was not interested in this activity and just stood around, too lazy to plant trees (AL12). She stood and watched for a bit but later took a leadership role (SW12).
SOCIAL	Eventually, we realised the purpose of the task. First, we were implementing our deepest feelings – then doing our duty to Nature (CB12). We believed that planting trees was a sign of being good citizens and it was a pleasure for us to do this (AL12). We came together to plant as many trees as possible. It showed us what we need to do to conserve our environment (SW12). We enjoyed the experience but did not much explore the meaning behind it (ME12). We found this is time-consuming and wondered if it is really worthwhile? We utilised each other's strengths to do the best job possible (JH212). We hoped that, in few years time, we will be able to revisit our trees and feel a sense of pride (KA12).	The task demonstrated a community working together to achieve great results (KA12). They tackled the task well, assigned roles according to their different strengths although the discussion side was neglected (ME12). Why are not they working? (JH12). Everyone worked together to get the job done and do it well (AL12). Everyone paired off in twos so that everyone could be directly involved in tree planting (SW12).

the trees they plant, and in reflecting upon the significance of this tree planting for themselves, their fellows and their tutors (Haigh, 2004).

Management of this exercise is, intentionally, "hands off"; learners are shown how and where trees are to be planted and left to sort themselves out, both to do their work and to observe what they do as a group. In every replication, gradually, the students organise themselves into teams and set to planting. Typically, some work enthusiastically while some avoid working, some do their planting carefully while some rush to get away quickly, some try to understand what the exercise is about while some do not, some take the task seriously while some clown around. In 2012, the class was introduced to AQAL and asked to use this tool to evaluate their experience in an assessed report. Table 3 is a composite from the work of eight learners.

Table 3 illustrates how AQ mapping encourages learners to construct different viewpoints on their experience and, especially, to reflect upon the differences between what was going on inside their minds and what actually happened. Their interior observations show their engagement with the meaning of the exercise, both for themselves and for their group. By contrast, their observations of their own behaviour show them both connecting with the practical and managerial aspects of the task and trying to determine what they might have seen as an outside observer. Long ago, Burns (1786) wrote: "O wad some Power the giftie gie us, To see oursels as ithers see us! It wad frae mony a blunder

Table 4. Four Quadrants of Integral Theory: ways of learning – styles of research (adapted from Wilber, 2006 & Marrero, 2007).

	INTERIOR	EXTERIOR
INDIVIDUAL	*Intentional, the Upper Left (UL) contains the Individual Interior.* The ways of learning within this "I think" quadrant are experiential learning, contemplative inquiry, phenomenology and critical reflection. Research addresses the mind, aspirations, ethics and intentions of the learner or instructor, attributes which are normally concealed, but which may be exposed by individual interview or by the analysis of the reflections within learning journals.	*Behavioural, the Upper right (UR) contains the Individual Exterior.* The ways of learning about "that person or object over there" include empirical observation, practical application and the demonstration of capabilities. In education, this is the view of the examiner and evaluator. It concerns how the learner actually performs. In pedagogic research, this is usually explored through assessment and empirical observation.
SOCIAL	*Communal – Cultural, the Lower Right (LR) or "We think" quadrant contains the Social Interior.* Ways of learning about "what we think" involves the exploration of our own social meme. Our inter-subjective ways of learning about "us" embrace empathy, participation, community engagement, cultural/multicultural (trans-personal) understanding and critical analysis of cultural presuppositions. Educational research ranges from connective practices, to service/workplace learning, ethical participation, etc. and may be tackled by questionnaires, focus groups and so forth.	*Systems, the Lower Left (LL) or "Many Its – They do" quadrant contains the Social Exterior.* Ways of learning about "what they do" include analyses of the functionality, performance, sustainability, creativity, adaptability and general ecology of the educational system. This is the realm of the educational manager who focuses on examination and evaluation scores, and whose concerns include the published curriculum, contracts, performance, targets and statistics as well as questionnaire returns. These data are explored interminably by Government and national agencies as well as by the managers of individual institutions.

free us, An' foolish notion". AQ mapping has the power to encourage learners, self-consciously, to develop this same valuable gift.

Using AQ mapping in pedagogic research

AQ data, like that collected by the AQ mapping of the tree planting exercise, can prove a valuable framework for pedagogic research. However, this approach creates complexity because, inevitably, each viewpoint implies different priorities and different ways of knowing and so each of the four quadrants suggests the need for particular types of research method (Table 4). Thus, while the subjective individual intentions (UL) of a learner might be explored by interviews or perhaps a reflective learning journal, those of the collective (LL) might be better explored by questionnaires and focus groups. By contrast, exterior behavioural aspects might best be explored by objective observation for individual case studies (UR) and by questionnaires and statistical analyses for the collective (LR).

Of course, as Esbjörn-Hargens (2007) points out, in pedagogic research, it is possible to recognise two distinct Types of investigation depending upon whether the researcher approaches the topic as an insider or, as common in Geography, an outsider (Table 5). The approach of the insider is typically subjective, personal and emotionally involved, while that of the outsider tends to be objective, scientific and emotionally detached. Please consider: even those studies of Ethical and Emotional Geographies, which deal with

Table 5. Two types (eight zones) of pedagogic research (after Esbjörn-Hargens, 2007 & Davis, 2008).

	INTERIOR	EXTERIOR
INDIVIDUAL	*Intentional, the Upper Left (UL) or " I think" subjective quadrant:* Researched from the i*nside* – me – in the first person OR researched from the *outside* – looking into his/her/ its feelings and thoughts. *Sample Research Questions:* Inside: How do I justify my way of teaching? Outside: What are individual teacher's beliefs about and aspirations their way of teaching?	*Behavioural, the Upper right (UR) or "It, S/he does" objective quadrant:* Researched from the *inside* – the third person view of my own experience OR researched from the *outside* – third person view of individual behaviour. *Sample Research Questions:* Inside: What actions do I take in order to advance my teaching? Outside: What strategies do individual teachers use to advance their teaching?
COLLECTIVE	*Socio-cultural, the Lower Right (LR) or "We think" inter-subjective quadrant:* Researched from the *inside* – the first person plural view of our social group's interior thinking OR researched from the *outside* – looking inside their collective mindset as they perceive it. *Sample Research Questions:* Inside: How can we best develop our Geography curriculum to meet the demands of the UK Geography Benchmark statement (Craig et al., 2007)? Outside: How does the culture of the Geography discipline affect the way teachers' conceive their Geography offer?	*Systems, the Lower Left (LL) or "They do" inter-objective " quadrant:* Researched from the *inside* – a view of our own system OR researched from the *outside* – a third person view of the system. *Sample Research Questions:* Inside: What are the results of this team's attempts to refine the curriculum to meet the Geography Benchmark statement? Outside: What strategies do Geography Staff teams adopt to match their offer with the Geography Benchmark statement?

personal morality and feelings, can be explored from a dispassionate, detached, outside perspective (e.g. Smith, 2000), while equally, the dynamics of Higher Education systems can be explored from the engaged perspective of an insider (Haigh, 2008).

Validity claims in the AQ quadrants

Geographers, more than most, work in cross-disciplinary contexts and tend to be familiar with the problems that arise from the different validity criteria of different disciplines and, indeed, of different tribes of Geography. Sometimes, this is as culture-bound as the necessity of citing a particular discipline's academic "Gods" and their disciplic chains in order to make a paper publishable (Haigh, 2012). However, frequently, it is a matter of creating results that match with disciplinary-specific notions of what validates a conclusion. Table 2 suggests that most traditional Geographical discourse maps into the exterior collective (LR) quadrant. Other disciplines place greater emphasis on other quadrants, especially those of the subjective interior. Of course, different validity criteria are implied by the perspectives contained by each of the four AQAL quadrants. Wilber (1997a) has mapped some of these typical validity claims. His work highlights the differences between the Interior Quadrants, where intentional "means" are emphasised, and Exterior Quadrants, where objective "ends" are key (Table 6).

Naturally, this has practical implications for ethics and decision-making about what is right and what is not. However, Wilber argues that the AQAL approach also highlights major inadequacies in those ethical systems that do not engage with its four perspectives

Table 6. Some types of validity claims in the AQ quadrants (after Wilber, 1997a).

	INTERIOR	EXTERIOR
INDIVIDUAL	Subjective Truthfulness: honesty, integrity, conscientiousness, candour, trustworthiness, enthusiasm, etc.	Objective Reality: correspondence of words and actions, demonstrated professionalism, skills and capability, outcomes, reliability, popularity, etc.
COLLECTIVE	Intersubjective Justness: best cultural fit, appropriateness, ethical, rightness, empathy – mutual understanding, inclusivity, community acceptance, etc.	Interobjective Reality – Functional Fit: team or systems performance, outcomes, statistical validity, process model, systems analysis, structural-functional constructive alignment, efficiency, effectiveness, systems mesh, popularity, etc.

(Wilber, 2005). Ultimately, Wilber, Leonard, and Morelli (2008) suggest that the Integral approach implies moving beyond Ethic's "Golden Rule" of "treat others as you would wish to be treated", to a "Platinum Rule", which is "treat others as they would wish to be treated" (Wilber et al., 2008, p. 259).

All Levels, Lines, States and Types

AQ mapping lies at the core of AQAL analysis. However, it provides only the first step in a five-step process (All Quadrants, All Levels, Lines, States and Types). Returning, briefly, to the tree planting exercise, the learner comments in Table 3 contain several layers of complexity, which they were invited to unpack for their end of semester project reports. These reports discussed different engendered Types of behaviour. They also reflected on their own and their group's transient personal emotional States at the time of the planting as well as the effects of the State of the weather. They recognised different Lines of interpretation that depended upon whether they saw themselves as environmentally or politically aware or more concerned with science or society (cf. Gardner, 2007). However, more critically, some of their key reflections concerned their interpretations of different Levels of motivation, comprehension and consciousness among their peers (Wilber, 2005, 2006).

In this case, the levels identified by learners were conditioned by their reading (Beck & Cowan, 1996) This source encouraged them to identify and AQ map the viewpoints of their peers at several predefined levels (cf. Haigh, 2011). Collectively, they identified (1) those who undertook the exercise simply because they were told that this was work they must do; (2) those who also did the work because it provided an opportunity for them to achieve a good grade; (3) those who also welcomed the task because of its communal and sharing aspects; (4) those who also recognised that it gave them a personal opportunity to improve their world and (5) those who also recognised its symbolic support for the whole planetary system.

Here, each AQ map exists within a multilayered structure (Koestler, 1978). Cook-Greuter (2005, p. 15) speaks of "telescoping" the AQ map as "quadrants within quadrants within quadrants, all the way up, all the way down". For example, Table 5's AQ map of pedagogic research, with its two Types (insider versus outsider), could be addressed at three levels. The first might identify the perspectives of the learner(s), the second the instructor(s) and the third the staff developer(s) and/or discipline. At each level, the discourse and the character of research questions would be different. Compare, for

Table 7. Levels in graduate attribute statements.

Level	Graduate Attribute	Goal
Trained employee	Proficient in skills and practice.	Good technician
Educated employee	Skilled, reflective and ethical practitioner, innovator, entrepreneur, good technician, self-developer/ manager, etc.	Good professional
Educated citizen	Widely knowledgeable, good neighbour, caring, cultured, wise, cosmopolitan, responsible trustworthy, ethical, reflective, empathic, good professional, etc.	Good citizen
Educated leader	Creative thinker, innovator, emotionally intelligent, good citizen, inspirational motivator, etc.	Good leader

example, the discourse contained by the Journal of Geography in Higher Education, which is mainly devoted to instructor views of student learning (Haigh, 2012), with that of the International Journal for Academic Development, which is mainly concerned with the perspectives of educational developers on instructors (Clifford, 2009).

Table 7 explores a series of levels that may be identified in curriculum design. The levels it describes will be well known to those who have served in Higher Education Institutes that have been upgraded from technical or vocational training institutions, like the former British Polytechnics. However, the same levels are also visible in the statements of those many universities in Australasia and beyond that have redefined their mission in terms of graduate attribute statements (Haigh & Clifford, 2011). Of course, such Graduate Attribute statements say a great deal about the aspirations of the institution and Table 7 outlines four levels of aspiration that may be observed in such statements. These rise from passive technical training towards active and informed engagement. Of course, ideally, an effective fully functioning educational system should develop at many levels – even if those different levels have to be developed in different ways. However, the key point is that each higher level transcends and includes, rather than replaces, the one below. Such a series, which resembles sets of Russian "Matryoshka" Dolls, is typical of AQAL Integral analyses (Wilber, 1998).

The recognition of these nested AQAL Levels can be empirical. It is not necessary to adopt any particular multilevel system and, indeed, Wilber addresses the construction of AQAL Levels in many different ways. However, his ideas always engage with Koestler's "transcend and include" concept of the Holon, which is an open system explored as a layer in a hierarchical system (Koestler, 1978), and with Levels being part of an evolutionary spectrum of consciousness (Da, 1995; Pandit, 1995; Wilber, 1993). Currently, Wilber favours the 8–9-level evolutionary values hierarchy contained by what Business Schools know as "Spiral Dynamics" (Beck & Cowan, 1996; Graves, 1974; cf. Haigh, 2011; Wilber, 2006). However, Wilber commonly focuses on just three basic Levels, which echo those of Deep Ecology (Naess, 1987; Wilber, 2005).

Wilber's three levels of human moral development are an egocentric "preconventional" level, an ethnocentric (social-group-oriented) "conventional" level and a "post-conventional" world-centred level (Wilber, 2005), which Naess (1987) develops as the "Ecological Self". Importantly, these same three levels have been detected in large-scale surveys of the attitudes of college students (Schultz, 2001). However, Wilber (2005, p. 34) conflates these levels with another basic triad, Body, Mind and Spirit, and this is his basic framework (Wilber, 2000).

Table 8. Fieldwork case study: three levels in the four AQAL Quadrants (after Wilber, 2005).

	INTERIOR	EXTERIOR
INDIVIDUAL	*Level 1: Body: how I felt* *Level 2: Mind: what I thought I should do* *Level 3: Spirit: what I aspired to do*	*Level 1: What she/he did* *Level 2: What she/he thought they did.* *Level 3: What she/he was trying to do.*
SOCIAL	*Level 1: About Me in My Group* *Level 2: My Group: what we agreed together* *Level 3: All of Us: what we believed as a class*	*Level 1: Team integration – behaviour of components* *Level 2: Team performance as a whole* *Level 3: Performance of Class as a whole.*

Case study: multilevel AQAL mapping of a foreign fieldtrip experience

The following case study adapts Wilber's three levels for an analysis of student participant reflections on, another typical Geographical Education context, a foreign fieldtrip (Table 8). Here, the field trip is the one that lies at the heart of Oxford Brookes University's compulsory Geography core module "Geographical Enquiry and Field Research". This two-semester course enrols around 50 learners in their second year of study. Its curriculum focuses on research training in the first semester and data analysis in the second. The field trip runs between the two semesters and, from 2009 to 2011, it addressed Malta. During the field trip, students divide into small teams of four to undertake research projects that they designed and prepared during the previous semester, and that they will write up after the field trip finishes.

Twelve months later, these learners were asked to reflect upon their experiences during the field course, both positive and negative, and to consider what happened from the perspective of the project group and its leadership. The context for this reflection is an assessed report for the final 'Education for Employment' section of that same final year Geography 'capstone module' called the 'The Ethical Geographer', which was mentioned earlier in connection with the tree planting exercise (Boyd et al., 2010).

Reading through these "one-year-later" reports grants a perspective on the impact of the field-study location. Of course, the focus of the exercise was project leadership and teamwork, so the subject of the field course receives much less attention. However, the Maltese context emerges strongly only from those groups that engaged directly with the local community through working with local conservation teams or sharing concerns with festival organisers. Others seemed to have remained at a detached touristic distance from the island's culture and people.

The student reports, almost uniformly, suggest a focus on grades but experience suggested that some participants were more committed to hedonism. Apparently, their approach generated much annoyance among peers and teammates, which was far from obvious at the time. It is also notable that, while these learners recognise their drive for good grades, many other motivations are described, including self-development and the benefits of doing something worthwhile. Table 9 unpacks the multiple levels of Table 8 into three AQ maps, which are populated with representative quotations from 15 student participant reflections on this exercise.

Again, Table 9 displays major contrasts between the commentaries in the interior and exterior columns. The interior columns emphasise self-development, empathy and critical self-awareness, while the exterior columns focus on behaviour, managerial factors and

Table 9. AQAL Levels from Table 8–Quotations from student reports on leadership issues in their Geography Fieldtrip Malta Project Team.

Level 1		
	Interior: *Body: how I felt*	Exterior: *What she/he did*
Individual	Of all my activities, this was, personally, the most significant to me (AP12). I thought it would be a fun trip to go on and easy to pass (JD12). I was worried by the project topic and resolved to find out as much as possible (AW12). I realised that "difficult" work is only truly difficult when I do not have the patience to work out the answer. Understanding this has been very helpful (RH12).	This experience showed that before one can comprehend where another person stands, one must have the ability to realise what they perceive (RH12). An autocratic order was given putting another person into our group. This person had undesirable attributes including a poor work ethic and poor organisation that put everyone's final result at risk (MJ12).
	Interior: *About Me in My Group.*	Exterior: *Team integration – behaviour of components*
Collective	Whilst I do not necessarily intend to exert control, I find it difficult to trust anyone other than myself and show little empathy towards people who fail to live up to my work ethic. As a result, I feel that I am prone to undermining their abilities (KL12). I felt annoyed in interviews when some people did not ask relevant questions (AW12); I was inexperienced as a leader and as a result our project suffered. I found myself in this role due to the lack of experience within the group (JH12).	On the whole, the group cooperated well. It did well to harness negative situations and adapt to change by focussing on working as a team (KA12). Some members of the team were not working to their full potential – so the task was not done as effectively as possible (AW12). A fail on the horizon instilled panic and immediately all of the members were interested only in themselves (JD12).
Level 2		
	Interior: *Mind: what I thought I should do*	Exterior: *What she/he thought she/he did.*
Individual	I am energetic, extroverted and creative. I would enjoy the chance to do something different in my studies (RH12). I was aware of the importance of the task not just for the selfish purpose of my course and gaining marks but to the community, the task was doing good in the world (JH12), When one of my team underperformed, I would help them and not punish them as if they underperformed to purposely spite me (NK12).	Collectively, she/he made sure that a strong team was selected to ensure the good team dynamics vital for good outcome to our project (MJ12). She/he tried to look to the future and be prepared to cope with undesired events (KA12). Each team member each took a leadership role when their particular strengths were required (MJ12). She/he was supposed to administer questionnaires in Valetta but because of procrastination, this never happened (DT12).

Table 9. (*Continued*)

Level 2		
	Interior: *My Group: what we agreed together*	Exterior: *Team performance as a whole*
Collective	In Malta, we adapted our project to cope with data access problems (DT12). Our competitive desire to achieve triggered a prompt and prepared attitude and a passion to collect data of a high quality (KL12). This involved working as a team, collaborating to reach a consensus by sharing our experiences and exploring our feelings. To do this we first had to accept everyone as equals and value their opinions (EL12). As a group, we found there was a conflict in our feelings about what ought to be done (AW12).	As a group, we had to organise our own project before, then carry it out whilst in Malta, and finally write it up during term time. A natural leader evolved in the group who focused on getting the best mark but less on the goal of the project. As a result, a important details were missed by the entire team (JH12). The team was full of energy and inquisitive due to finding themselves in a new and unfamiliar environment (AP12). The group-work was a success, they completed the task and got the marks desired (AW12). Ruled by lecturers who had the ability to punish deviance and reward conformity, and despite being teased by peers for working hard on the trip, our group played by the rules and, as a result, was rewarded by really good grades (JHo12).
Level 3		
	Interior: *Spirit: what I aspired to do*	Exterior: *What she/he was trying to do.*
Individual	Everything on earth is in the eye of the beholder, and so it is possible to alter my perceptions for personal development/benefit (NK12). I wish to carry on developing myself in my personal/work life and believe I can only achieve this by remaining open to learning new things and empathising with others. Empathy is my key, core principle (RH12).	Given strict instructions on how to conduct the survey, she did her best while realising that this type of activity creates follower drones rather than those who think and act independently (EL12). The procrastination, laziness and arrogance of individuals affected the experience. I am all up for people having a good time but found it hard to deal with those who panicked about getting the work done when they did absolutely nothing in the field (JHo12).
	Interior: *All of Us: what we believed as a class*	Exterior: *Performance of Class as a whole.*
Collective	The importance of the environment, sustainability and interculturality is becoming an ever more important concern... education to understand the problems and protect resources is essential (EL12). Our determination was to respect and be sensitive to each other and the local communities and their ways of life (KL12). Awareness of the self comes when the individual is conscious of the awareness of the Other (RH12). Even if we did not realise it then, this experience was representative of our future roles in the workplace (JD12).	Our group work was beneficial. We participated in local voluntary work for the benefit of the local community and wider world (KA12). At University, marks act as a substitute for money and material goods, so everyone wanted to be credited for their work (DT12). We barely knew each other and the whole class was working together in a completely unfamiliar environment, thus all were working to survive (AP12). Much of the class had a negative attitude due to not understanding the purpose of what was being done (JD12).

outcomes. The three levels parse out comments related to observations of self and others from analysis and critical observation, and from final reflection upon aspirations and outcomes (Table 8). Naturally, if the focus of the written exercise in Tables 8 and 9 had been "Malta" or even "the importance of foreign fieldwork" rather than reflection on teamwork and leadership, the outcomes expressed may have been very different and required different AQAL levels.

AQAL Integral in Geography research precedents

The AQAL *Journal of Integral Theory and Practice* (JITP) is subtitled a post-disciplinary discourse for Global Action (Esbjörn-Hargens, 2009). However, AQAL Integral commonly intersects with subjects of interest to the Geography discipline. Currently, the key intersection emerges through Future Studies and the arts of scenario planning. For example, Straatsma et al. (2009) apply Integral and Spiral Dynamic ideas to scenario planning for river management on the Lower Rhine. Integral theory commends itself as an approach to scenario planning, including environmental scanning, because it overcomes the problem that most scenario planning ignores subjective perspectives and focuses excessively on external, empirically based factors (Slaughter, 2004; Stewart, 2008). Exploring the design of sustainable habitats, Esbjörn-Hargens and Zimmerman (2011, pp. 242–243) add: "Technical information alone cannot persuade people to act. Motivation arises when we experience a given environmental problem through two additional perspectives – subjective and intersubjective". This is why Wilber (2005, p. 42) argues that "anything less than an Integral approach to environmental problems is doomed".

This contention is elaborated in Esbjörn-Hargens and Zimmerman's (2009) tome on "Integral Ecology", which concludes with three environmental management case studies that could form a part of any Geography programme. One deals with sustainable community development in Central America, the second with fisheries management in Hawaii and the third with rainforest conservation in Canada. Elsewhere, Integral thinking has been applied to sustainability issues in Malta and Lebanon (Bell & Morse, 2005) and for creating a geography of sustainability worldviews (Lynam, 2012). A special "Climate Change" issue of JITP includes papers on economics, social change, civic response and the search for global solutions (Esbjörn-Hargens, 2009), while in Urban Geography, Marilyn Hamilton's award winning work on the Integral City addresses urban managers and civic activists (Hamilton, 2008). Hamilton applies Integral concepts, including AQAL mapping, to "meshworking and managing change" in the "human hive" (Hamilton, 2006, 2008). In each case, using the AQAL Integral approach ensures that analysis engages with the roles of the self, the other and the world in human affairs.

By contrast, Brian Eddy, one of the pioneers of Integral Geography, uses an ecologically layered "ecoAQAL" model based on three layers that climb up from the physical environment, through biosphere to an anthroposphere that includes the noosphere (Eddy, 2005). He develops the concept of levels in terms of GIS map layers and aspires to show that biophysical and social realities can be explored by Geographers within "Integral places" (Eddy, 2005, 2008).

Further afield, AQAL Integral has supporters in Deep Ecology and also Peace Studies because it emphasises holism, pluralism and spirituality, while Integral Education has its own thriving literature (Esbjörn-Hargens, Reams, & Gunnlaugson, 2010; Gidley & Hampson, 2008). Miller (2003) argues that AQAL Integral Education is important for the peaceful resolution of conflicts and problems due to cultural differences because it is education for an expanded, evolutionary, transcendent consciousness within which such tensions become subsumed.

Sri Aurobindo's Integral Education: the other school

In theory and practice, there are two, largely independent, species of Integral Education – although both are internationalist and both aspire to Universalism (Esbjörn-Hargens et al., 2010). The other Integral Education is that being developed by the followers of Sri Aurobindo. Jailed by the British as "the most dangerous man in India", anti-imperialist activist Aurobindo Ghosh (like Gandhi, Vinoba and many other of his contemporaries) used his time in prison for creative thinking (Sharma, 2002). He emerged with a new idea, a "Yoga of the Earth", which would do more than seek escape, as in many traditional systems. It would transform the Earth by changing the consciousness of its people (Anderlini-D'Onofrio, 2006; Anon., 2005). Sri Aurobindo reasoned: "The world knows three kinds of revolution. The material can bring strong results, the moral and intellectual are infinitely larger and richer in their fruits, but the spiritual are the great sowings" (Anon., 2005, p. 1; cf. Table 7).

Sri Aurobindo's system, based in the Vedic notion of the five sheaths, develops learners through five levels (Ranande, 2006). It begins with perfection of the body. It then addresses the "Vital" energy, which involves controlling the sense and channelling desires. Next, it aims to attune the "Psychic" mind through building powers of concentration, observation and attention as in Yoga. Neeltje Huppes' "Psychic Education" teacher training workbook employs a three-stage transformative learning process that includes information collection, assimilation and the shifting of meaning perspectives along "a sunlit" evolutionary path (Huppes, 2001). At higher levels, this system aspires to unfold higher transpersonal visions of the union of the Self-within-all-things and to develop the deep inward intuition of Truth Consciousness.

Instruction is organised around three core principles: (1) nothing can be taught – only unfolded from within; (2) each learner must be allowed to develop at their own pace and (3) instruction should proceed from the close and personal towards things that are more distant both in space and time (Das, 1997). Sri Aurobindo's chief disciple, The Mother, offers 11 goals for learners: to live mindfully, to aspire to self-improvement, to expect to progress, to live beyond one's own ego, to be true to one's self and choose one's destiny, to overcome the tyranny of selfish desires, to live by the rule of enlightened reason, to be self-disciplined, sensitive to others and respectful of genuine individuality and to try to develop as a whole person (Mukherjee, 2005).

An Aurobindo Integral Education emphasises the constructive alignment of a quiet and harmonious habitat, a learning community involving students, teachers and other stakeholders, as well as conscious recognition by all of their need for spiritual growth and personal transformation (Partho, 2007). As Geographer Anderlini-D'Onofrio (2006) reports, Integral Education is (and should be) a life changing experience.

Appropriately, this system's learning strategies and research methodologies all stress UL Individual Interior attributes; its developers employ experiential practitioner-based participant enquiry, "Integral Inquiry, Intuitive Inquiry, Phenomenological Inquiry, Organic Research, and approaches to the study of Exceptional Human Experience" (SACAR, 2009, p. 1). Its key research areas include a narrative mode of assessment, which aspires to replace grades (SACAR, 2009) and a tutor-facilitated, "Free Progress", instructional technique (Das, 1997). Sri Aurobindo's ideas hold a high place among India's Philosophers of Education (e.g. Chandra & Sharma, 2004; Sharma, 2002), but have made few inroads into Geographical Education (Anderlini-D'Onofrio, 2006).

Discussion

The strength of the AQAL system is that it embeds the subjective interior and so helps scope and contextualise research by setting an investigation within its larger (many perspective and many-layered) context. In Geographical Education research, it may have especial value as an exercise for the training of young researchers, who will find themselves stretched intellectually by the task of AQ mapping at different levels, and who, as a result, may find themselves with a deeper, more plural and hence more rounded, understanding of their research problem.

Supporters of the AQAL Integral Education framework agree that it helps learners develop a more holistic vision. In addition, its many-layered, "holarchical", structure facilitates both indexing and comparison of different viewpoints (Filipsone, 2009). Filipsone (2009, p. 104) concludes that AQAL's combination of "quantitative pluralism and developmental structuralism" offers a quasi-universal dimension to "merely hermeneutic, phenomenological or symbolist realist approaches" and advances AQAL as an approach to more transformative modes of education. However, after reviewing more than 100 papers on Integral Education, Fisher (2007, p. 15) concludes: "... the more I learn what we have done, the more I know ... how inadequately educators have handled ... integrating [AQAL Integral] with the field of Education".

Naturally, AQAL receives critical attention, albeit perhaps not yet sufficiently (see Wilber, 1997b; Visser, 2006). Much of the extant criticism is aptly described as "immature emotional rebellion within the Integral Movement" (Fisher, 2007, p. 15). In fact, AQAL has many problems. For a start, its inherently spiritual and progressive view of reality is problematic, especially for Western materialists (Markus, 2009). Even for those inclined towards Eastern spiritual worldviews, Integral Education's structuring of the evolution of consciousness may seem partial or slanted. As Meyerhoff (2010) points out, Wilber shares the Da-ist, Advaita–Vedanta influenced, vision of there being "One Truth" (Da, 1995, 2007). AQAL also suffers from "Ken Wilber's ... ambition to [explain] how everything came to be, fits together and what it all means ... a true Theory of Everything: matter, life, mind and spirit .... this absurdly ambitious attempt cannot immediately be dismissed .... But the fact that it does not succeed is not surprising given the enormous diversity of knowledge it must integrate ..." (Meyerhoff, 2010, p. 11.1). As Meyerhoff (2010, p. 11.1) continues: "the deep dualities which Wilber's integration seeks to remedy are not resolved. Subject and object, science and spirit, individual and social are mapped and juxtaposed, but not integrated". The quadrants (Lines, Types and sometimes Levels and States) remain forever separate. Ultimately, the main benefits of Wilber's work may lie in its applications to research training and to life as much as to academic discourse (cf. Volkmann, 2011; Wilber et al., 2008).

AQAL Integral, however, is notable for its syncretic approach to the world's cultural and philosophical plurality and for its attempt to create "a composite map, a comprehensive map, an all-inclusive Integral Map that included the best elements from all of them" (Wilber, 2005, p. 1). The system is cosmopolitan in its attempt to unify the complexity of the human worldviews, albeit cosmopolitan with both an Aquarian and an American tinge, but from its melting pot emerges something that is new, influential and powerful. The approach distils human world-making and experience into five "essential elements or keys", namely Quadrants, Levels, Lines, States and Types (Wilber, 2005, p. 1). The outcome, as Wilber adds, "is just a map but it is the most complete and accurate map that we have at this time" (Wilber, 2005, p. 2). However, Wilber extends his argument beyond the map metaphor. AQAL Integral is both a methodology and a way of world-making in its own right, and in Wilber's words, it suggests a new operating system for the human mind (Wilber, 2005).

Conclusion

AQAL, which means "all quadrants, all levels, lines, states and types", is a framework developed by Ken Wilber and colleagues for mapping perspectives, worldviews and globalised knowledge. The approach, which is widely applicable across disciplines, aims to provide a deeper, more holistic, understanding of social processes and institutions and its key contribution is its emphasis on the interior, subjective perspective. The approach may have especial value in pedagogic research training as well as in scenario-based futures studies and environmental management and planning.

Proponents of AQAL Integral Education describe themselves as "map-making in a landscape in flux" (Esbjörn-Hargens et al., 2010, p. 2). AQAL provides a framework for the systematic mapping of multiple perspectives across a flux of different worldview levels. Its four quadrants represent four perspectives: individual versus collective, interior versus exterior. Its "Levels" represent different holons (Koestler, 1978), different system levels of organisation or conceptualisation, while its "Lines" represent different developmental directions (Gardner, 2007), its "States" unusual, often time-limited, circumstances, and its "Types" different categories, such as gender (Esbjörn-Hargens, 2007).

AQAL Integral differs from the Integral Education system of Sri Aurobindo in its emphasis on the "transcend and include" thinking of Wilber and its pluralism. AQAL Integral espouses an array of possible mappings of educational levels, where Aurobindo's Integral admits only one.

AQAL's is missing a thorough and effective external critique, which means that users have to deconstruct its claims to universality for themselves. However, it is a partial, pluralistic, worldview that is slanted towards Dharmic theories of spiritual evolution and the existence of One universal consciousness. In addition, it is misnamed because it less integrates than helps juxtapose different viewpoints and worldviews. Despite this, its greatest strength and potential for pedagogic research in Geography emerges from its provision of a framework that encourages researchers to describe education research problems from multiple perspectives and to consider the differing worldviews of protagonists in educational processes.

Acknowledgements

I thank Lukas Scherak, University of Exeter, UK, for his comments on an earlier draft of this paper.

References

Anon. (2005). *A new education for a new consciousness: Sri Aurobindo and the mother on education* (2nd ed.). Puducherry: Sri Aurobindo International Centre for Education.

Anderlini-D'Onofrio, S. (2006). A city in the forest: Gaia in the postmodern contact zones of Auroville's wider intentional community in Tamil Nadu, India. *Spaces of Utopia: An Electronic Journal, 1*, 56–89. Retrieved August 20, 2011, from http://ler.letras.up.pt

Beck, D., & Cowan, C. C. (1996). *Spiral dynamics: Mastering values, leadership, and change*. Oxford: Blackwell.

Bell, S., & Morse, S. (2005). Holism and understanding sustainability. *Systems Practice & Action Research, 18*(4), 409–426. Retrieved September 5, 2009, from http://oro.open.ac.uk/148/1/Holism__understanding_Sustain.pdf

Boyd, W., Healey, R., Hardwick, S., Haigh, M., Klein, P., Doran, P., Trafford, J., & Bradbeer, J. (2010). 'None of us sets out to hurt people': The ethical geographer and geography curricula in higher education. *Journal of Geography in Higher Education*, *32*(1), 37–50.

Burns, R. (1786). *To a louse: On seeing one on a lady's bonnet at church*. Retrieved December 23, 2008, from http://www.robertburns.org/works/97.shtml

Chandra, S. S., & Sharma, R. K. (2004). *Philosophy of education*. Delhi: Atlantic.

Clifford, V. A. (2009). Engaging the disciplines in internationalising the curriculum. *International Journal for Academic Development*, *14*(2), 133–143.

Cook-Greuter, S. R. (2005). AQ as a scanning and mapping device. *AQAL: Journal of Integral Theory and Practice*, *1*(3), 1–17.

Cosgrove, D. (2008). *Geography and vision: Seeing, imagining and representing the world*. London: I.B. Tauris.

Craig, L., Simmons, I., & Souch, C. (2007). *Subject benchmark statements: Geography*. Cheltenham, UK: Quality Assurance Agency for Higher Education – QAA 155 02/07. Retrieved July 2011, from https://www.qaa.ac.uk/academicinfrastructure/benchmark/statements/Geography.asp

Da, A. (2007). *Not two is peace*. Middletown, CA: IS Peace 723.

Da, A. (1995). The seven stages of life. In A. Da (Ed.), *The hymn of the true heart-master* (pp. 215–229). Middletown, CA: The Dawn Horse Press.

Das, M. (1997). *Sri Aurobindo on education*. Delhi: National Council on Teacher Education. Retrieved August 31, 2009, from www.ncte-india.org/pub/aurobin/auro_0.htm

Davis, N. T. (2008). Integral methodological pluralism in educational research. Integral Theory Conference, Mount Pleasant, California, August, 2008, Paper Presented. Retrieved January 4, 2013, from http://nextstepintegral.org/wp-content/uploads/2011/04/Integral-Methodological-Pluralism-Nancy-Davis.pdf

Eddy, B. G. (2008). AQAL topology: An introduction to integral geography and spatiality. *AQAL Journal of Integral Theory and Practice*, *3*(1), 184–198.

Eddy, B. G. (2005). Integral geography: Space, place, and perspective. *World Futures: Journal of General Evolution*, *61*(3), 151–163.

Esbjörn-Hargens, S. (2009). Climate change (Part 1). *Journal of Integral Theory and Practice*, *4*(4), 1–140.

Esbjörn-Hargens, S. (2007). Integral teacher, integral students, integral classroom: Applying integral theory to education. *AQAL: Journal of Integral Theory and Practice*, 2(2), 72–103. Retrieved January 4, 2013, from http://nextstepintegral.org/wp-content/uploads/2011/04/Integral-Education-Esbjorn-Hargens.pdf

Esbjörn-Hargens, S., Reams, J., & Gunnlaugson, O. (2010). The emergence and characteristics of integral education: An introduction. In S. Esbjörn-Hargens, J. Reams & O. Gunnlaugson (Eds.), *Integral education: New directions for higher learning* (pp. 1–17). Albany: State University of New York.

Esbjörn-Hargens, S., & Zimmerman, M. (2011). Integral ecology: Design principles for sustainable human habitats. In E. C. Mare & M. Lindegger (Eds.), *Designing ecological habitats: Creating a sense of place* (pp. 240–249). East Meon, Hants: Permanent Publications – Gaia Education. Retrieved January 4, 2013, from http://www.gaiaeducation.org/docs/4Keys%20-%20Designing%20Ecological%20Habitats.pdf

Esbjörn-Hargens, S., & Zimmerman, M. (2009). *Integral ecology: Uniting multiple perspectives on the natural world*. Boston, MA: Integral.

Filipsone, A. (2009). The integral philosophy of Ken Wilber and religious education. In M. De Souza, G. Durka, K. Engebretson, R. Jackson & A. Mcgrady (Eds.), *International handbook of the religious, moral and spiritual dimensions in education* (pp. 93–106). New York: Springer International Handbooks of Religion and Education.

Fisher, R. M. (2007). *A guide to Ken Wilber and the education literature: Annotated bibliography* (Technical Paper 27). Vancouver, BC: In Search of Fearlessness Research Institute. Retrieved September 5, 2009, from http://integralecology.org/integralresearchcenter/sites/default/files/Fisher._A_Guide_to_KW_and_the_Education_Literature.pdf

Gardner, H. (2007). *Five minds for the future*. Cambridge, MA: Harvard Business School.

Gidley, J. M., & Hampson, G. P. (2008). Integral approaches to school educational futures. In M. Bussey, S. Inayatullah & I. Milojevic (Eds.), *Alternative educational futures: Pedagogies for an emergent world* (pp. 285–304). Amsterdam: Sense Publishers.

Gold, J. R., Haigh, M. J., & Jenkins, A. (1993). Ways of seeing: Exploring media landscapes through a field-based simulation. *Journal of Geography*, *92*(5), 213–216.
Graves, C. W. (1974). Human nature prepares for a momentous leap. *Futures*, *8*(2), 72–87.
Haigh, M. (2004). Planting hope: Supporting the Greener curriculum. *Planet*, *13*, 22–25. Retrieved June 8, 2012, from http://www.gees.ac.uk/planet/p13/p13_7.pdf
Haigh, M. (2008). Internationalisation, planetary citizenship and higher education Inc. *Compare: A Journal of Comparative Education*, *38*(4), 427–449.
Haigh, M. (2011). Transformative learning for global citizenship: Turning to turquoise. *Learning and Teaching in Higher Education (LATHE)*, *5*, 7–16. Retrieved August 31, 2011, from http://www2.glos.ac.uk/offload/tli/lets/lathe/issue5/Lathe_5_M%20Haigh.pdf
Haigh, M. (2012). Writing successfully for the Journal of Geography in Higher Education. *Journal of Geography in Higher Education*, *37*, 1–19, iFirst Article published online 14 June 2012. Retrieved January 4, 2013, from http://www.tandfonline.com/doi/abs/10.1080/03098265.2012.692158
Haigh, M., & Clifford, V. (2011). Integral vision: A multi-perspective approach to the recognition of graduate attributes. *Higher Education Research & Development*, *13*(5), 573–584.
Haigh, M. J., Revill, G., & Gold, J. R. (1995). The landscape assay: Exploring pluralism in environmental interpretation. *Journal of Geography in Higher Education*, *19*(1), 41–55.
Hamilton, M. (2008). *Integral city: Evolutionary intelligences for the human hive*. Vancouver, CA: New Society (D&M Publishers).
Hamilton, M. (2006). Integral metamap creates common language for urban change. *Journal of Organizational Change Management*, *19*(3), 276–306.
Huppes, N. (2001). *Psychic education: A workbook*. New Delhi: Sri Aurobindo Education Society.
Koestler, A. (1978). *Janus: A summing up*. London: Hutchinson.
Lynam, A. (2012). Navigating a geography of sustainability worldviews: A developmental map. *Journal of Sustainability Education*, *3*, 1–14.
Markus, T. (2009). *Two roads diverging: Integral theory and contemporary science*. Retrieved January 4, 2013, from http://www.integralworld.net/markus2.html
Meyerhoff, J. (2010). *Bald ambition: A critique of Ken Wilber's theory of everything*. LaVerne, TN: Inside the Curtain Press. Retrieved January 4, 2013, from http://www.integralworld.net/meyerhoff-ba-toc.html
Miller, R. (2003). Education for a culture of peace. *Encounter: Education for Meaning and Social Justice*, *16*(1), 1–15. Retrieved January 4, 2013, from http://www.pathsoflearning.net/articles_Education_Culture_Peace.php
Mukherjee, J. K. (2005). *Principles and goals of integral education*. Puducherry: Sri Aurobindo Ashram.
Naess, A. (1987). Self-realization: An ecological approach to being in the world. In Glasser & A. Drengson (Eds.), (2005) *The Selected Works of Arne Naess, Volume 10: Deep Ecology of Wisdom* (pp. 515–530). Dordrecht, NL: Springer. Retrieved January 4, 2013, from http://www.springer.com/philosophy/book/978-1-4020-3727-6
Pandit, M. P. (1995). *A summary of Savitri*. Puducherry: Dipti, Sri Aurobindo Ashram.
Partho (2007). *Integral education: A foundation for the future*. New Delhi: UBS Publishers.
Pepper, S. C. (1942). *World hypotheses: A study in evidence*. Berkeley, CA: University of California.
Ranande, S. (2006). *Introduction to integral education: An inspirational guide*. Auroville, Puducherry: Sri Aurobindo Institute of Educational Research.
Reynolds, B. (2006). *Where's Wilber at? Ken Wilber's integral vision in the new millennium*. Saint Paul, MN: Omega Books – Paragon House.
SACAR (2011). *SACAR-IGNOU Programmes*. Puducherry: Sri Aurobindo Centre for Advanced Research. Retrieved January 4, 2013, from http://www.theuniversityoftomorrow.org
Schultz, P. W. (2001). The structure of environmental concern: Concern for self, other people, and the biosphere. *Journal of Environmental Psychology*, *21*(4), 327–339.
Sharma, R. N. (2002). *Neo vedanta education*. Delhi: Subhi.
Sima, Y. (2010). Geography and vision: Seeing, imagining and representing the world. *In-Spire Journal of Law, Politics & Societies*, *5*(1), 99–100.
Slaughter, R. (2004). *Futures Beyond Dystopia: creating social foresight*. Abingdon: Routledge.
Smith, D. M. (2000). *Moral Geographies: Ethics in a world of difference*. Edinburgh: Edinburgh University Press.

Stewart, C. C. (2008). Integral scenarios: Re-framing theory, building from practice. *Futures*, *40*, 160–172.
Straatsma, M., Schipper, A., van der Perk, M., van den Brink, C., Leuven, R., & Middelkoop, H. (2009). Impact of value-driven scenarios on the geomorphology and ecology of lower Rhine floodplains under a changing climate. *Landscape & Urban Planning*, *92*(2), 160–174.
Thrift, N. (2009). 5 space: The fundamental stuff of human geography. In G. L. Gill Valentine & S. L. Holloway (Eds.), *Key Concepts in Geography* (2nd ed., pp. 85–96). London: Sage.
Visser, F. (2006). *A spectrum of Wilber critics*. Retrieved January 4, 2013, from http://www.integralworld.net/visser11.html
Volkmann, R. (2011). Book review: Jeff Meyerhoff, Bald ambition: A critique of Ken Wilber's theory of everything. LaVerne, Tennessee, inside the curtain press, 2010. *Integral Leadership Review*, *11*(3), BR2. Retrieved January 4, 2013, from http://integralleadershipreview.com/2706-book-review-5
Wilber, K. (2006). *Integral spirituality*. Boston, MA: Integral Books.
Wilber, K. (2005). *The integral operating system, version 1.0. IOS users guide*. Boulder, CO: Sounds True Inc.
Wilber, K. (2000). *A theory of everything: An integral vision for business, politics, science and spirituality*. Boston, MA: Shambhala.
Wilber, K. (1998). *The essential Ken Wilber: An introductory reader*. Boston, MA: Shambhala.
Wilber, K. (1997a). An integral theory of consciousness. *Journal of Consciousness Studies*, *4*(1), 71–92. Retrieved August 31, 2011, from http://www.imprint.co.uk/Wilber.htm
Wilber, K. (1997b). *The eye of spirit: An integral vision for a world gone slightly mad*. Boston, MA: Shambhala.
Wilber, K. (1993). *The spectrum of consciousness* (2nd ed.). Wheaton, IL: Quest Books Theosophical Publishing House.
Wilber, K., Patten, T., Leonard, A., & Morelli, M. (2008). *Integral life practice*. Boston, MA: Integral.

SYMPOSIUM

Is Q for you?: using Q methodology within geographical and pedagogical research

Paul N. Wright

Maritime and Technology Faculty, Southampton Solent University, Southampton, UK

Q methodology is receiving growing interest as a research technique. Its use of factor analysis and a well-defined methodological approach may appeal to quantitative researchers who need to investigate issues more commonly associated with qualitative techniques. Such issues frequently arise when investigating student learning. This paper suggests that, appropriately used, Q is a methodology to be embraced by quantitative and qualitative researchers alike. Drawing on work within the field of education, as well as two case studies, the method's utility is illustrated, and Q is suggested as another potential method for investigating the subjective views of our students.

Discovering Q – a personal story

"Qualitative data are sexy" (Miles & Huberman, 1994, p. 1)

Eden, Donaldson, and Walker (2005) identified that the growth of qualitative methodologies within human geography, such as focus groups and in-depth interviewing, sometimes conflicted with the desires of research end-users and funding bodies, particularly in the area of policy development. Here, they commented, those who wanted to use the results of research for further development often sought to legitimize the work through the insistence of the presence of numerical data. In many ways, this observation echoed an earlier geographical debate, prompted by Massey (1999), which identified the role of "physics envy" within certain sections of the geographical community. Here, she focused upon the "soft" versus "hard" debate amongst geographers, usually conceptualized as the human/soft versus physical/hard divide. At the heart of this debate, it would seem, is the way in which human and physical geographers often work with and collect data, partly enshrined within the quantitative/qualitative "paradigm war" (Denzin, 2010).

So, if your background, like mine, is one of quantitative science, describing environmental and geographical phenomenon in terms of quantities, variables, numerical descriptions of state, then you might, like me, find an assertion such as Miles and Huberman's uncomfortable. In particular, faced with a pedagogical research question, you may feel that the use of a range of qualitative techniques you need to collect information to explore your question is disconcerting, problematic, but, in the end, essential. Let us face it, fellow quantitative researchers, we are unsexy!

Conflicted, I chose to carry on working with qualitative data, trying to understand the processes that researchers in the field used and, in particular, how they worked through their analysis of data. This was brought into sharp relief when I worked on a project looking at the conceptions of active learning, with a colleague well versed in using personal narrative as a form of enquiry (Wright & Romer, 2008). It took me some time to come to terms with the fact that issues of validity, precision (in the sense of repeatability), representativeness – in fact all of the things someone with an alleged reductionist/ positivist approach to quantitative data felt were important – looked and felt different to a qualitative researcher.

Whilst doing some background research into enquiry-based learning (EBL), I came upon a paper by Tim Deignan (2009), which mentioned Q methodology. Inquisitive, I made for the Internet, and discovered a technique that was at once described as both quantitative and qualitative. However, most of all, it appeared to solve a number of my qualitative research dilemmas. There seemed to be a more imaginative use of what, seemingly, appeared to be Likert-type data, beyond the "8 out of 10 were satisfied or greater" analysis I usually saw. At the same time, the technique seemed to give structure to the analysis of qualitative data, since the interpretation was driven by the statistics rather than my reading of the text. Timewise, and philosophically, it seemed to make more sense to me.

Due to its relatively slow diffusion across the Atlantic, and its birth in the realms of the human sciences, Q and geography have intersected less often than they might. The emphasis in the literature is often around environmental management and stakeholder analysis research, most regularly demonstrated in studies from North America. There is evidence that the methodological landscape is changing towards a more geographical diverse community of practitioners, and studies on place attachment (Hutson & Montgomery, 2011), footpath development (Hawthorne, Krygier, & Kwan, 2008), forest management (Steelman & Maguire, 1999), tourist experiences (Fairweather & Swaffield, 2002), attitudes to genetically modified crops (Hall, 2008), the environmental knowledge and politics of local communities (Robbins, 2006) and the views of science in conservation (Cairns, 2011) show the wide number of contexts in which Q has been used.

Q Methodology – a brief overview

Q methodology (or Q) was developed by William Stephenson in the 1930s. Stephenson was a holder of a PhD in Physics but was working in the laboratories of psychologists and statisticians like Spearman and Burt. Spearman's work was concerned with trying to establish patterns in data on individuals' performances on a range of tests of intelligence, attempting to derive a single factor, "g", which might describe human intelligence (Gould, 1996, p. 286). Stephenson's work, however, was not concerned with the measurements of people, but of people's measurements of themselves. Like the repertory grid work of Kelly (1955, cited in Jankowicz, 2004), Stephenson became much more interested in the personal constructs and subjective values of test subjects rather than their performance in "objective" tests. Thus, in Q, the subjects themselves become less important (as in more traditional, so-called, "R analysis"), it is their relative subjective views on a question that is of interest.

Q follows a series of relatively distinct phases, arriving at a set of factors that are judged to represent a variety of participant viewpoints. These methodological phases are the development of the concourse (defined below), a sifting of the concourse into a set of statements for survey use (called the Q sample), identification of participants (called the P-set), sorting of the Q-set by participants, data analysis and interpretation of factor solutions. There appears by a relative paucity of "how to" texts available for anyone interested in using Q; the oft-quoted book appears to be Steven Brown's "Political Subjectivity" (1980), which is out of print, but now available online. Fortunately, there are a number of primers and critiques (e.g. Dziopa & Ahern, 2011; van Exel & de Graaf, 2005; Webler, Danielson, & Tuler, 2009) that are freely available through the Internet. Whilst these are often strong on the practical side of the technique, there is less information available around the use of factor analysis, which, for the uninitiated, is perhaps the trickiest area of Q.

Concourse development and arrival at the Q sample

The concourse has been referred to as the "stuff of life" (Brown, 1991). It is, at its simplest and broadest, a collection of statements that encompass all views about the subject under scrutiny. As such the concourse is large, and the provenance of the statements can come from all walks of life: scholarly articles, blogs, face-to-face interviews, direct questions and personal opinion. These statements are usually sorted and thematically grouped. Repetitive statements are removed, and confusing statements can be reworded for clarity. In particular, it is useful here to make sure that statements that express more than one view are split into different statements, so that participants do not feel that accepting a statement is conditional upon more than one value judgement. Other authors tend to conform to a more rigid experimental design, identifying a range of dimensions and make sure that each dimension is represented equally (McKeown & Thomas, 1988).

Once these statements have been sifted into a range of themes, and final edits made, a set of statements is arrived at (called the Q sample). It is often useful to share this Q sample with a number of individuals unlikely to act as participants and to act as a pilot group.

The P-set and the Q sort

The P-set is simply the name given to the group of participants asked to provide a Q sort. The selection of the P-set is entirely consistent with the kinds of decisions made regarding the formation of sample groups in questionnaire design, in as much as the P-set membership should reflect a body of participants who are "theoretically salient" to the issue under study (Brown, 2010). Some studies may use rigorous sampling techniques, stratifying the P-set according to a range of factors that may have salience with the research aim. Other studies may use convenience samples, perhaps where the research question is framed in a more general, exploratory way. van Exel and de Graff (2005) suggest that the formation of P-set is not random; it is purposive such that there are enough participants to define a factor which can be readily compared with other factors extracted from the data. They suggest that having four or five participants associating with each factor (or viewpoint) is sufficient, and the number of the participants rarely is greater than the number of statements in the Q sample. Thus, the total number of individuals within the P-set only serves to give breadth so as to give the researcher confidence that the major viewpoints within the study have been obtained. This does lead to a methodological limitation, in that the viewpoints expressed through a Q study can only be applied to that Q sample, rather than the population at large. Baker, van Exel, Mason, and Stricklin (2010) provide an interesting methodological twist on combining small *n* studies, using Q, with a questionnaire technique to understand conceptions in a much larger population.

The statements are often sorted onto a grid with a quasi-normal distribution, with the distribution ranging from two poles representing points of most agreement or "most like I think" to least agreement or "least like I think". This technique shows some similarities to the Likert scales often used in questionnaires, where participants can make duplicate choices under each of the Likert headings. However, the grid system often used in Q studies tends to follow a quasi-normal distribution which only allows some duplication in choice (see Figure 1). This, at least, prevents the casual sorter from regarding everything as "neutral" or "most important". This feature of Q is often the one which sorters most dislike (Dziopa & Ahern, 2011). However, the issue here is to force sorters to make a value judgement of one statement over another. Thus, a sort could ask for participants to rank order the whole Q sample, but with large Q samples this could prove to be rather testing, and the grid frames the choices that a participant has to make in a more manageable way.

A short post-sort interview is often conducted to establish some qualitative data regarding the reasons behind positioning statements at the two extremes of the distribution. The statement numbers related to each score of +5 through to −5 are recorded along with any qualitative statements given as reasons for each distribution.

Data analysis

Analysis of the data is usually performed using a range of different software products, although the most recommended and used is the freeware PQMethod (Schmolck & Atkinson, 2002). This code is still MS-DOS based, so some computers may need a downloaded emulator in order for the MS-DOS code to run. One such emulator is also included with PQMethod. Following the manual, users input statements and sorts and can edit these easily. Once the scores against each statement have been entered, on a participant-by-participant basis, correlations are calculated between sorts. Factors are then extracted from this correlation matrix, an analysis that helps rationalize the various value choices made by participants based upon them placing statements against similar scores.

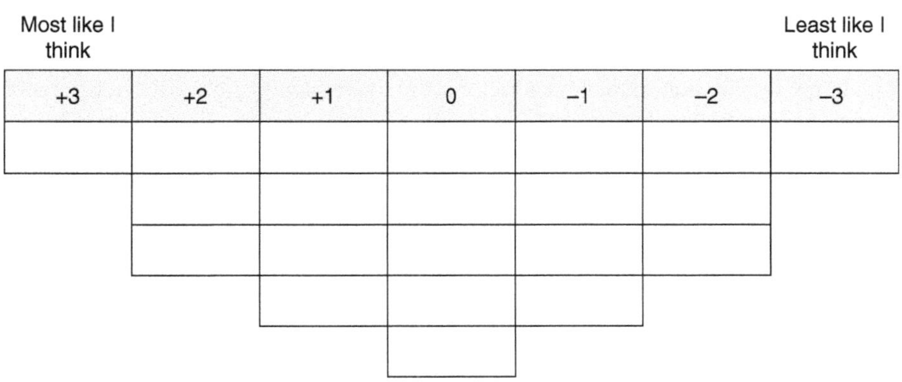

Figure 1. Example grid for Q sorting. Each cell of the grid should be filled with one statement from the Q sample, i.e. this Q sample has 21 statements.

Factor analysis can take two forms within Q, centroid factor analysis (CFA) and principal components analysis (PCA). There is still some debate over which method is preferable, with some authors (e.g. Watts & Stenner, 2005) suggesting CFA, which allows for decisions to be made on a theoretical basis (i.e. looking for factors that may chime in with a theoretical standpoint that the researcher or literature may identify), as opposed to PCA, where factors are determined purely by mathematical considerations. However, Webler et al. (2009) consider that in environmental work they have undertaken, PCA and CFA elicit very similar factors.

Even so, it must be stated that a factor solution does not represent an exact science. Webler et al. (2009) identify four criteria which the researcher must use to judge the best solution from the analysis performed. These are simplicity, clarity, distinctness and stability (i.e. the same participants loading to the same factor, or the same discourses coming from factors irrespective of the solution chosen). Thus, there are often many factors, possibly as many as there are participants, but some are seen as statistically non-significant, and others may be seen to make no qualitative sense at all. They are, in the end, mathematical constructs.

Thus, the researcher's task is to analyse many possible solutions and, paying attention to Webler et al.'s (2009) criteria, to choose the solution that gives an analysis that explains as much of the variance as possible, whilst attempting to "load" as many participants onto the chosen number of factors as possible, whilst rendering a coherent narrative of various viewpoints within the P-set. "Loading" refers to participants who have statistically significant agreement with one of the factors derived by the analysis. Here, the researcher is using decision-making processes much more akin to those used in qualitative research but is using the quantitative analysis to guide these decisions. Some practitioners also use an eigenvalue of more than 1 as a heuristic mechanism for assessing significance of factors (Deignan, 2009), whereas others use the scree plot technique. However, it is sometimes that a solution with a smaller number of factors, which yield only small changes in total variance, and where as many participants load to these factors as is possible (even though eigenvalues were just greater than one), is considered the chosen final factor solution.

To yield differences between the factors, rotation is used. Q utilizes hand rotation (also called judgemental rotation) or Varimax rotation. As in the decision between CFA and PCA, the choice of hand or Varimax rotation appears to be often down to whether the study is exploratory, and, therefore, that there are no theoretical constructs that might be tested. Where the study is exploratory, Varimax rotation is often used, rather than hand rotation, as this is a purely mathematical rotation of the factors that maximize the differences between them, rather than something based upon a theoretical or judgemental consideration. Hand rotation could be used to test a variety of ideas by physically manipulating the rotation so that other factor score moves away or towards viewpoints one may wish to measure against.

Once a solution has been chosen, software such as PQMethod identifies the relative scores for each statement in the Q sample. Therefore, you can identify the statements that are most valued, or in most disagreement with each factor. The software also allows the user to see which member of the P-set loads to which factor. Further analysis provides the researcher with a list of statements that most distinguish each factor and statements that are areas of greatest disagreement and consensus. From this analysis, the researcher interprets the factors to provide a qualitative view of each of the viewpoints identified by the analysis (sometimes referred to as "discourses" in Q). By placing these individual discourses together, the research provides a full narrative account of the qualitative views of the individuals studied.

Using Q in higher education

The benefits of Q in other educational studies

Eden et al. (2005) suggest that their analysis of Q within human geography leads them to identify Q as having benefits to the research community. They list a number of benefits: it is efficient and time/cost effective, it provides flexibility to the researcher, it has utility in the field of policy-making and its novelty requires researchers to use it thoughtfully and reflexively. Briefly reviewing where Q has been used in educational research suggests that these positive aspects are seen within this body of writing too.

Since Q studies work on small samples, and yet still derive rich data, it appears an ideal time and cost-effective addition to the arsenal of methods that researchers can use to understand the subjectivities of groups of students. Firstly, as sorts take less time than a formal interview, such that they can be done many at a time, sometimes online, the time saved in data collection is significant. Secondly, there is little of a need for lengthy transcription and analysis time. The selection and interpretation of factors is time consuming, but probably no more than any other method, quantitative or qualitative. Thirdly, sorts can be done as part of a classroom experience (e.g. Jurczyk & Ramlo, n.d.; Ramlo, 2008). In a recent non-educational Q study, Wright, Wink, and Pike (2011) conducted 25 sorts and analysed the data within 2 working days. To reach the same stage of the research process using interviews, even short ones, would take considerably more time and effort. This is not to say that Q offers a "silver bullet" that allows time and money poor researchers to establish a useful body of data; Q needs to be used thoughtfully and correctly as for every methodology. However, should the methodology be appropriate and the efficiencies are clear.

In terms of flexibility, studies have shown that Q can be used to explore perceptions around both specific and general educational questions. Susan Ramlo (2008), for instance, has used Q to study her students' perceptions around key concepts in physics, namely force and motion. She feels that Q gives her an advantage, when compared to other Likert-type inventories, in the study of the depth of understanding that her student show with these ideas. Similarly, recent work by Sparks (2011) has shown a range of approaches to learning in the arena of physical geography. As such, Q may represent an interesting technique that could be used in the blossoming field of geocognition (Petcovic, Libarkin, & Baker, 2009). More generally, Q has also been used to address research questions around pedagogic approaches. Recent examples include perceptions around initiatives in e-learning (Khatri, 2010; Kim, 2012), EBL (Deignan, 2009) and enterprise education (Deignan, 2011). In each of these cases, authors have used Q to investigate user subjectivities, to better inform inclusion or development of that educational practice within a curriculum. For example, in Deignan's study of EBL, he identifies that whilst EBL might represent a positive intervention in the student's learning experience, Q illuminates a range of potential sticking points in relation to the intervention and its applicability across all subject areas and for all learners. The research points to the training needs of staff and students in its effective use, and potential problems caused by a change in the division of labour in the classroom. It also suggests potential conflicts between university QA and finance considerations. Whilst such useful interpretations may be garnered through other, non-Q, methods, the utility of such a technique, along with the efficiencies described above, again point to a useful research methodology.

Whilst Q is used in the environmental arena to help give shape to policy discussions, in an educational context it has been used to elucidate views of stakeholders in much broader contexts than the classroom. In areas such as the adjustment of international students to a "home" educational culture (Yuan, 2010) through to "home" students learning about the diverse nature of the student body (Taylor, 2001), Q has been used to identify a range of viewpoints which authors have used to inform the Higher Education sector about the ways in which students deal with cross-cultural issues. In a study by Bradley & Miller (2010), the authors investigated much wider perspectives on attitudes to going to university, adding to the debate on widening participation, particularly from areas of social disadvantage. Studies such as that those on educators in citizenship (Anderson, Avery, Pederson, Smith, & Sullivan, 1997), health (Barker, 2008) and outdoors leadership (Hutson & Montgomery, 2006) have also taken place. The characteristic of all of these studies from an educational context shows that, although based upon small samples, Q can be used to broaden and promote debate about the implications of a wide range of important educational activities and move this debate outwards to a much wider audience than just one classroom. Eden et al. (2005) suggest that this is due to Q enabling policy-makers and stakeholders to see real differences in discourse, as well as points of consensus. Importantly, too, some researchers see Q as giving voice to marginal points of view (Wolf et al., 2011). However, David Okwell (2011, cited by Wolf et al., 2011) suggests caution, in which Q studies tend to involve too few people that proper consultation can really said to have taken place. Thus, whilst Q may be useful in shaping educational policy, by informing the debate, it cannot replace the debate itself.

Finally, Eden et al. (2005) suggest that Q's relative novelty promotes a thoughtful and reflexive use of the methodology, presumably whilst researchers accustom themselves to its use. They have some sympathy with Q critics that suggest that there is a claim in Q of objectivity and the removal of the researcher from the interpretation. Whilst there is no sense, this has been explicitly dealt with in educational research, there is an ongoing and very current debate within the Q community regarding the researcher sorting statements for themselves, and being included within the analysis (Gough, 2012). Proponents of this approach suggest that by doing so, the researcher can use Q to position themselves within the various discourses of the P-set, which may provide a degree of reflection upon their own subjectivities which could inform their interpretation.

Thus, it appears that the positive benefits seen to the human geography community were earlier established; Eden et al. (2005) have some resonance with the educational research community. In order to illustrate this further, two brief case studies follow, which show how Q can be used both to investigate students' attitudes and to let students investigate attitudes of the people around them.

The use of Q in pedagogical research – student perceptions of assessment

Many institutions have spent some time and effort attempting to decipher the messages from the results of the National Student Survey (NSS). Sectoral trends suggest that one of the areas in which many universities have not performed as well in is the section investigating satisfaction around "Assessment and Feedback" (HEFCE, 2010). Whilst there had been much work taking place internally within institutions on furthering colleagues' practice around these key areas, the responses from students were consistently lower than elsewhere within the NSS data-set.

There have been numerous studies identifying "good practice" in terms of the quality and quantity of feedback and the impact this has on the quality of student learning (Gibbs & Simpson, 2005). However, clearly, even if we were satisfied that our practice was guided by both good example and research findings, the perception of our students was still not as positive as it could have been. Thus, this example (Udall & Wright, 2009) was aimed at trying to understand what values students attached to a range of different characteristics of feedback; some characteristics we might want to encourage and some we may not.

The concourse was based on a literature study of good feedback practice and semi-structured interviews with students and teachers. The Q-set of 34 statements covered areas that involved who provided the feedback, when and how it was provided, what the purpose of feedback was and what it might tell both tutor and learner. The P-set involved 30 computer studies students from across all the three levels of their respective courses.

Three factors were identified, which accounted for all of the participants, and around 60 per cent of the total variance. The first factor, which was predominantly loaded with Level 4 students, appeared to have a developmental expectation of feedback. They were interested in how they were doing, and they liked their feedback plentiful and often with it being motivating and encouraging.

The second factor had echoes of the first factor within it, highlighting some developmental characteristics. However, there was a definite sense that this factor also wanted to see grades, and some indication of how best to put things right. There was less of a sense of a developmental "journey" seen in the first factor. This factor included a smattering of all levels but seemed most loaded to by Level 5 participants.

The third factor, which had some students from all factors but was most connected with Level 6 students, was much more adversarial. Students here valued direct instruction about what was wrong and how to put it right. There was no sense that knowing the "gap" (Sadler, 1989) between what the tutor expected and what was submitted was particularly useful. Each assessment took on an "episodic" feature, so that whilst each grade counted, there was little connection of feedback into developing better work in the future.

So, it appeared that as students progressed through the levels of their studies they moved away from a developmental to a corrective expectation of feedback. They became more interested in what they had done wrong – but more importantly they expected tutors to explicitly tell them how to put it right.

Interpreting this analysis, it felt that whilst "good practice" suggests that feedback should be timely, plentiful and encourage students on a developmental journey where they become autonomous and can self-assess, students in the latter stages of their university life wanted "better grades" and they expected us to tell them explicitly how to get them. Since the stakes are raised between Levels 4, 5 and 6 of a course, it might not be that much of a surprise that students expect different things from their feedback at each level. The data seemed to suggest that our quest to make them into independent learners, with the ability to assess their own performance, was one that became less important to students as they moved into levels where grades, and our judgement of that work, became more important to their final success.

The study left a difficult question hanging in the air. If this was their expectation at later levels, yet we continued to try and push the "developmental" nature of feedback, would students ever be satisfied, from an NSS viewpoint, with what we gave them?

The use of Q in the classroom – using Q within a class on survey techniques

As part of a Level 6 class on surveying techniques in environmental studies, qualitative methods were investigated, including questionnaires and interviews. As a class where active learning is the main pedagogical approach, it was important that students experience these methods and put them into practice. What was apparent that student struggled to develop good questionnaires, with insightful analysis and interpretation, and fitting in experiences of running, transcribing and interpreting interview data within the timetable was problematic. Thus, it was decided to run a 2-week session, investigating the use of Q as a way of establishing viewpoints on environmental problems.

Using a topic on which the class felt that all participants would hold a view, climate change was singled out as the research area. As an act of preparation, each student was asked to bring in 10 statements around the current debate on climate change, along with an indication of its provenance. Students were asked to source all material, both academic and popular, introducing them to the formation of a concourse. In the first class, students were introduced to qualitative methodology and, in particular, Q. At the end of this session, students were asked to pool their statements, and then thematically sort these viewpoints. Mimicking the methodology mapped out in earlier sections of this paper, students sifted out duplicates, rewrote statements, simplified and then checked to see whether all predominant views were covered. As a basis for this decision, the class used a report published by the Institute for Public Policy and Research (IPPR) in 2006, which identified 10 "attitudes" to the climate change debate, ranging from catastrophist to rhetorically sceptic. The result of this sifting, choosing and rewording resulted in a Q sample of 34 statements. After this session, the tutor then created an appropriate grid and list of statements, and students were asked to perform a sort with two members of their family or group of friends, emailing the data 24 h before the next session. This gave a P-set of 20 participants.

Before the next session took place, the tutor performed analysis on the data. However, students in the class were shown how to use the software and how to approach reaching a factor solution. The session closed with students developing a qualitative description of a factor, using the data supplied by PQMethod.

Using the Q sample, this work identified four factors by PCA with Varimax rotation. These accounted for 68 per cent of total variance, with all sorts loading significantly to one of these factors.

The first factor, which the class named the "Earth Mothers", as this appeared to be a predominantly female view, saw climate change as a global problem, which we all needed to work together to combat. These individuals felt that governments needed to act more responsibly and did not see climate change as something that can be left for future generations to change. They were optimistic that things can be done, especially in regard to reducing carbon dioxide emissions. This view was deemed to most closely match with the IPPR's "David and Goliath" discourse, which, it says, can be seen by opponents "...as 'long-haired hippies out to change the world.'" (IPPR, 2006).

Participants who loaded significantly on to the second factor expressed a view that, whilst they did not believe in a conspiracy, they were the only group that believed that there was a degree of uncertainty within the models. They did not believe that, if real, we could leave the problem for the future, but distrusted the government's motivation for action, feeling that it was another revenue-raising exercise. This view did not disbelieve that climate change was happening but felt that it was a natural process that we bore little responsibility for. The class felt that this view matched most closely to the IPPR's (2006) "Rhetorical sceptics", which characterize environmentalists and issues such as energy efficiency as the domain of "liberals, tree huggers and cissies".

Those who significantly loaded to a third factor were judged as "pragmatic" by the group. There was a concern from this group about the impact that climate change would have on poorer nations and that those impacts should be remediated with help from richer countries. There was a sense that these individuals felt a degree of confidence about their knowledge of the science behind climate change. They did not accept that people are confused, and they were the only ones who questioned whether the UK would necessarily become warmer under a warming global climate.

The final, fourth, factor was significantly loaded to by two sorts. The predominant view in this factor was rather catastrophist. It felt that climate change is unavoidable, and there is nothing that can be done to stop it. It is an irreversible fact, which is going to radically affect everyone. This group felt that this view most closely matched the "Alarmist" view found in the IPPR research of 2006. It viewed climate change as something "... awesome, terrible, immense and beyond human control". The IPPR report suggested that such views become so concerning that there is some thrill involved in the inevitable doom that awaits, a view that they defined as "Climate Porn".

As a final reflection, students were asked about the method, comparing and contrasting with what they already knew about questionnaires and interviews. Many students professed to like the method from the pragmatic view of small sample size. Some felt that questionnaires were impersonal and did not like their prior experiences of using them. This seemed to be around the issue of awkwardness and rejection, especially in the sense of taking questionnaires out to the public. The fact that participants "bought in" to the sort was important, and they felt that this created a more engaging atmosphere between sorter and researcher. Students were certainly interested enough to revisit this method when it came to final year dissertations. The major struggle with this independent work was around the interpretation of factors, especially its complexity. As one student wrote on their feedback:

> I used Q... everything went well until the interpretation. The information I got back was not as clear cut as the example we did in class.

However, this reflection has led to interesting discussions with students around the complexities of "real-world" data and how the researcher makes sense of it. Eden et al. (2005) use the word "engaging" to describe their experiences of using Q, and the interest and feedback received from students who have used the method or sorted statements for themselves certainly confirms this.

Conclusion

Frustrated by questionnaires? No time to run or transcribe interviews or focus groups? Qualitative research methods, which are predominantly used by many of us studying the HE student-learning experience are sometimes time consuming, methodologically complex in terms of interpretation and analysis, and often run counter to the training of many geographers and environmental scientists. At the same time, I have found teaching these techniques to students, as part of their development as fledgling researchers, to be problematic and frustrating.

Q methodology, developed by researchers within schools of psychology, is a technique that offers perhaps some degree of objectivity through statistical analysis. However, more importantly, it uses a well-defined structure for establishing a range of viewpoints from a group of individuals. Given thorough *a priori* research, a well-designed set of statements, some careful use of statistics, the Q user ends up with an interpretive task not unlike that confronting a qualitative researcher, but whose decisions and interpretations may be guided through use of the statistical output, a concept that many scientists and, often, students seem more comfortable with.

It is not a panacea to the problems associated with attempting to research student learning, but, carefully used, these case studies illustrate that it can be practiced in a number of contexts and quickly picked up by students. It is just another weapon in our arsenal of techniques to discover more about our students, and like any research methodology it has its appropriate and inappropriate uses. The use of Q appears to be growing, and it appears to be growing as a technique used by geographers of many persuasions. As such, maybe it is time to ask whether it is a technique that you might find useful.

References

Anderson, C., Avery, P. G., Pederson, P. V., Smith, E. S., & Sullivan, J. L. (1997). Divergent perspectives on citizenship education: A Q-method study and survey of social studies teachers. *American Educational Research Journal*, *34*, 333–364.

Barker, J. H. (2008). Q-methodology: An alternative approach to research in nurse education. *Nurse Education Today*, *28*, 917–925.

Baker, R. M., van Exel, J., Mason, H., & Stricklin, M. (2010). Connecting Q & surveys: Three methods to explore factor membership in large samples. *Operant Subjectivity*, *34*, 38–58.

Bradley, J., & Miller, A. (2010). Widening participation in higher education: Constructions of "going to university". *Educational Psychology in Practice*, *26*, 401–413.

Brown, S. R. (1980). *Political subjectivity: Applications of Q methodology in political science*. New Haven, CT: Yale University Press.

Brown, S. R. (1991). *A Q methodological tutorial*. Retrieved from http://facstaff.uww.edu/cottlec/QArchive/Primer1.html.

Brown, S. R. (2010, February 1). *Re: "Saturation."* Retrieved from Q-METHOD@LISTSERV.KENT.EDU.

Cairns, R. (2011). *Understanding science in conservation, a use for Q methdology*. Paper presented at the 27th annual ISSSS Q methodology conference, University of Birmingham, Birmingham, 7–9 September 2011.

Deignan, T. (2009). Enquiry-based learning: Perspectives on practice. *Teaching in Higher Education*, *14*, 13–28.

Deignan, T. (2011). Understanding and improving transitions in education for enterprise. In R. Kill & K. O'Rourke (Eds.), *Inspiring enterprise: Transforming enterprise education at Leeds metropolitan university* (pp. 56–69). Leeds: Leeds Met Press.

Denzin, N. K. (2010). Moments, mixed methods, and paradigm dialogs. *Qualitative Inquiry*, *16*, 419–427.

Dziopa, F., & Ahern, K. (2011). A systematic literature review of the applications of Q-technique and its methodology. *Methodology*, *7*, 39–55.

Eden, S., Donaldson, A., & Walker, G. (2005). Structuring subjectivities? Using Q methodology in human geography. *Area*, *37*, 413–422.

Fairweather, J. R., & Swaffield, S. R. (2002). Visitors' and locals' experiences of Rotorua, New Zealand: An interpretive study using photographs of landscapes and Q method. *International Journal of Tourism Research*, *4*, 283–297.

Gibbs, G., & Simpson, C. (2005). Conditions under which assessment supports students' learning. *Learning and Teaching in Higher Education*, *1*, 3–31.

Gough, H. (2012, March). *Re: "Can researchers include their Q sort in the analysis?"* Retrieved from Q-METHOD@LISTSERV.KENT.EDU.

Gould, S. J. (1996). *The mismeasure of man*. Harmondsworth: Penguin.

Hall, C. (2008). Identifying farmer attitudes towards genetically modified (GM) crops in Scotland: Are they pro- or anti-GM? *Geoforum*, *39*, 204–212.

Hawthorne, T., Krygier, J., & Kwan, M. -P. (2008). Mapping ambivalence: Exploring the geographies of community change and rails-to-trails development using photo-based Q method and PPGIS. *Geoforum*, *39*, 1058–1078.

Higher Education Funding Council for England. (2010). *National Student Survey-Findings and trends 2006 to 2009*. Retrieved from http://www.hefce.ac.uk/pubs/hefce/2010/10_18/10_18.pdf.

Hutson, G., & Montgomery, D. (2006). How do outdoor leaders feel connected to nature places? A Q method inquiry. *Australian Journal of Outdoor Education*, *10*, 29–39.

Hutson, G., & Montgomery, D. (2011). Stakeholder views of place meanings along the Niagara Escarpment: An exploratory Q methodological inquiry. *Leisure/Loisir*, *34*, 421–442.

Institute for Public Policy and Research. (2006). *Climate porn turning off people from action*. Retrieved from http://www.ippr.org.uk/pressreleases/?id=2240.

Jankowicz, D. (2004). *The easy guide to repertory grids*. Chichester: Wiley.

Jurczyk, J., & Ramlo, S. (n.d.). *A new approach to performing course evaluations: Using Q methodology to better understand student attitudes*. Retrieved from http://uakron.academia.edu/SusanRamlo/Papers/514651/A_New_Approach_to_Performing_Course_Evaluations_Using_Q_methodology_to_better_understand_student_attitudes.

Khatri, R. (2010). Investigating students' perceptions of e-learning within an occupational therapy program. *Operant Subjectivity*, *33*, 84–100.

Kim, J. -Y. (2012). A study on learners' perceptional typology and relationships among the learner's types, characteristics, and academic achievement in a blended e-Education environment. *Computers & Education*, *59*, 304–315.

Massey, D. (1999). Space-time, "Science" and the relationship between physical geography and human geography. *Transactions of the Institute of British Geographers*, *24*, 261–276.

McKeown, M., & Thomas, D. (1988). *Q-methodology*. Newbury Park, CA: Sage.

Miles, M. B., & Huberman, A. M. (1994). *Qualitative data analysis* (2nd ed.). London: Sage.

Petcovic, H. L., Libarkin, J. C., & Baker, K. M. (2009). An empirical methodology for investigating geocognition in the field. *Journal of Geoscience Education*, *57*, 316–328.

Ramlo, S. (2008). Student perspectives on learning physics and their relationship with learning force and motion concepts: A study using Q methodology. *Human Subjectivity*, *2*, 73–90.

Robbins, P. (2006). The politics of barstool biology: Environmental knowledge and power in greater Northern Yellowstone. *Geoforum*, *37*, 185–199.

Sadler, D. R. (1989). Formative assessment and the design of instructional systems. *Instructional Science*, *18*, 119–144.

Schmolck, P., & Atkinson, J. (2002). *PQ method*. Retrieved from http://www.lrz.de/~schmolck/qmethod/downpqx.htm

Sparks, K. (2011). *Students' conceptions of learning and approaches to learning in an undergraduate physical geography course*. Paper presented at the 27th annual international Q conference, Birmingham, UK, 7–9 September 2011.

Steelman, T. A., & Maguire, L. A. (1999). Understanding participant perspectives: Q-methodology in national forest management. *Journal of Policy Analysis and Management*, *18*, 361–388.

Taylor, T. (2001). *A Q-method study describing the diversity awareness of college students in theme housing* (Unpublished Ph.D. thesis). Oklahoma State University.

Udall, M. D., & Wright, P. N. (2009). *What students value about feedback – A study using Q methodology*. Paper presented at the International Society for the Scholarship of Teaching and Learning Annual Meeting, Bloomington, 2009.

van Exel, J., & de Graaf, G. (2005). *Q methodology: A sneak preview*. Retrieved from http://www.qmethod.org/articles/vanExel.pdf.

Watts, S., & Stenner, P. (2005). Doing Q methodology: Theory, method and interpretation. *Qualitative Research in Psychology*, *2*, 67–91.

Webler, T., Danielson, S., & Tuler, S. (2009). *Using Q method to reveal social perspectives in environmental research*. Greenfield, MA: Social and Environmental Research Institute. Retrieved from http://www.seri-us.org/sites/default/files/Qprimer.pdf.

Wolf, A., Good, J. M. M., Brown, S. R., Cuppen, E. H. W. J., Ockwell, D., & Watts, S. (2011). Q methodology and its applications: Reflections on Theory. *Operant Subjectivity*, *34*, 48–71.

Wright, P. N., & Romer, W. (2008). *Indivisible or intersecting? A cross institutional view of active learning*. International Society for the Scholarship of Teaching and Learning Annual Meeting, Edmonton 2008.

Wright, P. N., Wink, B., & Pike, K. (2011). *Stakeholder perceptions of the intangible benefits of Chichester Harbour Area of Outstanding Beauty: An investigation using Q methodology*. Unpublished consultancy report.

Yuan, W. (2010). Investigating international student perceptions of adjustment through Q methodology. *Intercultural Communication Studies*, *19*, 235–252.

RESOURCES

Using Observational Methods to Research the Student Experience

DEBBY R.E. COTTON*, ALISON STOKES** & PETER A. COTTON†
*Educational Development, University of Plymouth, UK, **Experiential Learning CETL, University of Plymouth, UK, †School of Marine Science and Engineering, University of Plymouth, UK

ABSTRACT *Much pedagogic research undertaken in geography and other disciplines relies on post-hoc methods such as surveys or interviews to investigate the student experience of higher education (often based on self-reports of behaviour). However, observation of students provides a far more direct route to obtain information about their behaviour, and there are many situations where observational data—collected by a researcher or by students themselves using video, audio or written diaries—may provide a deeper insight into their experience. This paper outlines some basic methods of observation, describes the potential advantages and disadvantages of collecting observational data, and provides examples of relevant research which draws on these methods. We conclude with a recommendation that observational methods should be used more widely in pedagogic research.*

Introduction

What Are Observational Methods?

Many observational methods have their roots in Anthropology, in which researchers visit little-known 'tribes', and record their activities for subsequent analysis and interpretation. Field notes are used to provide a 'thick description' (Geertz, 1973) of events and actors which can subsequently be drawn together to provide a detailed account of little-known cultures. Anthropological methods (including observation) have been adopted by education researchers and other social scientists, primarily through the development of ethnography as an approach to social research (see Hammersley & Atkinson, 1983). Ethnography might involve the researcher participating in events, observing activities, taking notes and collecting unstructured data to produce an account of the situation being studied (perhaps in a school or university rather than a little-known culture). A crucial part of observation—particularly important in educational research—is to 'make the familiar strange'. This may be problematic given that most researchers have a wealth of experience within educational settings, both as students and as tutors, yet it is crucial—in research terms—to look beyond what is overtly being 'taught' in any specific context. When using ethnographic approaches in educational settings, a range of alternatives to field notes for recording observed events have been developed (including audio or video recordings), and these may be of use to the pedagogic researcher. Recordings enhance accuracy, in that it is possible to have a more or less complete transcript of what was said, but may add to the researcher's workload, especially if a professional typist is not available to transcribe them.

Observers may take a 'participant' role, that is they are part of the situation under study (an example would be a lecturer doing research on their own teaching sessions), or a 'non-participant' role, where the researcher acts as a neutral observer of events (so far as this is possible). Protocols developed for peer observation of teaching, such as those in use in many UK universities, may provide an initial framework to structure observation of teaching. These protocols may encourage the observer to consider the physical environment and layout of teaching space; the tutors' organization and preparation; supporting materials and resources; student activities and contributions; questioning techniques and communication skills, and these could provide a useful starting point for classroom observation. Whilst this paper focuses largely on qualitative data collection, there is a long tradition of classroom observation in secondary schools, much of which relies on quantitative data, gathered using structured observation systems (Wragg, 1994). The best known of these is the Flanders Interaction Analysis Categories (FIAC) (Flanders, 1970) which focuses on teacher–student interaction, and codes events every few seconds according to a pre-ordained set of responses (e.g. teacher lectures, asks questions, gives directions, accepts ideas, praises, etc., student initiates or responds to teacher; silence). Although they may be helpful in some contexts, FIAC have been used far less frequently in higher education than in schools.

In the context of pedagogic research, observational methods might be used in situations such as:

- Undertaking a holistic, ethnographic study of an institution or department to investigate the daily activities and interactions therein;
- Participant observation of a lecturer's own teaching and interaction (using video or audio recordings of teaching sessions for example);
- Non-participant observation of student interaction and activities in the laboratory or field environment;
- Observation and analysis of online discussion forums or website usage, in order to judge the level of learning taking place.

Specific projects discussed in this paper include the use of: observation and audio recording of interactions in geography A-level classes; non-participant observation of geography undergraduates doing fieldwork; and video, audio and written diaries by student participant observers on an overseas field course.

Why Use Observational Methods?

Much pedagogic research undertaken in geography and other disciplines is based on post-hoc methods such as surveys or interviews, which ask students about their experiences and/or behaviours after the event. These methods have associated risks including the following:

- Selectivity—students may report only those aspects of their experience which are felt to 'fit' with the researcher's perspective;
- Memory limitations—students may not be able to remember details about their experiences, and more recent events take on a stronger resonance;
- Post-hoc rationalization—participants in the study provide a rational explanation for the actions after the event, rather than a report of what actually guided their actions at the time;
- Stereotyping—research has shown that (at least in certain circumstances) when asked about what behaviours they undertake, respondents provide a more stereotyped view than is actually the case if their behaviour is observed.

Therefore, instead of using a post-experience questionnaire, geographers doing pedagogic research might choose to use observational methods in order to see, first-hand, the kinds of interactions which take place in a class or field setting. This has the advantage of enabling data to be collected on events occurring in real-time, in a natural situation, rather than through the more artificial context of an interview, focus group or questionnaire. The data will be less influenced by the researcher's own agenda and will be (at least in the raw form) relatively free from bias. Observations may also be used in conjunction with other data collection methods as a form of triangulation (using more than one data collection method to provide a greater understanding of the situation). Stimulated recall (Bloom, 1953) can be used as a way of combining observations with interviews, for example by using a transcript of a teaching session as a prompt or stimulus for discussion during an interview with students, or by getting tutors to review a video of their lecture.

What Are the Associated Ethical Issues?

Ethical clearance is crucial for any research involving human participants, and observational research raises a number of specific ethical issues which need to be considered. In particular, it is important to gain informed consent from all those who might be affected (in a group situation, this includes all of the students who may be observed, whether or not they are directly involved in the research or not). If the research is being undertaken by a member of staff with teaching responsibilities for the student group, particular care must be taken that students are aware that participation in the research is voluntary and that non-participation will not affect their assessment. With groups of students, there may be limits to the claims of confidentiality which can be made since it is impossible to guarantee that an entire group of participants will not share information with others. In addition, if making recordings, these are more difficult to anonymize, therefore it is essential to have tight control over who has access to such data-sets. Finally, if observing a colleague's teaching, it is essential that researchers give careful consideration to the effect of a negative evaluation, and the potential impact this might have on the future career of participants were this to become more widely known. Many universities have regulations for research involving human participants, and this may involve gaining written consent from participants prior to undertaking any video recordings. For general guidelines on research ethics, see BERA: British Educational Research Association (2004).

What Are the Potential Pitfalls and How Can They Be Avoided?

A major difficulty with observational research is the tendency to collect a great deal of rather unstructured data presenting practical difficulties as for every hour of data collected, it is likely to take at least 4–6 hours to transcribe. It is crucial, therefore, to be clear about the research questions which guide data collection, to give explicit, written guidance to participants if they are acting as observers (for example, in video-diary methods). It may also be necessary to make decisions about transcribing only parts of the data-set, based on strict criteria relating to the research focus. Some qualitative researchers use a process known as progressive focusing (Agar, 1980), sometimes described as funnelling (e.g. Cousin, 2009). This process enables the focus of the research to be progressively refined over time through the study, and enables ideas or theories developed through initial data collection and analysis to be tested out on the same or different participants. Whilst this approach may seem appealing to busy academics who wish to 'shed data', significant time input is required early on in the project to ensure the emerging focus is underpinned by rigorous analysis.

Another difficulty in observational research is the potential for behaviour of participants to be influenced by the act of observation. This may be reduced in situations where the observer is known to, and friendly with, the research participant (unlike the more formal observations which might be used as part of a teaching review or inspection). However, even the act of participant observation, in the form of reflection on one's own teaching, is likely to incur some changes to normal behaviour. Audio or video recording is often recommended since this enables a more accurate and detailed record of events to be obtained than using field notes, and may reduce the impact of prior biases of the observer. However, there is a greater risk of 'reactivity' to the recording device, wherein people may act differently, or be less direct and honest when being observed. Whilst it may be tempting to undertake covert (undercover) observation to avoid this problem, it is rarely ethically acceptable! Reactivity may be greater with video recording than audio recording (although Example 3 below illustrates a case where this does not appear to apply)—but the video enables access to non-verbal communication, and information about the context which might not otherwise be available. It is possible to reduce reactivity by gradual familiarization to the equipment (for example, by recording a number of teaching sessions prior to the one which is of interest for the research).

How Can Observational Data Be Analysed?

Observational data are often highly unstructured and therefore difficult to analyse. Whilst a survey may involve pre-coded responses, and an interview can be based around themes of interest to the researcher, observational data are by their very nature 'naturalistic' (see Lincoln & Guba, 1985). The interpretative aspect of the analysis may therefore be harder than usual, simply because the researcher will need to make inferences about the key concepts of interest from naturally occurring conversation or descriptions. There is a wide range of ways of analysing qualitative data and for more detailed advice see Dey (1993). One popular approach is that of 'grounded theory', a model developed by Glaser and Strauss (1967) and used (at least in a loose form) by many qualitative researchers. This approach involves the identification of themes or categories, using the 'constant comparative method' whereby each new piece of data is compared to others already coded in order to identify similarities and differences. Further analysis attempts to clarify the meaning of categories, identify sub-categories and describe potential links between them. Data analysis may be aided by the use of computer software such as NVivo (QSR International), enabling easier organization and management of codes. However, useful as this is in terms of data organization and retrieval (especially when dealing with large data-sets), the intellectual work of interpreting the data generally has to be done by the researcher themselves.

How Can Validity and Reliability of Observational Data Be Enhanced?

Reliability, according to Hammersley (1992, p. 67), "... refers to the degree of consistency with which instances are assigned to the same category by different observers or by the same observer on different occasions". This is potentially problematic in qualitative research where interpretation plays a key part in the analysis of data. However, reliability can be enhanced by collecting data using more than one method (e.g. interviews and observations), or set of respondents (e.g. lecturers and students) and 'triangulating' the results (perhaps by comparing what respondents say with what they do when observed, or comparing lecturers and students' reports of the same event). In team research, it is also possible to use multiple observers and calculate inter-observer reliability (the degree of consistency between different researchers, interpretations of the same event or coding of the same data-set). Observational methods do have a strong claim to validity, described by Hammersley (1992, p. 62) as "the extent to which an account accurately represents the social phenomena to which it refers". This is because they are generally used to investigate actions in a natural setting, rather than an artificially induced context (such as an experiment), and they give a first-hand view of events, rather than a report of actions after the event (such as collected in an interview or questionnaire). In projects where it is important to understand accurately another person's perspective, validity can be enhanced by using some form of 'respondent validation', whereby the research findings are shared with participants, and refined in the light of their comments. The limitations of approaches such as triangulation and respondent validation are discussed by Silverman (2005), and Cohen *et al.* (2000) provide further helpful guidance on validity and reliability in observation.

Example 1: Observation of Interactions in Geography A-Level Classes

In this study (published as Cotton, 2006a, 2006b), teachers and students were observed taking part in class discussions about controversial environmental issues. Interactions were audio-recorded using a high-quality lapel microphone and transmitter, and the observational data were used as a point of reference for subsequent interviews. Accompanying notes were taken by the researcher including: a brief summary of events through time to enable a temporal dimension to be included in the transcripts; copies of any writing or diagrams made on the blackboard; occasional comments on the actions of teacher or case study students where these would not be clear on the tape; and any interpretative ideas occurring during observation. In addition, student contributions to whole class discussions (which were often picked up poorly on the radio microphone) were transcribed by hand in as much detail as possible, noting the name of the student involved. This enabled subsequent interviews with the students to include questions about specific comments they had made in class. Other details that would not appear on the tape, such as the seating position of students, were also noted.

A number of steps were taken to try and reduce the possibility of the teacher or students being influenced by the presence of a researcher. These included the use of a two-week 'lead-in' observation period prior to the start of formal data collection; the use of non-participant observation, whereby the aim was to be as inconspicuous as possible in the classroom; and the development of good relationships with both teacher and students to reduce the potential stress of being observed. In order to avoid being perceived by the students as part of the teaching staff, stringent efforts were made to avoid involvement in any teaching aspect of the lessons and communication with the tutor was minimized. However, a brief talk was given to the students during the lead-in period explaining the nature of the research and inviting comments or questions. This aimed to pre-empt any possible concerns about the purpose of the lesson taping, and to avoid the kind of queries which arose in the pilot study (such as, "Why are you wearing that microphone, sir? Is it a Walkman?"). Other than this initial interest in the research equipment, there were few occasions when there was evidence that students were particularly conscious of the researcher's presence. There was evidence that the students were more conscious of the tape, earlier on in the study, and became noticeably more relaxed as the observation period proceeded.

Observation, in the context of this study, was viewed as providing an essential reference point with which to compare the data obtained though interviews (triangulation). The observational data provided a background of shared experiences between researcher and researched, and enabled specific incidents to be discussed with participants (through stimulated recall, sharing with interviewees extracts from the transcript of the discussion). As Brown and McIntyre (1993, p. 36) point out, this constrains respondents to "concentrate on real and shared events rather than imagined reconstructions" and provides a "means for a limited check... on the teachers' accounts of what had happened". By focusing interviews on specific events, it was hoped that the discussion would focus on the reality of teaching about controversial environmental issues, rather than upon an ideal view held by the tutor: 'theory-in-use' rather than 'espoused theory' (Argyris & Schön, 1974).

Whilst this study generated unusually rich data on in-class discussions, the use of unstructured observation produced an overwhelmingly large volume of data. Structured observation systems designed for use in the classroom such as FIAC (Flanders, 1970) can assist with this process, since they involved researchers in scoring events at fixed points in time. However, they limit the researcher to questions which can be answered using quantitative data, and were deemed unsuitable for this study. Eschewing structured observations, and choosing to transcribe lessons in their entirety in the pilot study placed major limitations on the time available for reflection and analysis. In the subsequent research therefore, the entire teaching session was recorded (even those sessions in which the tutor stated that 'nothing interesting will happen!'), but only selected parts were transcribed in full, the remainder being indexed to enable them to be retrieved easily if required. This is likely to be the most practical course of action in many pedagogic research projects; however, the rationale for selection of parts of the recording for transcription should be clear and explicit.

Example 2: Non-Participant Observation of Undergraduate Geography Fieldwork

This study involved the use of non-participant observation as part of a mixed-methods approach to investigating the student experience of residential geography fieldwork. Two researchers participated in a week-long field course to Western Ireland attended by 190 stage two students following BA and BSc Geography programmes, together with 12 academic members of staff and four technical staff. Observation data collected over three separate days were triangulated with both quantitative data (from pre- and post-fieldwork surveys) and qualitative data (from individual and group interviews conducted over the duration of the field course) in order to identify and describe the factors most significant to the students' experiences, and those most likely to influence learning. Particular attention was paid to those aspects of fieldwork known to impact on students' affective responses, i.e. their attitudes, motivation and feelings (Boyle *et al.*, 2007; Stokes & Boyle, 2009).

The researchers were first introduced to the students during pre-fieldwork briefing sessions, during which the aims and purpose of the fieldwork were outlined, and volunteers for longitudinal data collection (involving participation in multiple interviews over the course of the fieldwork) identified. In accordance with the study's Ethics Protocol, students were provided with written information about the nature of the data being collected, their rights as participants and the obligations of the researchers in terms of confidentiality and anonymity. It was also made explicit that, by agreeing to participate in the study, students were consenting to their data being used in accordance with the research objectives. Participation in all or any aspect of data collection was entirely voluntary and students were free to decline or withdraw their information at any time. Once in the field, two types of observation were undertaken:

(1) *Direct observation of student activity.* Observations of students actively engaged in fieldwork were documented in the form of written notes using a pre-designed pro-forma. This semi-structured approach enabled the researchers to focus on observing specific aspects of the fieldwork, e.g. interactions between students and their peers or academic staff, the level of engagement with the learning activity; whilst also allowing wider observations to be recorded and thus allowing further themes to emerge (Cohen *et al.*, 2000). Working independently and accompanying different groups meant that the researchers were able to observe a reasonable cross-section of the overall field activity, although it reduced the opportunity to test the reliability of their observations through cross-checking their data against each other.
(2) *Stimulated recall of field activity.* Brief instances of field activity were recorded using a digital camera, and used as a basis for stimulated recall activities during interviews with individual students. These were recorded and later transcribed. After watching the recordings, participants were asked to describe what was happening, what they thought they were learning and how they had been feeling. Encouraging students to talk about their actions and behaviours in the field provided valuable information about the ways in which they perceived the learning situation and environment. Interestingly, it was clear that, despite being shown footage on the same day that it was taken, some students had difficulty accurately recalling what was happening or what they were learning:

> OK, I can't even remember what he [the lecturer] was talking about there to be honest. Was he talking about the till?

Combined with the interviews, these observation exercises generated a substantial amount of very rich data which was first coded in NVivo, and then subjected to content analysis (Patton, 1990) to identify key themes. The combination of observation and interviews provided a powerful means of validating qualitative findings; interviews were used to check inferences arising from observations, whilst observation was used to check information given in interviews. Furthermore, observation provided an insight into the students' external behaviours, whilst interviews allowed some exploration of their internal feelings and perceptions (Patton, 1990). Pre- and post-fieldwork surveys are valuable for identifying changes in, for example, attitude or perceptions (e.g. Boyle *et al.*, 2007), but say little about the factors influencing these changes, the students' experiences or the learning process (Taber, 2000). By triangulating the qualitative data with quantitative methods, a more holistic insight into the student experience was obtained. Direct observation coupled with *in situ* interviews carried out whilst students were immersed in field activity provided insight into the students' 'lived experiences', whilst out-of-field interviews and stimulated recall enabled students to provide reflective accounts of their thoughts and actions.

In terms of the researchers' experience, observation was felt to be a valuable means of recording student activity, particularly in terms of identifying topics worthy of greater exploration (e.g. during interview) and providing context to information collected outside of the field environment. A key strength of the observation was its relative 'detachment' from the students. Although it is possible that the students' behaviour was to some degree influenced by the researchers' presence, this technique required no direct interaction between the observer and the observed. The extent to which an observer can remain personally 'detached' from what is being observed is an interesting question, however, and it is important that the researcher recognizes how their own viewpoints and biases can impact on the data collected. In this instance, both researchers explicitly 'tried not to read too much into the situation', conscious of how their own field experiences might influence their perceptions and interpretations of what was being observed.

Example 3: Video, Audio and Written Diaries by Student Participant Observers in the Field

This research formed part of a study investigating the impact of novelty (Falk *et al.*, 1978, Orion & Hofstein, 1994) and prior preparation on the student experience of fieldwork (Cotton & Cotton, 2009). Twelve student volunteers were provided with video cameras, voice recorders and notebooks and were asked to keep a diary of their experiences whilst on a field course in South Africa. The students were requested to spend at least one hour in total on this activity during the seven-day field course. Small incentives (textbooks) were provided to students to encourage their engagement with this, largely autonomous, data collection. The project was authorized by the Faculty of Science Human Ethics Committee prior to commencement. All students were given a research information sheet which provided some information about the aims of the project, data collection procedures, possible benefits of the research, and assurances about confidentiality and their right not to participate or to withdraw from the study. Students were also asked to sign a consent form which indicated that they understood the objectives of the research that they were free to withdraw at any time and that anonymity was guaranteed unless they expressly waived this right.

The three methods of data collection were highly successful in eliciting accounts of the students' experiences whilst on the field course. However, student responses to the different data collection methods varied and there was considerably more enthusiasm for the video-diary method than for the other approaches. The students engaged strongly with the idea of video-diaries and produced lengthy accounts of all aspects of the field course (fieldwork on the beach, lectures at the field centre, social events and even videos taken in bed in the evening!). Videos were often something of a communal effort—other students commented, were interviewed or even took charge of the camera at times. Whereas Cohen *et al.* (2000, p. 281) talk about video recording being associated with a "connotation of surveillance", students in this study appeared immediately comfortable with the cameras and related well to this medium, perhaps because the students themselves were in charge of the cameras. They captured very successfully the first-hand 'lived experience' of the field course, with an exuberant, 'as-it-happens' style. When asked to rate their experience of data collection (using a choice of three paired words such as 'easy' or 'difficult'), the video-diarists were unanimously positive about their experience, describing it as easy, enjoyable and interesting:

> Never used one before, never wanted one—but found it quite addictive! Didn't use camera much—felt like video captured experience much better. (Q2.4)

However, the recordings were extremely time-consuming to transcribe and analyse (some took more than a day's work simply to transcribe). Also popular with the students were the written diaries or field course logs, which ranged from 10 to 27 pages in a small notebook provided by the research team. These provided more intimate, reflective accounts of the experience (perhaps because of the increased anonymity of this method). There were also differences between different diarists—some were very chatty and described the parties and socializing, others were very serious and factual "we searched around in rock pools recording organisms within a piece of equipment called a 'quadrat'". (F1:1). The field course logs provoked a mixed response as a form of data collection, with two students rating it as an entirely positive experience but two others included 'difficult', or 'unpleasant' and 'boring':

> Difficult: Because I didn't always find time to write in it. Enjoyable and interesting because it is a good way of reflecting your day and telling someone you don't know, how you're feeling and dealing with things. (Q2.6)

However, as real-time observational accounts, the written diaries were limited, being mostly—though not entirely—written in the evening after the event. The impact of memory issues, even within the same day, has been noted earlier and applies equally in this context. The audio-diaries were the least popular data collection approach, proving brief and somewhat constrained (they were frequently halting with lots of pauses and 'ums and ers'). They tended towards formality, with lots of detail about work schedules which was lacking from the other data collection methods, giving the impression that they might have been using briefing notes as an aide-memoire. It appeared that students were not accustomed to capturing their thoughts in this way and found it a hard technique to adapt to. They seemed self-conscious in a way which did not seem to happen with the video-diaries. The audio-diaries also tended to be much less personal and reflective than the written diaries (perhaps because they were spoken out loud making it harder to express uncomfortable emotions). They did not capture as much of the first-hand 'lived experience' of the field course as the videos did, and lacked the reportage-style approach of many of the video-diaries. An advantage is that they were much less time-consuming to transcribe (30 minutes to 1 hour on average)—but the richness of data was substantially reduced by using this method. Only two formal ratings for this method were received, of which one was positive, the other neutral. Both mentioned a degree of uncertainty over what to say:

> Difficult to find time sometimes, found it a bit awkward to speak into it, not sure what to say, I was only repeating what I had done during the day. (Q2.9)

Here again, it is clear that the digital recorders were used largely at the end of the day, rather than for direct observation and reporting of events, as we had hoped. Data were analysed by the member of the project team who was not involved in teaching of the module and who was not known to the students. Video- and audio-diaries were transcribed prior to analysis (in the case of the video-diaries with some additional notes about the visual content); written diaries were analysed directly from the notebooks returned by students. Analysis entailed an iterative process of reading and re-reading data, looking for similarities and differences between accounts, and noting any references to the various novelty domains identified by Orion and Hofstein (1994).

Conclusions

These three examples demonstrate possible ways in which observational methods might be used in pedagogic research in geography, and illustrate some of the advantages and disadvantages of these approaches. Whilst the examples are drawn from a range of disciplines and levels, and are at very different scales, all have potential for use in this context. There are relatively few examples of direct observation in pedagogic research in geography reported in the literature, which is perhaps surprising given the potential value of this technique in gaining insight into learning processes. Further useful examples are provided by Orion and Hofstein (1994) and Lai (1999), who describe the use of structured observation and participant observation respectively in the investigation of fieldwork. Unquestionably, observational methods can be complex and time-consuming; however, they provide such a rich data-set that they deserve more widespread use, even in quite small pedagogic research and development projects. In particular, student video-diaries can enable large amounts of data to be collected with relatively low input from the researcher and some basic thematic analysis can be undertaken in real-time whilst playing back the recordings. Other observational methods (such as the use of stimulated recall) are more time-consuming, but they may serve the dual purpose of encouraging teaching staff to reflect upon their actions in the light of video or audio recordings. In many cases, observational data could supplement—or even replace—the ubiquitous questionnaire evaluation of teaching innovations, and provide greatly increased potential for understanding teaching, learning and the student experience.

References

Agar, M. (1980) *The Professional Stranger: An Informal Introduction to Ethnography* (London: Academic Press).

Argyris, C. & Schön, D. A. (1974) *Theories in Practice* (San Francisco: Jossey Bass).

BERA: British Educational Research Association (2004) Revised ethical guidelines for educational research. Available at http://www.bera.ac.uk/files/2008/09/ethica1.pdf (accessed July 2009).

Bloom, B. S. (1953) Thought processes in lectures and discussions, *Journal of General Education*, 7, pp. 160–169.

Boyle, A., Maguire, S., Martin, A., Milsom, C., Nash, R., Rawlinson, S., Turner, A., Wurthmann, S. & Conchie, S. (2007) Fieldwork is good: The student perception and the affective domain, *Journal of Geography in Higher Education*, 31(2), pp. 299–317.

Brown, S. & McIntyre, D. (1993) *Making Sense of Teaching* (Buckingham: Open University Press).

Cohen, L., Manion, L. & Morrison, K. (2000) *Research Methods in Education*, 5th edn (Abingdon: RoutledgeFalmer).

Cotton, D. R. E. (2006a) Implementing curriculum guidance on environmental education: The importance of teachers' beliefs, *Journal of Curriculum Studies*, 38(1), pp. 67–83.

Cotton, D. R. E. (2006b) Teaching controversial environmental issues: neutrality and balance in the reality of the classroom, *Educational Research*, 48(2), pp. 223–241.

Cotton, D. R. E. & Cotton, P. A. (2009) Field biology experiences of undergraduate students: The impact of novelty space, *Journal of Biological Education*, 43, pp. 169–174.

Cousin, G. (2009) *Researching Learning in Higher Education* (London: Routledge).

Dey, I. (1993) *Qualitative Data Analysis: A User-friendly Guide for Social Scientists* (London: Routledge).

Falk, J., Martin, W. & Balling, J. (1978) The novel field trip phenomenon: Adjustment to novel settings interferes with task learning, *Journal of Research in Science Teaching*, 15(2), pp. 127–134.

Flanders, N. A. (1970) *Analyzing Teaching Behaviour* (Reading, MA: Addison-Wesley).

Geertz, C. (1973) Thick description: Toward an interpretive theory of culture, in: C. Geertz (Ed.) *The Interpretation of Cultures: Selected Essays*, pp. 3–30 (New York: Basic Books).

Glaser, B. G. & Strauss, A. L. (1967) *The Discovery of Grounded Theory: Strategies for Qualitative Research* (Chicago: Aldine).

Hammersley, M. (1992) *What's Wrong with Ethnography?* (London: Routledge).

Hammersley, M. & Atkinson, P. (1983) *Ethnography: Principles and Practice* (London: Tavistock Publishers).

Lai, K. C. (1999) Freedom to learn: A study of the experiences of secondary school teachers and students in a geography field trip, *International Research in Geographical and Environmental Education*, 8(3), pp. 239–255.

Lincoln, Y. S. & Guba, E. G. (1985) *Naturalistic Inquiry* (Beverley Hills: Sage Publications).

Orion, N. & Hofstein, A. (1994) Factors that influence learning during a scientific field trip in a natural environment, *Journal of Research in Science Teaching*, 31(10), pp. 1097–1119.

Patton, M. Q. (1990) *Qualitative Evaluation and Research Methods*, 2nd edn (Newbury Park, CA: Sage Publications).

Silverman, D. (2005) *Doing Qualitative Research*, 2nd edn (London: Sage).

Stokes, A. & Boyle, A. P. (2009) The undergraduate geoscience fieldwork experience: Influencing factors and implications for learning, in: S. J. Whitmeyer, D. W. Mogk & E. J. Pyle (Eds) *Field Geology Education—Historical Perspectives and Modern Approaches*, 461, Geological Society of America Special Paper, pp. 313–321.

Taber, K. S. (2000) Case studies and generalizability: Grounded theory and research in science education, *International Journal of Science Education*, 22(5), pp. 469–487.

Wragg, E. C. (1994) *An Introduction to Classroom Observation* (London: Routledge).

SYMPOSIUM

Visual methodology as a pedagogical research tool in geography education

Bryan Wee, Amy DePierre, Peter Anthamatten and Jon Barbour

Department of Geography and Environmental Sciences, University of Colorado Denver, Denver, CO, USA

This article offers an overview of visual methodology and its applications to geography in higher education. Drawings from an undergraduate mapping and map analysis course are used to concurrently highlight and deconstruct students' sense of place in the context of teaching and learning. As a pedagogical research tool, visual methods nurture self-reflection in learners, inform curriculum and instruction for diverse classrooms, and support the construction of meanings in geography.

Introduction

Sense of place in geography education

Educational research has shown that students bring prior knowledge to learning (Driver, 1989). These cognitive schemas (we call them foundational ideas) are existing frames of reference which are used to make sense of new concepts; they are constructed from past experiences situated in specific socio-cultural settings and they influence how and what is learned (Vygotsky, 1962). For example, a student from an urban environment may view water systems differently from one who lives in a rural area due to differences in their foundational ideas (Shepardson, Wee, Priddy, Schellenberger, & Harbor, 2009). In geographical terms, these different ways of interpreting interactions between human and natural systems are used to describe *sense of place*. According to Agnew (1987), sense of place has a relationship to humans and "the human capacity to produce and consume meaning" (Agnew, 1987 in Creswell, 2004, p. 7). Creswell (2004) adds that, "when we look at the world as a world of places we see different things. We see attachments and connections between people and place. We see worlds of meaning and experience" (p. 11).

Tuan (1977) distinguishes space from *place* based on experience, where "what begins as undifferentiated space becomes place as we get to know it better and endow it with value" (p. 6). He adds that these human experiences (mediated by symbols) provide insight into how and what people know about the world that they inhabit, e.g. a cabdriver finds her/his way around a city differently from a geographer who studies the city conceptually (Tuan, 1977). Critically, for geography in higher education, an appreciation of place frees us from thinking of students as merely empty vessels awaiting the transfer of information from expert to novice. Instead it forces us, as educators and researchers, to ask ourselves what attachments students have to places and how these connections inform students' understandings of geography. Consequently, our study focused on the following two questions: How do undergraduate geography students conceptualize home and where they live? Do these students' conceptions vary by gender, ethnicity and year in higher education? The clarification of pedagogical research focused on sense of place in geography education recognizes students as active learners who inhabit spaces in unique ways and who reside in communities with different socio-cultural norms.

Methods

Instructional context and data collection

A total of 38 undergraduate students in a geography course on mapping and map analysis (2011) at the University of Colorado Denver participated in this study. The course is designed to use visual methods to communicate spatial information, and it typically enrolls geography students in their first, second, or especially their third year (Table 1). Students are introduced to basic concepts of geospatial tools while developing map reading and analytical skills, as well as their understandings of mapping concepts such as scale and projection. Data presented in this paper were collected with the hope that it would help the instructor relate students' sense of place to curriculum and pedagogy in meaningful ways. Specifically, students were asked to (a) draw a map of their home and (b) draw a map of where they live. We were interested in using visual methods (drawings) to ascertain the different scales at which students conceptualized these ideas – home and where they live – in the context of this course. The limitations of this study include a small

Table 1. Characteristics of students in the study.

Total	38
Gender	
Men	18
Women	18
Not specified	2
Ethnicity	
White/Caucasian	33
Hispanic/Latino/a	2
Mixed	2
Not specified	1
Year in School	
First/freshmen	1
Second/sophomore	7
Third/junior	23
Fourth/senior	6
Not specified	1
Age	
19–22	17
23–26	10
27–30	6
>30	4
Not specified	1

sample size as well as a relatively large proportion of White/Caucasian, third-year students; hence, the findings should be interpreted and contextualized with these constraints in mind.

Significance of visual methodology and methods in geography education

Drawings represent a unique assessment for students in formal classroom environments at universities; as an image-based research approach situated within the broad framework of visual methodology, drawings provide valuable insight into students' sense of place (Wee, 2012). A visual methodology is an approach to framing social reality through visual methods (Banks, 2007), and has its origins in anthropology, sociology, and human geography, where photographs, diagrams, and maps have been key research tools used in these disciplines (Prosser, 1996). A visual methodology provides an underlying set of epistemological assumptions for pedagogy and research; it assumes that people "see" the world differently because of individual as well as shared experiences in unique socio-cultural settings. Visual methods, such as drawings, are the tools that elicit foundational ideas in ways that encourage creative expression while reinforcing the notion that students are not all the same with regard to teaching and learning. Rose (2007) uses the term "visual culture" to describe how the world is rendered in visual forms on an everyday basis. Indeed, the organization of human space is largely dependent upon sight (Tuan, 1977). Seeing, however, does not only refer to what the eye sees but also the ways by which meanings are constructed from seeing (Rose, 2007), e.g. how we see, what we are allowed to see, or what we are made to see are examples of a "selective process in which environmental stimuli are organized into flowing [cognitive] structures" (Tuan, 1977, p. 10).

One of the strengths of visual methodology, therefore, is that it has the potential to highlight a range of visualities emanating from an individual and collective sense of place. Visual methodology has been used successfully in academic fields such as health (Radley & Taylor, 2003), childhood (Rasmussen, 2004), teacher education (Taylor, 2002), and science education (Wee, 2012). Researchers have urged the continued application of visual methodologies in the social sciences (Banks, 2007; Prosser, 1996). For example, Anthamatten, Wee, and Korris (2012), Pearce et al. (2009) and Holt, Spence, Sehn, & Cutumisu (2008) used visual methods such as drawings and photographs to understand children's conceptions of play and physical activity. In terms of applying visual methodology to educational reform, Wee (2009) encouraged the contextualization of geography curriculum to reflect students' lifestyles and preferences, as well as the teaching of complex geographical concepts at an early age to minimize the formation of misconceptions. When images are used in the context of research, the questions have primarily focused on visual forms constructed by researchers rather than participants; there is "a concern for the power relations in which those images are embedded" (Rose, 2007, p. 6). In other words, when participants are asked to comment on images produced by researchers, they are essentially being asked to construct meanings from an object that was not produced by them. Hence, their voice (in a figurative sense) is absent in the research process. In their work on children's geographies, Aitken, Lund, and Kjørholt (2007) as well as Holloway & Valentine (2000) stress the need for a recognition of children' lives in the context of their respective cultures, and they reiterate the need for research to embrace voices that are still largely missing from academic debates. Another strength of visual methodology, therefore, is its ability to empower participants by changing the relationship between researcher and subject, where disparate ways of knowing are expressed through images created by the participants themselves (Wang,

Kun, Wen Tao, & Carovano, 1998). Within the context of teaching and learning in university courses, encouraging students to create their own images makes pedagogical research a more humanistic and equitable endeavor.

This approach is not without its limitations. Fundamentally, there is always some level of subjectivity involved in the interpretation of images. What we "see" in an image is filtered through our own foundational ideas; when we look at an image, we are looking (consciously or otherwise) at the relationships between ourselves and the image such that there can never be a true understanding of what an image is "saying." Rose (2007) cautions that "taking an image seriously involves thinking about how [the image] positions you, its viewer, in relation to it" (p. 10). In the same way, no two images are alike because the production of images inevitably includes both intentional (subject) and unintentional (personal) views of the world. For example, Burgin (1999) describes photography as a complex interaction of signs, where the content of images are produced deliberately by the photographer who relies upon "knowledge that is cultural and in the final analysis, ideological" (p. 45). From a pedagogical viewpoint, images do not necessarily point to *how much* a student knows; instead, they illustrate *what* or *how* a student knows. Each has its place within the broader goals and outcomes of learning in higher education.

Data analysis

Drawings in our study were analyzed using mixed methods to provide a deeper and broader understanding of students' sense of place. Specifically, qualitative open coding (see Miles & Huberman, 1994) allowed thematic ideas to emerge from the drawings while descriptive and analytical statistics helped to ascertain if students' conceptualizations of places differed by gender, ethnicity, and year in higher education. After a preliminary review of all drawings, a coding scheme was developed by the authors to systematically categorize how students portrayed their home and where they lived (Table 2).

These ratings were reviewed and agreed upon by all members of the research team prior to the initiation of data analysis. If multiple scales were present, the drawing was coded to the largest geographic scale. All drawings were rated and coded by the same person to ensure consistency in data interpretation.

Results and discussion

We have purposefully selected a few examples of student work representing a cross section of the qualitative data to highlight key themes that emerged from our data.

Table 2. Rating scheme for scale.

Rating	Drawing scale
1	Items or a specific room
2	Floor plan or specific building
3	Multiple buildings or a single street
4	Multiple streets or a neighborhood
5	Single town or locality
6	Multiple towns or a single state/province
7	Multiple states/provinces or a national region
8	Single entire country
9	Multiple countries
10	Global or universal

This is followed by a broader discussion of all students' responses and their sense of place in relation to pedagogy in geography.

Sense of place: a qualitative description

Figure 1 shows two drawings demonstrating students' sense of place and different interpretations of "home." Both drawings in Figure 1 consist of symbols that reveal what these students know about the world they inhabit as well as the different scales in which these ideas are situated. One student exhibits a more conceptual idea of home (as part of earth), whereas the other views home as a specific location on a map. Mountains, which are a central geographic feature in Colorado, figure prominently in both drawings. These differences and similarities in students' sense of place can function as learning and assessment tools in geography education. For example, having drawn these maps by hand, students may have a heightened awareness of their spatial thinking skills in a geographical context. The drawings also depict the varying levels of knowledge that these students bring to map reading and analysis (cardinal directions, for example, are present in one drawing and absent in the other), thereby offering valuable insight into how the course curriculum might be revised to nurture a deeper understanding of geography content.

These drawings also point to the fact that the same word in a geography class can have different meanings for different students. First, when instructed to draw a map, students drew both conceptual- and direction-oriented types of maps. Neither drawing is "right" or "wrong." Given that universities continue to reflect an increasingly diverse student population, the importance of a classroom pedagogy that is responsive to students' unique sense of place should be recognized.

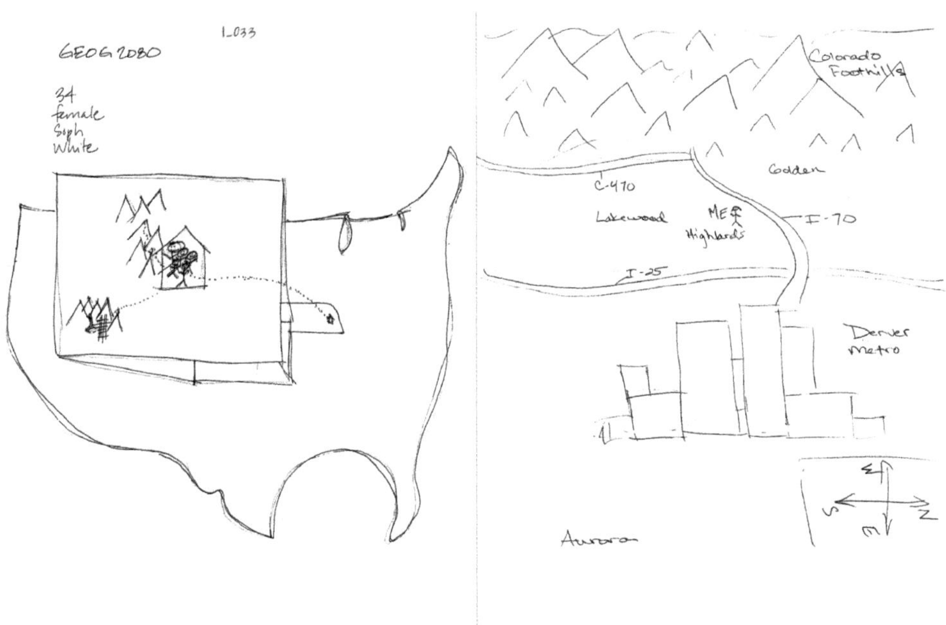

Figure 1. Two students' responses to the question: Draw a map of your home.

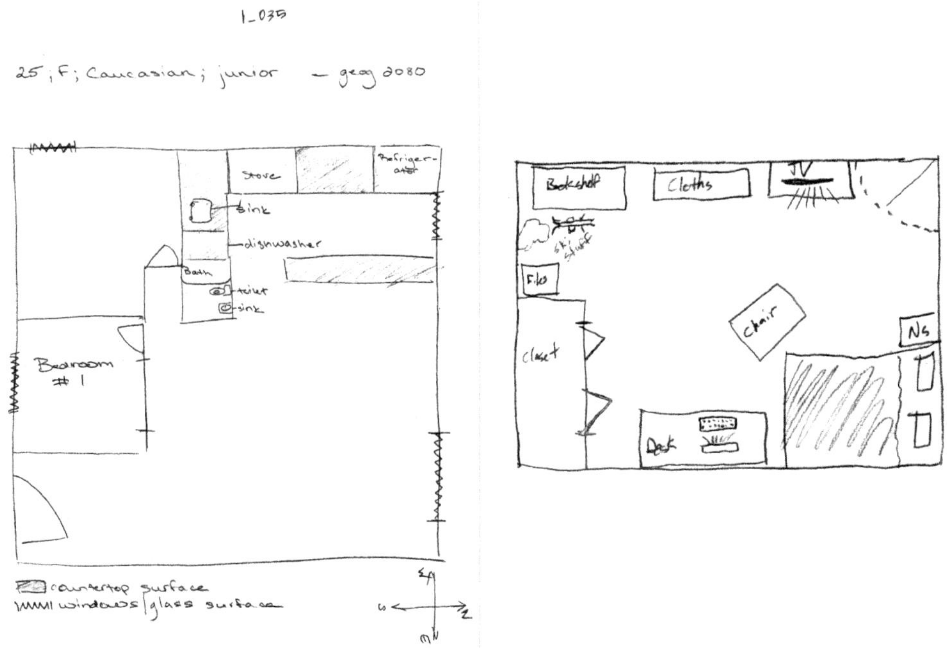

Figure 2. Two students' responses to the question: Draw a map of where you live.

Additional drawings (Figure 2) underscore the ability of visual methodology to provide a nuanced understanding about sense of place. These students drew maps (of where they live) with high levels of detail, likely a reflection of their familiarity with these places. The drawings in Figure 2 resonated with our own experiences living in urban environments. For example, a couch forms part of a person's identity when it has endured countless spills, sleepless nights, or years of scratching by pets. In other words, evidence of tangible materials such as a kitchen and bed reflects the feelings, thoughts, and emotions embodied in a person's sense of place (Tuan, 1977). For example, the "ski stuff" located in the corner of a room (Figure 2) suggests that one student is an avid skier with a particular connection to the winter landscape in Colorado. According to Tuan (1977), place has geometric properties in the form of objects, where

> Neither the newborn infant nor the man who gains sight after a lifetime of blindness can immediately recognize [an object] such as a triangle. The triangle is at first "space," a blurred image. Recognizing the triangle requires prior identification of corners – that is, places. Objects and places are centers of value. To attend to them even momentarily is to acknowledge their reality and value (p. 17).

The careful attention given by students to these objects in their drawings is both informative and insightful in that it discloses the meanings associated with (and within) places.

Sense of place: a quantitative description

Place is a construct that can be described in several ways and at different levels. In addition to our nuanced interpretation of individual drawings (above), we also used descriptive and analytical statistics to characterize students' sense of place collectively as a class.

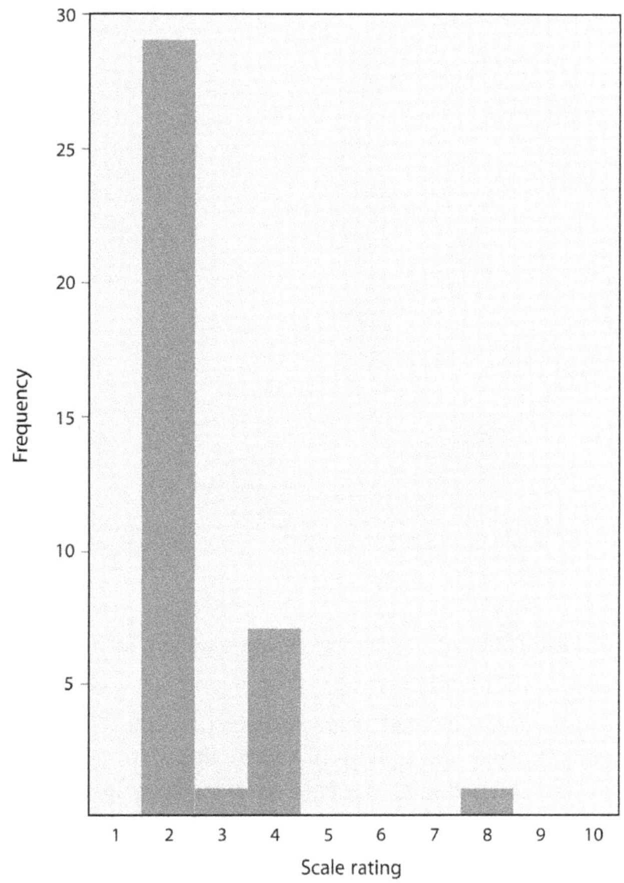

Figure 3. Frequency distribution of scale ratings for the question: Draw a map of your home.

Specifically, we used a coding scheme (Table 2) to characterize the scale at which students conceptualized "home" and "where they live." There were some notable differences in the ways that students depicted these concepts. A majority of students, 29 out of 38 (76%), drew "home" as a single building or as rooms in a single building (Figure 3). Several students drew streets or a neighborhood to describe "home," but there was comparatively little variation in the scale of the drawings.

When asked to depict where they live, the scale shifted "upwards" to encompass more area. Most students drew streets and neighborhoods, but the scale of the drawings ranged from depictions of specific items to global regions (Figure 4).

Both "home" and "where you live" are social constructs representing different sets of ideas in a geographical context. In this case, students' conceptions revealed a deeper connection to, and familiarity with, the former (home). The diversity of scale around which students depict these different notions of place begs the question of how individual differences mediate these concepts. Hence, we compared the drawings by gender, ethnicity, and year (Table 3).

It is interesting to note that the non-White/Caucasian students in this class depicted broader scales in their drawings of "where they live" and narrower scales in their drawings of "home," relative to their White/Caucasian peers. For different students, the notion of

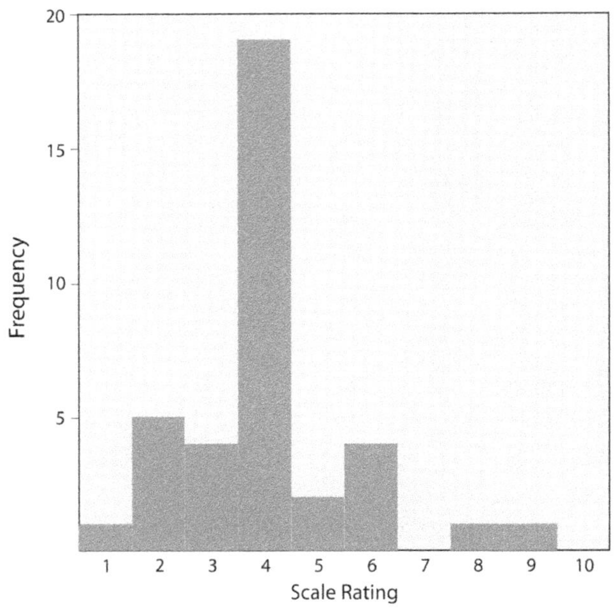

Figure 4. Frequency distribution of scale ratings for the question: Draw where you live.

home may take on a more personal characteristic (e.g. the Chinese word for home translates into "family"), while conceptions of "where they live" may reflect a more global perspective if they are international students or were born in a different country. In addition, college seniors in this sample had a relatively holistic view of where they lived, perhaps reflecting four years of academic training and a better ability to appreciate "the big picture" in higher education.

We performed basic statistical tests in order to determine whether the differences observed were statistically significant. Non-parametric tests were used when appropriate, because the data were not normally distributed, and the specific test was chosen to reflect the type of data we analyzed. As reported in Table 4, while there appear to be some differences in the ratings between groups, none of the differences were statistically

Table 3. Average scale rating of drawings by individual characteristics.

	n	"draw your home"	"draw where you live"
Gender			
Men	18	2.3	3.9
Women	18	2.9	3.9
Ethnicity			
White/Caucasian	33	2.6	3.8
Hispanic/Latino/a	2	2.0	4.0
Mixed	2	2.0	6.5
Year in school			
First/freshmen	1	2.0	4.0
Second/sophomore	7	2.9	3.6
Third/junior	23	2.6	3.8
Fourth/senior	6	2.2	5.2

Table 4. Statistical tests for differences in scale of drawing by individual characteristics.

Characteristic	Statistical test for difference	p-value, "draw your home"	p-value, "draw where you live"
Gender	Mann–Whitney	0.19	0.68
Ethnicity	ANOVA	0.63	0.09
Year in school	Spearman's ρ	0.93	0.24

significant ($p < 0.05$). Future work, using a larger sample size, might identify some statistical differences in the scales at which students conceptualize space and place.

Nonetheless, these variations in students' conceptions of "home" and "where they live," as evidenced by their drawings, demonstrate the integral role that individual and collective experiences play in developing one's sense of place. It also reiterates the contextual nature of meanings that are, in geographical terms, relative to locations and distances between spaces.

Implications for research and teaching

Without the ability to converse with people regarding the images they produce, it is difficult to infer what the presence or absence of certain features and objects might mean to them. This is a challenge with visual methodology, where follow-up interviews are needed to ensure that images accurately represent the thoughts, emotions, and perspectives of the person(s) who produced them. This is particularly important when working with young children, whose world views are very different from that of adults (Holloway & Valentine, 2000; Wee, 2012). Rose (2007) concurs with this approach, writing that

> It is very unusual to encounter a visual image unaccompanied by any text at all, whether spoken or written. Even the most abstract painting in a gallery will have a written label on the wall giving certain information about its making... and this makes a difference to how [people] will see that painting. Visual images are multi-modal in this way – they always make sense in relation to other things, including written texts (p. 11).

From a pedagogical viewpoint, our findings highlight the wide range of foundational ideas that students bring with them to geography courses. Similarities and differences exist, for example, in how students' conceptions of home largely reflected a more intimate knowledge of specific places and how students of different ethnicities conceptualized space at different scales. Geography instructors need to be aware of the nuanced ways by which students make sense of content because knowledge is ultimately grounded in students' foundational ideas. Students learn better when they are able to relate new information to existing foundational ideas uniquely derived from human experiences (Bransford, Brown, & Cocking, 2000). For example, a geography course on urban farming might intentionally include language about "home" if the goal is to encourage students to learn more about sustainable food systems in local neighborhoods.

Conclusion

Sense of place is a construct that is simultaneously abstract and real; it is also (arguably) the cornerstone of geography education where human–environment interactions tend to be the focal point of teaching and learning. This article suggests that educators can effectively connect students' foundational ideas to course curriculum if they know what these ideas are or better yet, grasp the meanings that students attach to geographical concepts. The use of visual methodology in geography education supports a student-centered pedagogical approach by valuing the foundational ideas that are brought to learning environments in universities on an everyday basis. Images, in particular, can highlight a person's sense of place, that is, their interpretations of the world and the interactions that occur within it. As geography educators, it is critical that we recognize students' differing sense of place and leverage these ideas in ways that provide meaningful opportunities to support learning in our classes.

References

Aitken, S., Lund, R., & Kjørholt, A. (2007). Why children? Why now? *Children's Geographies*, *5*, 3–14.

Agnew, J. (1987). *Place and politics*. Boston, MA: Allen & Unwin.

Anthamatten, P., Wee, B., & Korris, E. (2012). Exploring children's perceptions of play using visual methodologies. *Health Education Journal*, doi: 10.1177/0017896912443119.

Banks, M. (2007). *Using visual data in qualitative research*. Thousand Oaks, CA: Sage.

Bransford, J. D., Brown, A. L. & Cocking, R. R. (Eds.). (2000). *How people learn*. Washington, DC: National Academy Press.

Burgin, V. (1999). Art, common sense and photography. In J. Evans & S. Hall (Eds.), *Visual culture: The reader* (pp. 41–50). Thousand Oaks, CA: Sage.

Creswell, T. (2004). *Place: A short introduction*. Malden, MA: Blackwell Publishing.

Driver, R. (1989). Students' conceptions and the learning of science. *International Journal of Science Education*, *11*, 481–490.

Holloway, S. L., & Valentine, G. (2000). Children's geographies and the new social study of childhood. In S. L. Holloway & G. Valentine (Eds.), *Children's Geographies* (pp. 1–26). London: Routledge.

Holt, N., Spence, J., Sehn, Z., & Cutumisu, N. (2008). Neighborhood and developmental differences in children's perceptions of opportunities for play and physical activity. *Health & Place*, *14*, 2–14.

Miles, M. B., & Huberman, A. M. (1994). *Qualitative data analysis*. Thousand Oaks, CA: Sage.

Pearce, A., Kirk, C., Cummins, S., Collins, M., Elliman, D., Conolly, A.M., & Law, C. (2009). Gaining children's perspectives: A multiple method approach to explore environmental influences on healthy eating and physical activity. *Health & Place*, *15*, 614–621.

Prosser, J. (1996). What constitutes an image-based qualitative methodology? *Visual Sociology*, *11*(2), 25–34.

Radley, A., & Taylor, D. (2003). Images of recovery: A photo-elicitation study on the hospital ward. *Qualitative Health Research*, *13*(1), 77–99.

Rasmussen, K. (2004). Places for children, children's places. *Childhood-A Global Journal of Child Research*, *11*, 155–173.

Rose, G. (2007). *Visual methodologies: An introduction to the interpretation of visual materials*. Thousand Oaks, CA: Sage.

Shepardson, D., Wee, B., Priddy, M., Schellenberger, L., & Harbor, J. (2009). Water transformation and storage in the mountains and at the coast: Midwest students' disconnected conceptions of the hydrologic cycle. *International Journal of Science Education*, *31*(11), 1447–1471.

Taylor, E. W. (2002). Using still photography in making meaning of adult educators' teaching beliefs. *Studies in the Education of Adults*, *34*(2), 123–139.

Tuan, Y. F. (1977). *Space and place: The perspective of experience*. Minneapolis: University of Minnesota Press.

Vygotsky, L. S. (1962). *Thought and language*. Cambridge, MA: The M.I.T. Press.

Wang, C., Kun, Y., Wen Tao, Z., & Carovano, K. (1998). Photovoice as a participatory health promotion strategy. *Health Promotion International*, *13*, 75–86.

Wee, B. (2009). Understanding children's connections to the environment in the U.S. and Singapore: Implications for geography educators. *Research in Geographic Education*, *11*(2), 79–93.

Wee, B. (2012). A cross-cultural exploration of children's everyday ideas: Implications for science teaching and learning. *International Journal of Science Education*, *34*(4), 609–627.

RESOURCES

Using student interviews for becoming a reflective geographer

Hanne Kirstine Adriansen[a] and Lene Møller Madsen[b]

[a]*Department of Education, Aarhus University, Tuborgvej, Copenhagen N, Denmark;* [b]*Department of Science Education, University of Copenhagen, Øster Voldgade, Copenhagen K, Denmark*

This paper presents a case for interviewing students as an effective yet complex way to integrate reflexive practice into teaching and research. Even though many human geographers are accustomed to conducting qualitative interviews in various contexts, it is not straightforward to interview one's own students. This paper addresses three issues: implications of doing insider interviews; ethical issues of interviewing students where power relations are at stake and using visual co-constructions as a means of levelling the analytical power of the insider interviewer. We show how student interviews have enhanced our reflection-on-action and give recommendations for prospect student interviewers.

1. Introduction

In this paper, we examine how student interviews can be employed as a means to become reflective geography scholars in relation to our teaching obligations. It deals with both the opportunities that interviewing students open, as well as some of its key challenges. We are inspired by the idea of the reflective practitioner (Schön, 1983). While the concepts and ideas underlying reflective practice date back to researchers such as Dewey (1933) and Piaget, Ackermann, and Berthoud-Papandropoulou (1977), it was Donald Schön's seminal book from 1983 *The Reflective Practitioner* that brought attention to the idea that professionals, such as teachers, should develop a reflective practice. According to Schön, professionals can become better at their job by developing reflection-in-action and reflection-on-action. The former refers to the ability to reflect on challenges in the situation where and when they unfold, whereas the latter involves systematic reflection after the situation, e.g. teaching a class. Schön has also been of inspiration to geographers: In an edited volume by Ashley Kent directed at teachers of geography at secondary/high school level, the editor writes that the book "attempts [. . .] above all to encourage reflection by experienced practitioners" (Kent, 2000, p. vii).

The question is how we can develop this reflective practice? In this paper, we will show how a method that is well known to many human geographers, namely qualitative interviews, can be used in a manner that is unusual to many geographers, namely by interviewing one's own students. Qualitative interviews can provide in-depth understanding of the interviewee's life world (Kvale, 1996); hence by interviewing one's own students, we may gain valuable knowledge about the "how" and "why" of students' learning thereby adding to our reflection-on-action as geographer teachers. Unfortunately, interviewing (one's own) students is not as straightforward as it may seem. At least we were surprised when we began using our extensive interview experience *inside* academia (Madsen & Adriansen, 2006; Madsen & Winslow, 2009). Based on these surprises, we will address three main themes: the implications of doing insider interviews; ethical issues when interviewing one's own students and using visual co-constructions as a means of interviewing. A visual co-construction is an artefact, a paper on which the interview is written and drawn by the interviewer and interviewee. The three themes are interrelated in so far as insider interviews raise special ethical concerns, especially when interviewing one's own students where power relations are at stake, and the creation of visual co-constructions during interviewing can be a means of levelling the analytical power of the interviewer.

The paper offers close reflection on subtle nuances of the interview experience. It gives voice to the concerns that qualitative researchers may have when they interview their own students as a means of getting feedback on teaching and reflecting-on-action. It is based on our own experiences of creating reflection-on-action through student interviews as seen in Example 1 below. With ever increasing demands on productivity and measurable outputs in Academia, it may be difficult to find time for developing a reflective practice. We have found time by combining our reflection on teaching with research, which means we have used our teaching as empirical material for research (e.g. Adriansen, 2010; Adriansen & Knudsen, 2013). Furthermore, we find developing a reflective teaching practice a more productive way to work towards better teaching than the ranking of teaching and "competition fetish" introduced by management (Naidoo, 2008).

> Example 1: Some years ago, we made research interviews with students about their engagement in undergraduate course in Geographical Information Systems (GIS) (Madsen, Christiansen, & Rump, 2014). These interviews let us to insights about the importance of students' motivations and reasons for choosing a particular course. These interviews showed us in a very direct way how the students' approaches to a learning activity are related to their reasons for attending the course. For example, two students attending the same course in GIS had quite different perspectives: one was almost finished with his studies and wanted to be updated on GIS to get a job within the GIS environment while the other student wanted to learn GIS in order to use it in an assignment in another course to specialise within a certain geography topic. These two students engaged very differently with the learning activities and had different views on the learning objectives of the course. As teachers, we can use this knowledge in other courses to be aware of the importance of going into dialogue with the students about their motivations and intentions with the course, especially if we feel that there is a lack of motivation. By doing so, we create a possibility for ourselves to understand the relevance and perception of the learning activities we try to get the students engaged in.

2. Insider interviews: interviewing one's own students

"What do you want to ask me about?" is a common reaction when a researcher/teacher asks students to become interviewees. The question can occur in any potential interview situation, but when the interviewee is a student, there are extra layers to the question. "Is this some type of exam?", "Are there wrong/stupid answers which will haunt me afterwards?", "I like you as a person, but wasn't too impressed with your way of teaching, how can I tell you that?" Likewise the teacher/researcher[1] may be (and should be) concerned about the ethics involved in interviewing students. Also it can be challenging for the researcher to let students experience the researcher's capabilities as an interviewer. What is at stake here is positionality. The concept positionality refers to the multiple positions the interviewer and the interviewee have and can take on in relation to each other during an interview, such as gender, age, social position, ethnicity and profession. As noted by Chacko: "Positionality is a critical factor in framing social and professional relationships in the field; it sets the tone of the research, affecting its course and its outcomes" (Chacko, 2004, p. 52). A teacher–student interview may have different dynamics than a researcher–informant interview. The different positions of teacher and student will give raise to different power relations and legitimate ways of talking and behaving during an interview. Positions and power can be negotiated throughout the interview – often in subtle ways (Adriansen & Madsen, 2009).

The issue of positionality has been analysed widely among geographers (Chacko, 2004; DeVerteuil, 2004; McCleery, 2004). This discussion has especially been presented by feminist and post-colonial geographers (e.g. England, 1994; Larner, 1995; Rose, 1997). When interviewing one's own students, it is especially the insider/outsider and in-between position which demands attention. As pointed out by Kitchin and Tate, being an insider may make research more difficult: "You may fail to notice pertinent questions or issues because of the inability to step back from a situation and fully assess the circumstances" (Kitchin & Tate, 2000, p. 29). Conducting insider interviews with students demands an attentive interviewer who is conscious about pursuing answers where the interviewee says "you know" and other implicit expressions. Otherwise, the interviewer may end up with empirical material that is of little or no use, because it is full of implicit remarks. On the other hand, the insider position also provides advantages such as easy access to the field and close knowledge of the context, culture and language (Adriansen & Madsen, 2009).

The in-between position has been described by Chacko (2004) by the use of Entrikin (1991), as the position "between the objective pole of scientific theorising and the subjective pole of empathy and understanding" (Chacko, 2004, p. 54). As researchers interviewing our own students, we have experienced this in-between position ourselves and will add a third professional pole of interviewer curiosity. This curiosity often drives the questioning which can oscillate between the insider and outsider position. The distinction between being an insider and an outsider is not an easy one. Sikes and Potts (2008) have noticed that being an insider is common within educational research. They define an insider as somebody who is attached to or involved in the organisation or its social groups prior to commencing the study. By this definition, all teachers interviewing their own students are insiders. While we find this correct, it does not capture the dynamic aspect of the insider position. Therefore, we prefer to adopt Narayan's (1993) stance that we all belong to a number of communities simultaneously. Here also, the in-between position can be at play. To us, an insider is somebody who is considered an insider by the other members of a given community and/or who participates on par with the other members of that community. In this view, a teacher interviewing her own students can be both an insider and an outsider. The teacher is an insider in so far as she was present during teaching, but an outsider in the sense that she did not participate on par with the students. One might say that the interviewer and the interviewee were present at the same event but did not experience it from the same place all the time. This means that during an interview, the student can position the teacher in different ways from one minute to another. Likewise, the teacher can emphasise these different positions by asking questions in different ways, e.g. by drawing on the common knowledge from the classroom thereby emphasising the insider position or by asking from a distance to employ an outsider position.

Asking a colleague to interview one's students could be a way to avoid insider difficulties. We have tried interviewing a colleague's students about their approaches to learning GIS (both individual and focus-group interviews). Even though we were not insiders in Sikes' and Pott's (2008) understanding, it turned out we were insiders in relation to the subject matter. The students knew that we knew GIS and thereby it became difficult to ask outsider questions (reported in Madsen & Nielsen, 2013; Madsen & Rump, 2012). Being an insider both in relation to the subject matter and to the interviewees can be labelled "double insider". We have discussed the challenges this poses in Adriansen and Madsen (2009).

Interviews with students make power relations very present, but this is also the case in the opposite situation, i.e. when students interview researchers as seen – but not discussed – in Dwyer (2001). Student interviews of researchers may also be useful for developing a reflective practice, yet it is beyond the scope of the present paper. The power that a teacher can exercise over a student can inform the content of the interview. It goes without saying that the teacher has to think carefully about which questions to ask and how they should be asked. But the teacher should also consider her analytical power. Rose (1997) explains that the researcher holds a privileged position by deciding who is a relevant interviewee, by formulating the questions asked and when they are asked, by directing the flow of the discourse and by having the final power of interpretation. This is the analytical power of the researcher and it should not be neglected. Rose further links the analytical power of the researcher to the issue of distance. Even though the teacher and the students may be situated in the same landscape of power, there is still a distance between them because of the analytical power of the teacher. However, the students also have power in the sense that they hold the knowledge that the teacher wants. These power relations raise ethical questions some of which we will address in the next section.

3. Ethical issues when interviewing one's students

Researchers are required to consider ethical implications of their research (e.g. American Sociological Association, 2008). However, when interviewing one's own students, additional ethical issues are raised, e.g. regarding confidentiality and anonymity. When using interviews for becoming a reflective practitioner, confidentiality and anonymity only becomes a problem when the information from the interviews is shared with other people. In regard to using the interviews for analysing and publishing one's reflection on teaching practice, the usual aspects of confidentiality and anonymity in research apply. In the context of conducting student interviews in general, we want to point to three issues: The selection of interviewees, the interview location and the timing of the interview. These issues may seem banal, but in our experience it is important to reflect on them especially when interviewing one's own students.

The selection of interviewees: Besides being able to justify research design, the selection of students for interviews should be wise in regard to the signals it sends to students. Are the "clever" ones chosen, the "difficult" ones or those who seemed reflective in terms of pedagogy? While qualitative research design allows us to select critical cases or maximum variation (Flyvbjerg, 2006), doing so may send unfortunate signals to students. In one study, we asked the students to sign up themselves, thereby we did not have control over the selection. It worked well for reflection-on-action and we avoided unethical signalling, but we did not have the best research design (Adriansen & Madsen, 2013).

The interview location: As geographers we know that space matters, and this is also the case when interviewing (Elwood & Martin, 2000; Sin, 2003). Different interview locations have different benefits and drawbacks for interviewing students. The benefits of interviewing at the location where teaching took place are as follows: the discussed activities can be placed in relation to each other, and the materials used during teaching may be present. The drawback is that power relations may be more difficult to overcome. Likewise, interviewing outside the classroom or even outside campus has its advantages and disadvantages. On the positive side: the explicit power relations can be diminished, especially off campus. However, both the power relations and ethical issues are still at stake. On the negative side: it may be awkward for the student and the teacher to meet off campus, e.g. in a cafe. If choosing a location on campus for interviewing, it is important to choose a location that does not resemble an exam situation.

The timing of the interview: By timing, we refer to scheduling of other activities one may have with the students; primarily timing in relation to exams. Just like the location should not resemble an exam, we would usually not recommend interviews prior to examining a student because the student may want to please the teacher and the answers therefore reflect the students' idea of a "good" answer instead of her perception of teaching and learning. Nevertheless, if the latter bias can be overcome, valuable information can be obtained by interviewing students before, during and after a course. We have made individual interviews with students about their learning process during a course on GIS in planning and management (Madsen, Christiansen, & Rump, 2014). This study showed to us that interviewing the same students twice during the course provided valuable information on how the students' perceptions of the assignments and learning activities continuously were incorporated in the students' sense making of the course. A longitudinal setup with interviews during the course prior to the exam is necessary for constructing this type of reflection on students' learning processes.

Generally, we find that the unequal power relations make interviews with students more delicate than many other types of interviews. However, there are ways to level the analytical power of the teacher as discussed in the next section.

4. The creation of visual co-constructions: a means to interview one's students

In our experience, visual co-constructions can be used for balancing power relations. We focus on different ways of writing and drawing an interview (here called "pen and paper" interviews), but many of our considerations here can be translated to situations where other materials are used for creating visual co-constructions. The basic ingredients in a "pen and paper" interview are a large piece of paper and a number of coloured pens. The interview may be recorded (and transcribed); it depends on the type of analysis intended and on how the use of a recorder may affect the interviewee. Before we go into further detail with the pen and paper interview, we need to understand how visual co-constructions can affect power relations.

4.1 Visual co-construction: a way of balancing power relations

The benefit of using pens and (large pieces of) paper is that the student can participate in writing and drawing. After the teacher has explained the purpose of the interview and the process, she can encourage the student to take part in writing and drawing on the paper.

Kvale (1996) has explained that qualitative interviews are co-constructed. When the interviewer and the interviewee both participate in writing, the co-construction becomes physically explicit – a visual co-construction, as seen in Figure 1.

The paper serves as a "collective memory" where the content of the interview can be seen both by the interviewer and the interviewee. It is easy for both parties to return to an issue already discussed and this can be linked with other events along the way. Whether the interviewee is participating in writing or not, she is usually engaged in following how the interview unfolds on the paper. This is quite different from a "conventional" interview. The visualisation on the paper affects how the interview is conducted. While some researchers (e.g. Goodson & Sikes, 2001) recommend that the interviewer keeps eye contact with the interviewee, we find that it is culturally specific what constitutes appropriate eye contact. While eye contact can create an intimate atmosphere, it can also be intimidating. In our experience, it can be easier to talk about personal issues when there is no eye contact, but an atmosphere of presence – for instance because both the interviewee and the researcher are engaged in the same activity such as drawing. As an interviewer said after conducting a pen and paper interview: "The paper removes the focus from the slightly formal situation" (Adriansen, 2012, p. 50).

Furthermore, there are two important and intertwined points, which we consider benefits of visual co-constructions. These concern ownership and analytical power.

Ownership: In our experience, interviewees often take ownership of this type of interview. This was evident when the interviewee wanted a copy of paper afterwards – something we have experienced more than once with pen and paper interviews but interviewees have never asked for our notes or the digital recording after a "conventional" interview. Hence, using the paper as a collective memory can make the interview a collective process allowing the interviewee to take ownership. She has the possibility to steer the interview in a certain direction, although it is still the interviewer who ultimately holds the analytical power and can decide which issues are relevant for the interview.

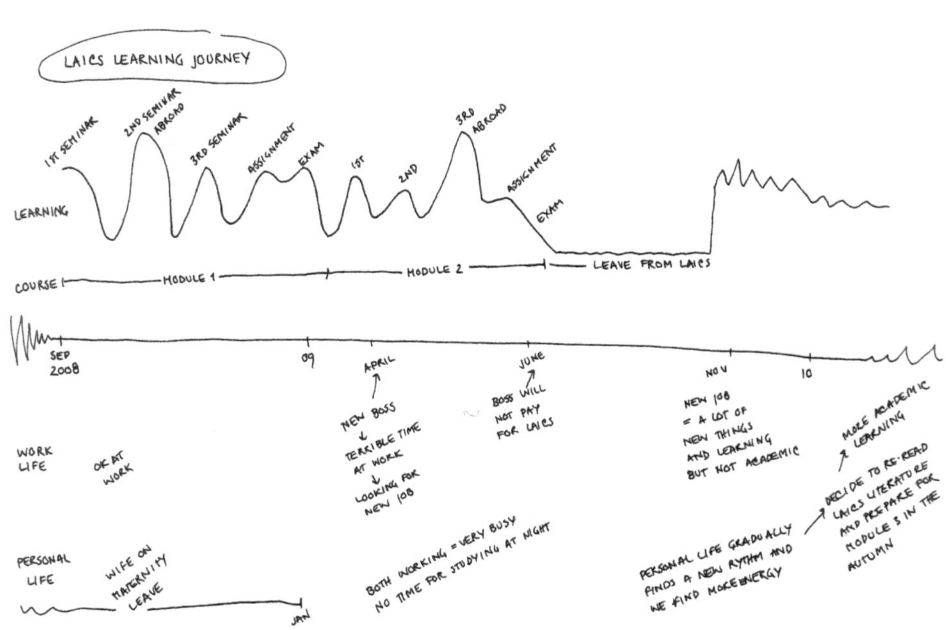

Figure 1. An example of a learning journey interview from the study of creativity and criticality.

Analytical power: In "conventional" interviews, the interviewer holds the analytical power. It is the interviewer who decides that the interviewee is a relevant person to talk to; it is the interviewer who constructs the interview guide and it is the interviewer who asks the questions (Kvale, 1996). When conducting a "pen and paper" interview, it is still the interviewer who decides that the interviewee is a relevant person to talk to, the type of questions asked and the way the interview is conducted. So far, the analytical power is held by the interviewer. When the "pen and paper" interview starts, however, the situation is different from a "conventional" interview because the interviewee is invited to participate in "constructing the interview". In this way, the analytical power is shared, although not equally, in the interview situation. The creation of visual co-constructions allows the interviewer and the interviewee to be situated closer in the landscape of power – to use the vocabulary of Rose (1997). However, when the interview ends, it is not different from other qualitative interviews; here the analytical power belongs to the researcher solely.

4.2 Using time as an organising principle: timeline interviews

We have analysed the use of timelines interviews elsewhere (Adriansen, 2012), here a few words on the technique suffices. The backbone of the method is the drawing of a timeline in the middle of the paper in the "pen and paper" interview. Ordering of events can be guiding the interview. An important issue is when the timeline should begin and end.

In an educational setting, it could begin with the interviewee's first university experience, enrolment in a specific course or a significant event prior to university important for the interviewee's choice of education. Likewise, the timeline could also continue into the future to capture how the student sees her future use of the competences she has gained. The paper is a physical layout which permits different lines representing different perspectives; these may be the "core" of the story, e.g. the course/degree programme/student life in the middle and family/outside towards the edges. Thereby, there is an emphasis on unfolding the context and its ascribed meaning for the student's choices and experiences. Usually, the interviewee begins with the most important events, which are noted on one side of the timeline and gradually "outside" events that have affected the events may come up as the story unfolds and are noted on the other side of the line. When trying to place events in order, the interviewee sometimes realises that these events took place in a different order from what she initially remembered; when conducting focus group interviews, disagreements among the group are also laid out very clearly. This may mean that they re-negotiate their story. In this way, the paper and the drawings along the timeline become important visual tools for seeing how the story is constructed.

4.3 Using time and space as organising principles: learning journey interviews

The timeline interview uses time as an organising principle. Likewise, space or location can be used in a "pen and paper" interview as an organising principle. Inspiration can be found within geographical research (e.g. Oldrup & Carstensen, 2012; Tani & Surma-aho, 2012).

A learning journey interview is a combination of the two methods mentions above. A learning journey is the learning process experienced during time and space. In principle, the journey can be any course or learning process at university or elsewhere. The benefits of using interviews of students' learning journeys are that it becomes possible to link learning activities to other activities in space and time. Example 2 below, together with Figure 1, provides a description of how we have used individual learning journey interviewes for our reflection-on-action.

Example 2: The chapter "How criticality affects students' creativity" (Adriansen, 2010) is an example of our own reflective practice. Higher education is met with increasing demand for providing students with creative and innovative competencies. How can these competencies be reconciled with critical thinking which is the hall mark of Academia? Does teaching the subject matters creativity and innovation mean that we have to be creative and innovative in our way of teaching? If yes, how do we practice what we preach? These questions were answered during a reflective process including student interviews. We also read relevant literature which was outside our field of research but provided valuable input for reflection. The process ended with a book chapter based on student interviews and reflection. In this way, the time spent on reflection provided useful output "for the system" as well as reflection-on-action.

The learning journey approach gave rise to detailed accounts of when the students felt motivated and experienced high learning and vice versa. One of the interviewees explained how she had experienced the first seminar as unstructured and very confusing, even frustrating, leaving the seminar with more questions than answers:

"I had a feeling of having wasted three days (Adriansen, 2010, p. 73)".

But already when preparing the first assignment, she became a little more positive:

"I could feel that I had some new thoughts, was thinking in a new way (Adriansen, 2010, p. 73)".

And in hindsight, she could see how the first seminar had made sense, how it had prepared her for the journey.

The drawings of the learning journeys (Figure 1) showed how the location of the seminar (whether in Denmark or abroad) also affected the intensity of their learning. This information was invaluable for understanding our role as teachers and for reflecting on the importance of location as well as the coherence of the degree programme.

5. Discussion and concluding remarks

In this paper, we have shown how to address challenges involved in conducting qualitative interviews with one's own students for the purpose of becoming reflective geographers. Being an insider in relation to one's interviewees presents difficulties and advantages as well as ethical dilemmas. All this calls for reflection. We have shown how reflection on the subtle nuances of the interview setting and experiences can be used. This paper may be seen as a "meta-reflection" on reflection-on-action, a reflection on how to reflect on our actions as geography teachers. The reason why we encourage student interviews as a tool for reflection-on-action is that qualitative interviews can provide us with in-depth understanding of a student's life-world, their meaning-making and ways of learning. These are issues we find important for any teaching professional wanting to develop her reflection on her teaching practice.

We have argued that we may find the resources for developing a reflective practice by using our reflections as input for research. But one question remains: why should our students spend their valuable time on being interviewed? The answer can be found in Kvale's (1996) notion of the interview as a co-construction where the interviewee is affected and also has learned something through the interview. Hence, participating in such an interview would hopefully lead to reflection-on-action for both the interviewer and the interviewee. Some scholars have argued that higher degree students should develop a reflective practice – usually through journal writing (Boud, 2001; Cunliffe, 2004; Morrison, 1996). We would argued that a reflective practice may not (only) be developed in solitude; it can be a collective endeavour developed through qualitative interviews.

While we find qualitative interviews beneficial for developing a reflective practice, teachers should be prepared for the challenges and ambiguities involved when interviewing one's own students. As an insider interviewer, one becomes an integral part of the process. We concur with Wilson (1997) that if researchers are not prepared for that, then they have better choose another method.[2]

Based on our findings, we have extracted the following recommendations for teachers who want to interview their students for reflection-on-action. The recommendations are also applicable when researchers want to interview their students for other purposes such as course evaluation, getting feedback and ideas for course development:

- Pay attention to positionality during the interview process – also before and after, i.e. when recruiting interviewees and when reporting back to them. There are three important (sets) of positions – the teacher/researcher–student position, the insider/outsider position and the in-between position.
- The teacher/researcher–student position has to do with power relations and one's respective identities. One's position as a researcher/teacher and the student's as learner can be at stake in the process. Being explicit about one's role and their roles may help both the students and the researcher/teacher to reflect upon them. As a teacher, one should be aware of and pay attention to the ethics involved. This can be done by carefully selecting interviewees, and mindfully choosing a time and place for the interview.
- Using visual co-construction for interviewing may affect ownership and analytical power in a positive manner.
- The insider position provides both benefits and drawbacks: Make the most of the benefits (easy access to the field and close knowledge of the context), and be aware of how to handle the drawbacks e.g. pursue implicit answers and presupposed shared understandings.

Notes

1. Both the teacher position and the researcher position are present and at stake when interviewing one's own students. In the following, we use the word teacher or researcher depending on which position is most important in the context.
2. Wilson made the remark in the context of focus group interviews in educational research, but we find it equally valid for the arguments made here.

References

Adriansen, H. K. (2010). How criticality affects students' creativity. In C. Nygaard & C. Holtham (Eds.), *Teaching creativity–creativity in teaching* (pp. 65–84). Faringdon, Oxon: Libri Publishing.

Adriansen, H. K. (2012). Timeline interviews: A tool for conducting life history research. *Qualitative Studies*, *3*, 40–55.

Adriansen, H. K., & Knudsen, H. (2013). Two ways to support reflexivity: Teaching managers to fulfil an undefined role 'A problem cannot be solved at the same level of thinking that created it'. *Teaching Public Administration*, *31*, 108–123.

Adriansen, H. K., & Madsen, L. M. (2009). Studying the making of geographical knowledge: The implications of insider interviews. *Norsk Geografisk Tidsskrift–Norwegian Journal of Geography*, *63*, 145–153.

Adriansen, H. K., & Madsen, L. M. (2013). Facilitation: A novel way to improve students' well-being. *Innovative Higher Education*, *38*, 295–308.

American Sociological Association. (2008). *Code of ethics and policies and procedures of the ASA committee on professional ethics*. Washington, DC: The Association.

Boud, D. (2001). Using journal writing to enhance reflective practice. *New Directions for Adult and Continuing Education*, *2001*, 9–18.

Chacko, E. (2004). Positionality and praxis: Fieldwork experiences in rural India. *Singapore Journal of Tropical Geography*, *25*, 51–63.

Cunliffe, A. L. (2004). On becoming a critically reflexive practitioner. *Journal of Management Education*, *28*, 407–426.

DeVerteuil, G. (2004). Systematic inquiry into barriers to researcher access: Evidence from a homeless shelter. *The Professional Geographer*, *56*, 372–380.

Dewey, J. (1933). *How we think: A restatement of the relation of reflective thinking to the educational process*. Lexington, MA: Heath.

Dwyer, C. (2001). Linking research and teaching: A staff-student interview project. *Journal of Geography in Higher Education*, *25*, 357–366.

Elwood, S. A., & Martin, D. G. (2000). "Placing" interviews: Location and scales of power in qualitative research. *The Professional Geographer*, *52*, 649–657.

England, K. V. L. (1994). Getting personal: Reflexivity, positionality, and feminist research. *The Professional Geographer*, *46*, 80–89.

Entrikin, J. N. (1991). *The betweenness of place: Towards a geography of modernity*. Baltimore, MD: Johns Hopkins University Press.

Flyvbjerg, B. (2006). Five misunderstandings about case-study research. *Qualitative Inquiry*, *12*, 219–245.

Goodson, I., & Sikes, P. J. (2001). *Life history research in educational settings: Learning from lives*. Milton Keynes: Open University Press.

Kent, A. (2000). *Reflective practice in geography teaching*. London: Sage.

Kitchin, R., & Tate, N. J. (2000). *Conducting research in human geography: Theory, methodology and practice*. London: Prentice Hall.

Kvale, S. (1996). *Interviews: An introduction to qualitative research interviewing*. Thousand Oaks, CA: Sage.

Larner, W. (1995). Theorising 'difference' in Aotearoa/New Zealand. *Gender, Place and Culture: A Journal of Feminist Geography*, *2*, 177–190.

Madsen, L. M., & Adriansen, H. K. (2006). Knowledge constructions in research communities: The example of agri-rural researchers in Denmark. *Journal of Rural Studies*, *22*, 456–468. doi:10.1016/j.jrurstud.2006.02.001

Madsen, L. M., Christiansen, F. V., & Rump, C. (2014). Students individual engagement in GIS. *Journal of Geography in Higher Education*, *38*, 251–265.

Madsen, L. M., & Nielsen, T. T. (2013). Learning to do geography? University students posing questions in GIS laboratory exercises. *Norsk Geografisk Tidsskrift–Norwegian Journal of Geography*, *67*, 157–161.

Madsen, L. M., & Rump, C. (2012). Considerations of how to study learning processes when students use GIS as an instrument for developing spatial thinking skills. *Journal of Geography in Higher Education*, *36*, 97–116. doi:10.1080/03098265.2011.576336

Madsen, L. M., & Winslow, C. (2009). Relations between teaching and research in physical geography and mathematics at research-intensive universities. *International Journal of Science and Mathematics Education*, *7*, 741–763. doi:10.1007/s10763-008-9134-y

McCleery, A. (2004). So many Glasgows: From 'personality of place' to 'positionality in space and time'. *The Scottish Geographical Magazine*, *120*, 3–18.

Morrison, K. (1996). Developing reflective practice in higher degree students through a learning journal. *Studies in Higher Education*, *21*, 317–332.

Naidoo, R. (2008). The competitive state and the mobilised market: Higher education policy reform in the United Kingdom (1980–2007). *Critique Internationale*, *39*, 47–65.

Narayan, K. (1993). How native is a "native" anthropologist? *American Anthropologist*, *95*, 671–686.

Oldrup, H. H., & Carstensen, T. A. (2012). Producing geographical knowledge through visual methods. *Geografiska Annaler: Series B, Human Geography*, *94*, 223–237.

Piaget, J., Ackermann, E., & Berthoud-Papandropoulou, I. (1977). *Recherches sur l'abstraction réfléchissante*. Paris: Presses Universitaires de France.

Rose, G. (1997). Situating knowledges: Positionality, reflexivities and other tactics. *Progress in Human Geography*, *21*, 305–320.

Schön, D. A. (1983). *The reflective practitioner: How professionals think in action.* New York: Basic Books.

Sikes, P., & Potts, A. (2008). *Researching education from the inside: Investigations from within.* Oxon: Routledge.

Sin, C. H. (2003). Interviewing in 'place': The socio-spatial construction of interview data. *Area, 35*, 305–312.

Tani, S., & Surma-aho, O. (2012). Young people and the hidden meanings of the everyday: Time-space path as a methodological opportunity. *International Research in Geographical and Environmental Education, 21*, 187–203.

Wilson, V. (1997). Focus groups: A useful qualitative method for educational research? *British Educational Research Journal, 23*, 209–224.

Using Focus Group Research to Support Teaching and Learning

Heather Winlow*, David Simm*, Alan Marvell** and Rebecca Schaaf*

**Department of Social Sciences, Bath Spa University, UK; **Department of Leisure, Tourism, Hospitality and Events, University of Gloucestershire, UK*

Using qualitative research methodologies for pedagogic research can provide informative insights into student experiences of the teaching and learning environment in Higher Education. This paper considers the practicalities of undertaking a pedagogic study using focus groups by outlining strategies for effective focus group organization, highlighting good practice from the social sciences and pedagogic literature and providing a critique of a recent geographical study, in which focus groups were used as the sole research method. Furthermore, the paper illustrates how a pedagogic research study can be used to enhance teaching and learning and to inform curriculum planning and course management.

Introduction

Involving students in qualitative research can be valuable for pedagogic study and can provide insights into learning and teaching issues, including the effectiveness of course delivery and the adequacy of support networks in place in Higher Education (HE) environments. Here we outline how focus group methodology can be applied to pedagogic research, discuss the key stages to be considered in designing the framework for analysis and provide critique of a recent geographical research project that we have undertaken. We build on the practical guidance given by Breen (2006) by critically reflecting on the logistics and practicalities of designing, undertaking and applying a focus group methodology to pedagogic research. The paper is intended to provide guidance to lecturers who are considering undertaking qualitative pedagogic research with students in HE contexts.

Project Summary: Supporting Foundation Degree Students

This paper incorporates discussion of the methodological design for a recent pedagogic research project (2008–2009) which focused on the effectiveness of support networks in place for Foundation degree students studying courses in Development Geography and

Tourism Management at three institutions: Bath Spa University (BSU); Weston College, Weston-super-Mare; and City College Plymouth. The results and recommendations of this research are discussed elsewhere (Simm *et al.*, 2011; Simm *et al.*, 2012). Our project aims were:

- To gain understanding of the needs, motivations and experiences of Foundation degree students, focusing on (i) their transition from school or further education (FE) into HE (either within a FE college or at university) and (ii) their transition by 'top-up' into the final year of an undergraduate honours programme.
- To investigate the experiences of Foundation degree students taking contrasting pathways.
- To enable managers and tutors to gain a better insight into the recruitment and retention of this group.
- To maintain and develop appropriate academic and pastoral support structures.

To meet the project aims we required a qualitative methodology which allowed for both expression of opinions and student cohort discussion of issues relating to their course, institution and the support networks in place. Focus groups were selected as appropriate to our project aims.

Focus Groups: An Effective Method for Pedagogic Research?

Pedagogic research may be designed with the goal of improving aspects of teaching and learning and educational support systems. When undertaking pedagogic research the appropriateness of the method must be considered in relation to your research aims. Focus groups are a participatory form of qualitative research and can be viewed as group-based interviews which can reveal a range of opinions and perspectives on a topic. Focus groups were originally used for market research and are now widely used within the social sciences to gather opinion-based data. Like other qualitative methods, focus groups allow the researcher to gather in-depth information from a small group of participants, but can be time-consuming to plan and undertake. The advantage of focus group research is that, like interviews, they allow informants a role in shaping the research (Bennett, 2002) and can be semi-structured in format, allowing dialogues, rather than interrogations (Valentine, 2005). Unlike one-to-one interviews, focus groups can be used to expose the differences, contradictions, unique experiences, views, perceptions and attitudes expressed by different group members (Bennett, 2002; Hydén & Bülow, 2003), allowing for a richer understanding of the issues. Cameron (2000, p. 101) observes that although focus groups require much planning and interpretation, they are "an exciting and invaluable research tool for geographers to use" and "can generate insights and understandings that are new to both participants and researchers." This approach also allows for observation of interaction between group members (Denscombe, 2010).

Cameron (2000) argues that focus groups can either be the sole research method for a project, or can be combined with quantitative methods. If undertaking pedagogic research, focus groups can be used as part of a medium or large scale research project, but the method can also be applied at a smaller scale, perhaps to analyse the effectiveness of teaching methods within a module and to aid in its future development. For example, focus groups can be combined with a quantitative survey, such as annual module evaluation or a questionnaire, in which summary information can be collected quickly from a large group. Volunteers can then be selected for participation in a focus group to provide additional depth to the findings. For a larger research project, it may be appropriate to incorporate quantitative and qualitative methods into a longitudinal study, in which the progression of particular groups of students is measured, or, alternatively, to take a 'snapshot' approach with students from different year groups. In some contexts, pedagogic research approaches may evolve into an integral part of the learning and teaching strategy of a module or programme. For instance, focus groups may be beneficial in situations requiring students to reflect on their experiences of undertaking research, overseas fieldwork (Simm & Schaaf, 2011) or work placements. Hopkins (2007) suggests that focus groups can also be combined with other qualitative techniques, such as one-to-one interviews—which can reveal individual opinions that may not be revealed in a group-based approach.

As demonstrated in the literature, focus groups have been used to explore a wide range of socio-geographic issues. For example, raising awareness of education and training needs in ethnic communities (Fallon & Brown, 2002), community reactions to the threat of natural hazards (Burgess, 1996), analysis of retail behaviours (Holbrook, 1996), identity construction among young Muslim men (Hopkins, 2007) and reactions to school regulations among teenagers (Raby, 2010). Geographers and other educationalists have also used focus groups for pedagogic research: firstly, to study institution wide changes, such as the effectiveness of personal development portfolios (Kneale, 2002), the development of IT resources (Breen, 2006) and student attitudes to plagiarism (Gullifer & Tyson, 2010); secondly, to compare the effectiveness of different teaching approaches, such as Fuller *et al.'s* (2003) cross-institutional comparison of fieldwork-based versus classroom-based approaches and thirdly, to reflect on the effectiveness of specific courses or modules, or on academic and pastoral support in place for specific groups of students, as in our study.

Focus Group Organization

If focus groups have been selected as appropriate for your pedagogic research project, several considerations need to be made in planning and implementing the research, including:

- Research ethics
- Sampling strategy and group size
- Focus group design
- Session management
- Recording, transcription and data analysis

Here, these elements are considered as stages in the process of carrying out effective focus group research, and while some of this advice is generic, we also suggest how the methods can be applied specifically to pedagogic research.

Stage 1: Research Ethics

Any research involving human participants must undergo an ethics assessment in advance of undertaking research. Most universities, or academic schools, have ethical guidance and procedures in place which must be followed (see The British Sociological Association (2002) for in-depth guidance on research ethics). Ethical questions include:

- Is the research covert or deceptive?
- Does the research involve working with vulnerable groups?

- Has informed consent of all participants been sought? (i.e. the aims of the research should be clearly explained and on this basis individuals may consent to be involved in the project).
- Will individual responses be anonymized?
- Will individuals have the right to withdraw from the project at any time (without penalty)?
- Will there be financial rewards for involvement in the project?

Some of these points are more applicable to a pedagogic environment than others. For example, working with vulnerable groups is more likely when working with the general public on social issues (e.g. homelessness and domestic violence). However, the power relations inherent in the lecturer–student relationship should be taken into account in research design and focus group management. Other aspects such as anonymization of student responses will allow for more liberal discussions about their experiences of their course and institution (without fear of repercussion) and overt research and informed consent will usually be required.

Ethical considerations also need to be made if pedagogic research methods are being used to evaluate and develop teaching. Can focus groups and other participatory approaches be integrated within the formal teaching of a course or module? If this is done, is the pedagogic research study advertised to students or are data surreptitiously collected? Do students have the choice to opt-out of the research, without missing out on key teaching input? Alternatively, focus groups may take place outside the taught curriculum, resulting in enhanced independence but presenting issues of representivity.

Stage 2: Sampling Strategy and Group Size

Focus groups can be selected in several ways: they may be arranged around a random sample drawn from a broader population, or could be selected on the basis of specific demographic characteristics (such as gaining a mixture of different genders, ethnicities or ages), or through 'purposeful sampling' (Cameron, 2000) with participants chosen on the basis of their shared experience in relation to the research topic. The latter provides a 'common communicative ground' (Hydén & Bülow, 2003, p. 311) around which discussions can take place. 'Natural groups'—where group members know each other—will provide different group dynamics to a group of strangers (Holbrook, 1996; Hydén & Bülow, 2003). In an educational context, a purposeful sample, or natural group, may be of greater value than a random sample. As Breen (2006, p. 466) observes "if the focus groups are designed to investigate students' learning experiences, they should consist of participants who have been exposed to similar experiences."

Guidance on focus group size suggests using between 5 and 12 participants (Cameron, 2000; Fallon & Brown, 2002; Hydén & Bülow, 2003; Longhurst, 2003). Group dynamics, and the contribution made by each person, will vary with group size. Crang & Cook (2007) note that 10–12 members is a large group, while 6–8 is 'lively.' Wibeck *et al.* (2007) suggest five for promoting small-group discussion, allowing each participant to play a prominent role, while Onweugbuzie *et al.* (2009) propose that 'mini' focus groups of three to four participants are feasible. There can be difficulties in recruiting participants to focus groups, so advanced planning, and keeping participants informed is vital. If the budget allows, a free lunch or other refreshments may help in terms of recruitment! If working with students from a particular course of study, it may be appropriate to use a scheduled timetabled slot to increase participation rates. A carefully structured letter or e-mail explaining the aims of the study should be sent in advance to potential participants, to ensure the informed consent of participants.

In our study, the student cohort for each year group on each award provided a purposeful sample of an appropriate size for carrying out focus group research, with 7 focus groups undertaken, and 32 participants (see Table 1). We used focus groups as the sole method for this study as we wished to gauge student opinions on the effectiveness of support mechanisms and to allow for group discussion of issues.

In this study, two focus groups had only two participants due to cohort size. The length of these sessions was adapted to 40 minutes, which allowed time for individual reflection. Our experience was that, rather than being a disadvantage, the small groups enabled a full discussion of participants' views, whereas in larger groups more management was required. Hopkins (2007) observes that focus group guidance can be overly formulaic and that small groups can yield useful data, and Longhurst (1996) observed similar advantages of (unintentionally) small focus groups. Therefore, if a very small focus group or dual interview occurs, this should still be regarded as useful data to be incorporated into your wider study.

Stage 3: Focus Group Design

Effective question or topic design includes, firstly, deciding on the key topics of discussion, which will be linked to research aims and, secondly, careful consideration of wording, useful key phrases and the sequencing of questions (Cameron, 2000). Some authors suggest the use of a topic or interview guide (if using a semi-structured framework) consisting of around 6–12 main points, with sub-points (Dunn, 2000; Crang & Cook, 2007). Dunn (2000, p. 61) notes that semi-structured interviews are organized around "ordered but flexible questioning", while Krueger & Casey (2009) note that the questioning route should flow in a natural, logical sequence. Dunn (2000) identifies 'funnelling' and 'pyramid' techniques, where 'funnelling' involves an initial focus on general issues, moving towards more personal matters or specific issues, with sensitive questions at the end (similar to Breen's (2006) suggestion that key questions should take place in the final third of the schedule), while the 'pyramid' technique begins with easy to answer questions at the start, deeper questions in the middle and generalized questions at the end. In terms of individual question design, Longhurst (2003) suggests starting with a question that puts participants at ease before getting on to more sensitive or thought-provoking topics. Krueger & Casey (2009) argue that, to maintain a balanced perspective, it is important to ask positive questions before negative questions. Guidance suggests a focus group length of between 1 and 2 hours (maximum). The length that a session takes will be linked to how fully participants engage in the discussion and how much they have to say about the topics raised.

For our research project, the focus group design followed a logical sequence, employing the 'pyramid' approach (following Dunn, 2000), covering the path of the student from entry to graduation and beyond. The topic schedule included five linked sections: (1) pre-arrival, (2) on-arrival, (3) at college/university, (4) looking back and (5) looking forward. We applied the pyramid approach because by following the sequential experiences of the students, the central part of the interview (peak of the pyramid) would relate to their experiences of their programme and the university environment. This section is where both positive and negative experiences, and potentially contentious issues, would be discussed. Questions were designed to ease participants into the

Table 1. Focus group composition

Institution	Programme of study	Year group	Number of participants	Gender (male/female)	Number of mature students	Sampling strategy
Weston College	Fd Tourism Management	1	7	1M/6F	2	Purposeful
Weston College	Fd Tourism Management	2	6	1M/5F	2	Purposeful
City College Plymouth	Fd Tourism Management	1	6	1M/5F	2	Purposeful
City College Plymouth	Fd Tourism Management	2	3	1M/2F	1	Purposeful
BSU	Fd Development Geography	1	2	0M/2F	1	Purposeful
BSU	Fd Development Geography	2	6	2M/4F	2	Purposeful
BSU	BSc Tourism Management (top-up from Fd Tourism Management)	3	2	1M/1F	0	Purposeful

discussions at the start including asking them about where they had applied and why, with more general questions about future aspirations coming at the end. Our topic schedule is shown in Table 2. Other aspects of our focus group design included:

- Prompts for all topics, to be used only if required. (Often key issues were raised by students, without prompting.)
- Repetition of Sections 3 and 4 for top-up students, in relation to their experience at their second institution.
- Same set of questions and prompts used for all seven focus groups, allowing comparison between data-sets.
- Focus group recordings were between 40 minutes (in which there were two participants) and 1 hour. 'Recorded' focus groups were preceded by *c.* 15 minutes session briefing.

Stage 4: Focus Group Management

Focus group management includes consideration of setting or environment, the role of the moderator(s), group dynamics and power-relationships. To help put participants at ease, it is generally recommended that focus groups take place in familiar surroundings. If you are working with students the focus group can take place on their campus, either in a seminar room, or in a more informal meeting area. Alternatively, as campus can be seen as a formal environment, another neutral, but accessible, venue could be chosen. A decision about the practicalities of on-campus sites versus the need for a neutral venue will need to be made. Crang and Cook (2007) note that the 'associational contexts' of a venue need to be considered, including whether it has links with particular groups of people or types of information, or could reinforce power relationships. Power-relationships between group members (for example friendship groups, gender, age and confidence levels) can also affect responses.

The role of the moderator(s) is a key consideration. Burgess (1996) suggests using two moderators, one 'interviewer' and one 'note-taker,' who acts as a participant-observer but does not contribute to discussions. The role of the 'interviewer' includes ensuring that all topics are covered, using prompts where required, ensuring that all participants have chance to comment on each topic, and picking up on issues raised by participants. The role is to guide discussions when needed, but not to dominate. Crang and Cook (2007) note that the moderator must be sensitive to power relations, and that the group may need to be managed to encourage participation from all members. Some participants may feel under-confident in expressing their views, and moderators must actively encourage inclusivity and participation. This may involve directly asking quieter participants what they think about an issue, and in some cases may require interrupting more vocal members to allow someone else to speak. Another role of the moderators is to outline verbally the aims of the focus group at the beginning of the session, so that this is clear to all participants. Participants should be reminded directly before the focus group takes place that they have the 'right to withdraw' from the process at any stage during the discussion. This is particularly important if sensitive topics are being discussed. If the aims and reasons for your study have been clearly explained to participants throughout the planning stages, the risks of withdrawal will be minimized.

Our pedagogic study was multi-institutional and involved travel to several different towns/cities, so logistically arranging focus groups on the college or university site was

Table 2. Topic schedule, based on 'pyramid' approach

Pyramid approach	Question type	Topic/question	Example prompts (only use if required)
1. Pre-arrival	Opening question to put students at ease	What attracted you to the Foundation Degree?	Why did you choose an Fd course, rather than a BSc?
2. On-arrival	Transition question	When you arrived, did you feel equipped for university/college life?	Did you find the transition into the first year of university/college easy or difficult?
3. At college/university	Key questions (always ask positive questions before negative questions)	What are/were your positive experiences of the course?	What has the level of support been like? (e.g. tutor availability, availability/access to hardcopy/online resources)
		What are/were your negative experiences of the course?	
		What were your positive experiences of the institution?	How have these issues been dealt with? (if contentious issues raised by students)
		What were your negative experiences of the institution?	
4. Looking back	Transition question	So far, has your FE/HE experience been as you expected?	Is there anything you would have changed about your experience as a whole?
5. Looking forward	Closing question	What are your plans after completion of the Fd/BSc?	Do you plan to top-up to year 3? (reasons?) What career are you planning to pursue?

preferable. While the power-relationships associated with the campus environment (e.g. staff–student hierarchies) may be reinforced through this approach, conversely the venue is familiar to students. Focus groups took place in seminar rooms, and to provide an informal setting the focus groups were set up in a circle around tables, and refreshments were provided. Before beginning each focus group, the research aims were outlined, and participants were advised that they could withdraw without penalty at any stage of the process. BSU staff acted as moderators due to familiarity with project research aims and of the courses involved. Following Burgess (1996) two moderators were used for each group, one note-taker and one interviewer.

Recording, Transcription and Data Analysis

In the planning stages you will need to decide whether to record the focus groups. Commentators acknowledge that the presence of a recording device may deter some people from full participation (as can taking extensive notes), but it is also recognized that recording discussions provides the fullest record of the conversation, enables eye contact and removes the potential biases of hand-written notes. Whether or not discussions are recorded (traditionally using a dictaphone, or alternatively a video-recorder to record both sound and body language), it should be made clear to all participants that individual responses will be anonymized—which may alleviate some concerns with being recorded. The alternative to recording is to take written notes, and make notes of body-language (Onwuegbuzie *et al.*, 2009), and the tone of the discussions. Even if discussions are being recorded it is recommended that written notes should always be taken, in case of equipment failure (Dunn, 2000; Crang & Cook, 2007), and also to help with clarity. If taking written notes only, these should be written in short hand and fully written up as soon as possible after the session. Recording discussions may raise tensions as this can affect the flow of discussions, and ground rules need to be put in place, such as asking participants not to talk over one another, to facilitate clear recording.

A decision needs to be made about whether to transcribe the focus groups before analysis. If you have the option of paying a research assistant to undertake this time-consuming task (or out-sourcing the work to an online company), this is probably worthwhile. Where a detailed record is required, recordings should be fully transcribed and transcripts checked for accuracy against the recordings before beginning data analysis. Transcripts should then be read through several times in order to identify key themes within and between groups. Coding systems can be used to identify themes, such as colours, number, letters or symbols (Cameron, 2000). If working on a large study, it may be appropriate to use a qualitative data package such as Atlas ti* or NVivo (which has largely absorbed the earlier Nu*dist package) making it easier to assign codes to emerging themes and to establish relationships between themes (Breen, 2006; Onwuegbuzie *et al.*, 2009). However, these packages can be cumbersome and time-consuming to use at least initially. The final write-up should be balanced between direct quotes from participants and interpretation of the discussions (Cameron, 2000). Quotes may be used to reflect important themes and also to illustrate new findings.

For our study, focus groups were digitally recorded to provide a complete record of discussions and to aid in data analysis. Transcription was undertaken by a paid research assistant, scripts were cross-checked against the original recordings and participant responses were anonymized. Transcripts were used as the basis for our analysis and have allowed identification of both general themes and specific points, and aided in the process of 'triangulation' or cross-checking of participants views (Holbrook, 1996; Denscombe, 2010), both within individual focus groups and between the seven focus groups.

Challenges of Method

Participation

Participation in terms of attendance at the focus group and the contributions of individual members can be problematic to organize. Whether a focus group is embedded into a scheduled teaching session or outside the timetable, participation in the research must be voluntary to align with the ethical principles of research. Motivation for students to participate might include a sense of obligation, peer-pressure, volunteering with a desire to improve others' learning based on their own experiences or as a platform to convey grievances. Thus, participants of focus groups may be self-selecting and need to be carefully managed, in particular when participants become too vocal or seek to promote individual agendas. In our study, we worked with quite close-knit cohorts, but in some instances there were quieter and more vocal participants. In some focus groups, it was necessary to directly address quieter group members to ensure feedback from all participants.

Power Relations

The relationship between participants and moderators needs consideration, with arguments on both sides for familiar or independent moderators. In our case study, we used moderators that were familiar to students. Inevitably, this means that power-relationships between staff and students will have had some effect on what individual students felt able to discuss. However, familiarity with staff appeared to encourage students to participate and this approach wielded much more in-depth information than relying only on written module evaluations. As Goss & Leinbach (1996) note, focus groups do not eliminate unequal power relations between researcher and research subjects, but they provide a more open forum for discussion and are collaborative. If more than one institution is used, ensuring representivity and comparability between institutions is also important.

Benefits of Method

Informal Environment

Focus groups allow for liaison with students in an environment which is different from a more 'formal' teaching session. Staff may act as moderators, but the onus is on gathering student opinions and reflections on their experiences (and staff act as facilitators and are not 'experts' in the area).

Student Empowerment

Students are given the opportunity to discuss in-depth issues relating to teaching and learning and support networks in HE, rather than just providing summary data in response to quantitative surveys.

Potential to Integrate into Modules

Focus group methodologies can be appropriately integrated into the evaluation of modules, and could be used in a seminar-based context.

Dissemination of Experiences and Good Practice

The qualitative nature of the data allows for reflection on student experience, further developing provision or modifying learning and teaching methods where required, and allows for wider dissemination of good practice.

Implementing Change

In our study with Foundation degree students, the focus group findings (Simm *et al.*, 2012) led both to increased awareness amongst staff of support issues, and to the development of a series of one-off sessions which students could participate in. These included group workshops, reading group sessions and social events for Foundation degree students, events which address two issues (i) assisting student understanding of key concepts and (ii) providing identity to the degree programme. For the on-site Foundation Degree in Development Geography, student mentoring of second year students by third year top-up students was adopted to support students in undertaking their overseas development placement which is a key component of the course (Winlow *et al.*, 2007; Simm & Schaaf, 2011). This meant that students gained wider perspectives on how to adjust to a different culture and to carry out a small research project overseas, beyond the advice that could be offered solely by academic staff (who do not have in-country experience of all destinations).

The facilitation of pedagogic or administrative change is often a key objective of pedagogic research. It is important that pedagogic research is not an ephemeral episode but that findings are translated into constructive outcomes for participant institutions, courses, modules or cohorts of students. Outcomes may include disseminating findings as a report or presentation to participant institutions and staff, publishing in relevant journals or books, conference presentations and reporting to students through staff–student committee or similar formal mechanisms. Focus group discussions may reveal negative points, and researchers must exercise tact in dealing with contentious issues (in particular complaints directed at individual staff or institutions). This includes ensuring anonymity, and exercising caution when findings may involve 'washing dirty linen in public', which may arise with external dissemination.

Focus groups can provide insights that quantitative approaches, such as module evaluations or the National Student Survey, cannot. They can be a vehicle for student empowerment, offering students a platform to voice their views and to influence decision making, such as periodic reviews of teaching programmes or courses, and providing formative feedback on new modules. Through participation, students may also develop greater understanding and empathy for the objectives and challenges of university teaching and learning.

Acknowledgements

The project explained in the case-study was funded by the HEA-GEES Subject Centre Small-Scale Learning and Teaching Research and Development Fund 2008–2009. We thank the staff and students at Weston College, City College Plymouth and Bath Spa University for participating in this study and Emma Temple-Malt who transcribed the audio recordings.

References

Bennett, K. (2002). Interviews and focus groups. In P. Shurmer-Smith (Ed.), *Doing cultural geography* (pp. 151–162). London: Sage.
Breen, R. L. (2006). A practical guide to focus group research. *Journal of Geography in Higher Education, 30*(3), 463–475.
Burgess, J. (1996). Focusing on fear: The use of focus groups in a project for the community Forest Unit, Countryside Commission. *Area, 28*(2), 130–135.
Cameron, J. (2000). Focusing on the focus group. In I. Hay (Ed.), *Qualitative Research Methods in Human Geography* (pp. 50–82). Oxford: Oxford University Press.
Crang, M., & Cook, I. (2007). *Doing ethnographies*. London: Sage.
Denscombe, M (2010). *The Good Research Guide for Small-Scale Social Research Projects* (4th ed.). Maidenhead: McGrawHill/OUP.
Dunn, K. (2000). Interviewing. In I. Hay (Ed.), *Qualitative Research Methods in Human Geography* (pp. 50–82). Oxford: Oxford University Press.
Fallon, G., & Brown, R. B. (2002). Focusing on focus groups: Lessons from a research project involving a Bangladeshi community. *Qualitative Research, 2*(2), 195–208.
Fuller, I., Gaskin, S., & Scott, I. (2003). Student perceptions of geography and environmental science fieldwork in the lights of restricted access to the field, caused by foot and mouth disease in the UK in 2001. *Journal of Geography in Higher Education, 27*(1), 70–102.
Goss, J. D., & Leinbach, T. R. (1996). Focus groups as an alternative research practice: Experience with transmigrants in Indonesia. *Area, 28*(2), 115–123.
Gulliver, J., & Tyson, G. A. (2010). Exploring university students' perceptions of plagiarism: A focus group study. *Studies in Higher Education, 35*(4), 463–481.
Holbrook, B. (1996). Shopping around: Focus groups research in North London. *Area, 28*(2), 126–142.
Hopkins, P. E. (2007). Thinking critically and creatively about focus groups. *Area, 39*(4), 528–535.
Hydén, L. C., & Bülow, P. H. (2003). Who's talking: Drawing conclusions from focus groups – Some methodological considerations. *International Journal of Social Research Methodology, 4*, 305–321.
Kneale, P. (2002). Developing and embedding Reflective Portfolios in Geography. *Journal of Geography in Higher Education, 26*(1), 81–94.
Krueger, R. A., & Casey, M. A. (2009). *Focus Groups: A Practical Guide for Applied Research* (4th ed.). London: Sage.
Longhurst, R. (1996) Refocusing Groups: pregnant women's geographical experiences of Hamilton, New Zealand/ Aotearoa, *Area, 28*(2), pp. 143–149.
Longhurst, R. (2003). Semi-structured Interviews and Focus Groups. In N. J. Clifford & G. Valentine (Eds.), *Key Methods in Geography* (pp. 117–132). London: Sage.
Onwuegbuzie, A. J., Dickenson, W. B., Leech, N. L., & Zoran, A. G. (2009). A qualitative framework for collecting and analyzing data in focus group research. *International Institute for Qualitative Methodology, 8*(3), 1–21.
Raby, R. (2010). Public selves, inequality, and interruptions: The creation of meaning in focus groups with teens. *International Institute for Qualitative Methods, 9*(1), 1–15.
Simm, D., & Schaaf, R. (2011). Providing specialist support to peers: Evaluating a student mentoring scheme for semi-independent overseas fieldwork. *Planet, 24*, 30–36.
Simm, D., Marvell, A., Schaaf, R., & Winlow, H. (2012). Foundation degrees in Geography and Tourism: A critical reflection on student experiences and the implications for undergraduate degree courses. *Journal of Geography in Higher Education*. DOI: 10.1080/03098265.2012.692075.
Simm, D., Marvell, A., Winlow, H., & Schaaf, R. (2011). Student experiences of Foundation degrees in Further and Higher Education. *Planet, 24*, 2–9.
The British Sociological Association. (2002). Statement of Ethical Practice for the British Sociological Association. Available at http://www.britsoc.co.uk/equality/Statement+Ethical + Practice.htm (accessed November 2011).
Valentine, G. (2005). 'Tell me about…': Using interviews as a research methodology. In R. Flowerdew & D. Martin (Eds.), *Methods in Human Geography a Guide for Students Doing Research Projects*, 2nd ed., (pp. 110–127). Harlow: Pearson Education.
Wibeck, V., Dahlgren, M., & Oberg, G. (2007). Learning in focus groups: An analytical dimension for enhancing focus group research. *Qualitative Research, 7*(2), 249–267.
Winlow, H., Simm, D., & Haslett, S. (2007). Extended work-related placements in developing countries: Supporting remote learning and assessment. *Planet, 18*, 28–32.

The Power of Debate: Reflections on the Potential of Debates for Engaging Students in Critical Thinking about Controversial Geographical Topics

RUTH L. HEALEY
Department of Geography and Development Studies, University of Chester, UK

ABSTRACT *Many controversial subjects characterize geography in the 21st century. Issues such as climate change, sustainability and social exclusion generate much discussion and often involve clear differences in opinion of how they might be addressed. Higher education is an important space for critical engagement with challenging issues. Preparing for and participating in debates enables students to develop critical thinking skills, alongside a variety of oral presentation and discussion skills. This paper reflects on the potential for teaching through debate in geography. The arguments are illustrated through a debate about whether asylum seekers should be allowed to work in the UK.*

Debates as a Method for Developing Critical Thinking and Oral Skills in Geography

Many controversial subjects characterize geography in the 21st century. Issues such as climate change, sustainability and social exclusion generate much discussion, particularly when explored through a geographical lens of who gains and who loses, when and where. Consequently, a student graduating from an undergraduate geography degree should be able to analyse critically the evidence on such topics from several different perspectives and be able to express their opinions on how these issues should be tackled orally as well as in writing (Quality Assurance Agency [QAA], 2007). Organized debates in class are one teaching method which supports students in developing such skills.

It has been proposed that we should "devote heightened levels of energy to activities that offer the potential to sustain universities as places for critical, activist and just scholarship" (Hay, 2001, p. 141). The debate activity discussed here supports the development of critical thinking skills, but also has the potential to support critical engagement, along the lines suggested by Hay (2001). Teaching, which offers students the opportunity for critical reflection on the wider world, may have long-lasting impacts on their views and the potential for social transformations (Castree, 2000; Cook, 2000; Wellens *et al.*, 2006).

This paper argues for the value of debate in teaching geography in higher education. As a pedagogic exercise, a debate offers the potential to engage students in critical thinking, analyse political issues and develop a range of transferable skills. The extent to which a debate delivers on all of these aspects depends upon the design. For example, it may provide opportunities to develop transferable skills, but the topic under discussion may not relate to critical thinking about a political issue. Here both critical thinking about controversial issues and the development of key transferable skills are brought together. This paper focuses upon an example of teaching through debate in a small class, at a post-1992 university in the UK.

Debates within Teaching

The use of debate to explore different perspectives and arrive at conclusions dates back to the Egyptians over 4000 years ago (Kennedy, 2009). Debates as a method of teaching began in ancient Greece with the 'father of debate', Protagoras of Abdera (481–411 BC) (Smith, 1918; Kennedy, 2009). Whilst academics in the 1980s highlighted debate as "a promising teaching device" (Lewin & Wakefield, 1983, p. 116), only certain disciplines, such as law, have taken the method on board to any great extent. The use of debates within higher education has generally remained restricted predominantly to extra-curricular debate teams outside the classroom (Bellon, 2000). This limited use may relate to the criticisms levied against the Socratic debate method, for being too adversarial and combative. Yet, research has indicated the effectiveness of debates in supporting learning in a variety of disciplines, including economics, education, history, marketing, medicine, nursing, psychology, sociology and social work (Dundes, 2001; Helenius *et al.*, 2006; Omelicheva & Avdeyeva, 2008; Kennedy, 2009). There is, however, a significant dearth of literature specifically reflecting on the use of debates in teaching geography, despite the range of topics in the discipline which offer potential for such a method. Research has discussed opportunities for exploring particular sides of debates through role playing activities such as in preparation for planning a public inquiry (Maddrell, 1994; Livingston, 1999). This offers the opportunity to focus upon particular perspectives of issues, but not the same level of 'thinking on your feet' which forms a central part of a debate. Learning to cope with the uncertainty associated with participating in a debate is a useful skill in future employment and is seen as an increasingly important function of higher education (Barnett, 2000).

A debate is defined as a formal discussion about an issue or a problem (Thomas, 1996). Historically debates have an oral tradition, although Web 2.0 technologies have the potential also to create a space for online debate in a written format (Selwyn, 2007), or virtual debates via *Second Life* or *Skype*. Other pedagogic options for developing oral communication skills include oral presentations and discussions. During oral presentations the students who are not participating in the presentation are often largely passive audience members, whereas in discussion activities and tutorials students rarely have specific sides that they have to argue for. Debates are therefore distinct as an active, argument form of oral pedagogy (Omelicheva & Avdeyeva, 2008; Kennedy, 2009), which can bring drama to the classroom and engage students in lively discussion (Green & Klug, 1990; Crone, 1997; Helenius *et al.*, 2006). Kennedy (2009, p. 225) argues that "students learn more effectively by actively analyzing, discussing, and applying content in meaningful ways, rather than by passively absorbing information". It is suggested that through debate students gain greater knowledge by reinforcing material already covered in the lecture (Crone, 1997; Kennedy, 2009). However, there is relatively little research demonstrating the value of this method to students and little scholarship comparing debates with other teaching methods (Omelicheva & Avdeyeva, 2008).

Although, as already noted a debate does not have to be conducted in an oral format, the opportunity to engage in immediate counter-arguments and rebuttals is likely to be enhanced through live oral communication. In whatever format a debate takes place, it offers the opportunity to develop enhanced written and/or oral communication skills; research skills; analysis, synthesis and evaluation skills; critical thinking skills; argument and persuasion skills (Green & Klug, 1990; Dundes, 2001; Kennedy, 2009). Significantly, these include the development of the higher level cognitive skills of Bloom's Taxonomy (Kennedy, 2009). Omelicheva & Avdeyeva (2008, p. 606) go as far as to argue that "debates appear to be more effective in developing students' comprehension of complex concepts and application and critical evaluation skills" than lecture-based teaching. The method makes it necessary for students to seek reasons to justify their viewpoint by developing such abilities as identifying value assumptions within arguments and judging whether data are misleading or absent (Green & Klug, 1990). This pushes students to evaluate critically the evidence on either side of a debate and creatively build counter-arguments. Debates offer such opportunities as they encourage students to question people's behaviour and perspectives, helping them to recognize the complexity of decisions and opinions.

The method of debate offers students the opportunity to stimulate development of their critical thinking through encountering students with views contrary to their own, and in doing so induce them to either change those views or learn to defend their own views with better logic and more substantial evidence (Green & Klug, 1990). Students rarely experience a direct challenge to their ideas, and as such they infrequently have the opportunity to defend their own arguments (Green & Klug, 1990). The highly engaged atmosphere of a debate supports students to focus on their attitudes and opinions from a particular perspective (Omelicheva & Avdeyeva, 2008). Yet, although they are focused on one side of an argument, the formation of successful counter-arguments requires an understanding of the opposing viewpoint.

Teaching for Critical Thinking through Debate

> Critical human geography must make radical pedagogy a central concern—and must recognize the classroom as a site of potential political engagement. (Heymann, 2000, p. 303)

Critical thinking skills can "give students the tools to understand what they are learning" (Wolf *et al.*, 2010, p. 43). This ability relates to analytical, interpretation, inference, explanation and evaluation skills (Facione, 2000). Through using these skills students may monitor and, where appropriate, correct their own reasoning, meaning that critical thinking is about "judging in a reflective way what to do or what to believe" (Facione, 2000, p. 61). Critical thinking is a complex term, meaning different things to different people. It may be considered as "the identification and evaluation of evidence to guide decision making. A critical thinker uses broad in-depth analysis of evidence to make decisions and communicate his/her beliefs clearly and accurately" (Critical Thinking Co., 2011, no page). Table 1 outlines some of the different meanings associated with the term critical thinking, alongside different types of activity which support the development of critical thinking and its subsets.

Table 1. Critical thinking and its subsets

Critical thinking and subsets of it—tools for the manipulation of knowledge	Forms of primary and secondary representation
Critical thinking—which includes critical: appraisal evaluation reflection understanding analysis review appreciation management awareness care	Oral representation— debate, discussion and other oral representation Written representation— critical reports, reviews, critique, satire, essays, metaphor— Graphic depiction—cartoon, pastiche, sketch
Critical incident analysis Problem-solving and decision-making are forms that can be broadly similar to critical thinking when there is no one fixed solution to be sought	Forms of action— assertive action; critical or professional practice, reflective practice, dramatic or the applicable representation, etc. Various forms of representation

Source: Moon (2008, p. 30).

Focusing upon oral representations specifically it is possible to recognize how students may engage with the different elements of critical thinking as they prepare for a variety of different types of oral activity. However, a debate provides engagement in a quite specific way by forcing a student to evaluate and reflect on materials and to produce a supporting argument for a particular position. In order to do this, it is necessary for the student to be aware of the strengths and weaknesses of the opposing position(s) and put forward a case for why their argument is the strongest. A discussion encourages students to explore different ideas depending upon how the conversation goes, but may lead to an exploration of the most interesting aspects of an issue; an oral presentation may lead students to focus on the arguments in the literature that support their point and ignore those that do not; in contrast in order to be successful in a debate students need to appreciate the issue from several different angles, and develop these views to form the most appropriate argument. Geography as a discipline is well positioned to contribute to teaching critical thinking as the subject often requires students to present informed opinions based on the analysis and synthesis of information from a variety of sources (Korkmaz & Karakuş, 2009). Preparing and conducting a debate requires such analysis and synthesis of the opposing sides of the argument. Fullan (2003, p. 23) suggests starting with the notion of a moral purpose, identifying key problems with that purpose, suggesting desirable directions, but allowing the students to find their own path through the topic. Through the preparation for a debate students are then able to be creative in deciding how they wish to argue for or against the topic. Jackson (2006) argues that people tend to be happier or more satisfied if they are able to be creative. In choosing a debate topic, it is necessary to make sure that the chosen topic has clear opposing sides. Through preparation, discussion and critical reflection it is possible for students to investigate and develop the nuances of individual arguments.

Case Study

This debate was an unassessed exercise as part of a third-year undergraduate degree module entitled: 'Excluded Peoples? Migrants and Refugees'. This is a 24-week unit in which the students have contact time of two hours a week. This module is open to students studying Single and Combined Honours Geography, and Combined Honours International Development Studies. The preparation and the debate ran over two consecutive teaching sessions (four hours), beginning around five weeks into the course. The moral purpose in this exercise was the right of asylum seekers to work in the UK. An asylum seeker is someone who seeks recognition as a refugee. A refugee is defined as a person

> who owing to a well-founded fear of being persecuted for reasons of race, religion, nationality, membership of a particular social group or political opinion, is outside the country of his nationality and is unable or, owing to such fear, is unwilling to avail him [or her]self of the protection of that country; or who, not having a nationality and being outside the country of his [or her] former habitual residence as a result of such events, is unable or, owing to such fear, is unwilling to return to it. (United Nations, 1951, 16, Article 1.2)

The majority of the students on the module were studying geography and had little, if any, past formal tuition about asylum. Although some of the students doing International Development Studies would have had a brief introduction in the first year of their degree.

There is much media interest in the experiences of asylum seekers, particularly as a consequence of perceived abuses of the asylum system (Schuster, 2003). Many governments have sought to create increasingly restrictive policies towards asylum seekers resulting, for example, in the right for asylum seekers to work in the UK being withdrawn in 2002. The right of asylum seekers to work draws upon particularly geographical moral issues. Smith (1995, p. 275) argues that geography is well placed to address 'moral issues':

> The most geographical of human practices may be to define a space or territory, whether by erecting fences or borders, or simply drawing lines on a map. In so doing, we may also be defining who belongs to a community or nation (e.g. who is a citizen).

The findings discussed here are two years worth of data based on student reflective responses to a series of points before and after the debate had been conducted. The 2009 class consisted of 12 students (5 male, 7 female) and the 2010 class consisted of 13 students (3 male, 10 female), all of the students were in their early to mid-20s. The majority of these students were present for the debate (11 in 2009 and 10 in 2010). In both cohorts, all except one student participated in the reflection pieces before the debate; however, those who did not attend the debate itself did not complete a second reflection.

Pre-Debate Views

Having established ways of thinking about exclusion and inclusion over the first few weeks, the students on the course were asked to reflect on and write down their initial views towards the following three points:

1. Do you think asylum seekers should be able to work in the UK?

2. Why do you think asylum seekers should or should not be able to work in the UK?
3. Comment on your views of the way asylum seekers are treated in the UK.

This exercise helped the students to clarify their initial views before being influenced by the positions put forward in the resource pack of academic and grey literature on the subject. This helped students to focus on their personal position in relation to the topic. When students take personally the issues they study, they draw these ideas into their everyday lives (Angus *et al.*, 2001).

This debate was, in part, designed to teach for social transformation. This is "teaching that aims to promote knowledge, skills and values amongst all students that, through critical thinking, encourages social justice and equity" (Wellens *et al.*, 2006, p. 118). The tutor had anticipated that with relatively little knowledge of the subject the students would believe that asylum seekers *should not* work in the UK. It was hoped that through the debate students would explore whether or not this was a socially just policy. However, of the reflections written prior to the first teaching session (9 out of 12 students—2009 cohort—and 12 out of 13 students—2010 cohort—participated in writing the reflections) the majority of the students entered the debate believing that asylum seekers *should* be allowed to work in the UK. This is perhaps an unusual reaction when compared to general attitudes towards immigrants in the UK (Crawley, 2005). Students who choose to take a module about migrants and refugees may be predisposed to be more positive about the rights of such individuals than students who do not. However, the students' justifications for why asylum seekers should be allowed to work varied from a focus upon moral responsibility towards individuals in need, to only if it would benefit the broader society (Table 2 illustrates the different types of responses, all student names are pseudonyms). The majority of students justified their beliefs that asylum seekers should be allowed to work in the UK on the basis of the benefits this would have to the UK as a whole. Even those students who demonstrate a moral position supported this further through how this was good for the whole country. Only two students across both cohorts entered the debate preparation believing that asylum seekers should not be able to work in the UK.

At the teaching session the students listened to a talk about the asylum system in the UK and the rationale behind the change in policy withdrawing of the right to work in 2002. After this brief introduction to the topic, the groups were divided into two opposing teams (in 2009 this was done randomly, and in 2010 the division was more strategic to separate individuals whose attendance had been variable at the teaching sessions). Each team was provided with the same pack of information containing material from both sides of the debate, alongside details of the debate format. One student commented how he was excited about the teaching method as it was different from his past experience, describing it as "a new way to learn" (*David*, 2009). They were given two hours in teaching time and two weeks between the introduction of the topic and the debate itself to prepare. The debate was then run as a panel debate.

The panel debate chosen was highly structured (see Appendix 1 for debate structure). Each of the teams chose to allocate members a specific role within the debate. These roles coincided with different stages within the debate so that individuals could prepare the opening argument, a specific argument or cross-examination as part of the main debate, or the closing argument in advance. Each section of the debate then had a set time in which

Table 2. Examples of students' initial views on the topic

Focus of viewpoint	Examples of student comments (quoted directly from students' written reflections)
Moral right of the individual	"From a personal and moral perspective I feel that the Asylum seekers should be allowed to work. From the limited knowledge I have learnt so far, asylum seekers live in terrible conditions not only in their country of origin, but also in the UK. They may spent years waiting for their application to be accepted in the meantime they have no money, little support and they have to cope with cultural differences, language barriers and so much more! From the video I realized how much asylum seekers want to work and support themselves ... I believe that asylum seekers should be given the chance and right to work but also to pay taxes and contribute to society" (*Sarah*, 2009). "I absolutely think that asylum seekers should be able to work in the UK, it seems illogical for them to not ... Excluding them from work just re-enforces the 'us and them' attitude portrayed by the media and creates an even bigger divide" (*Lisa*, 2009).
For the good of broader society	"I believe that asylum seekers should have the right to work in the UK, as surely it's more beneficial for them to provide towards taxes than to claim benefits" (*Susan*, 2009). "Yes I think that asylum seekers should be able to work particularly as it has not been proved that they cause unemployment for British citizens across the UK. If asylum seekers were given the opportunity to work, this would enable them to pay for food and accommodation themselves and by not sleeping on the streets this could also reduce crime rates. In employment they would also be paying National Insurance and Tax. As the UK is becoming less attractive due to the recession, major UK skills problems are being caused therefore allowing asylum seekers to work would 'fill these gaps' and employers would not have to spend the extra time and money training other people" (*Larissa*, 2010). "An asylum seeker who is able to work after obtaining key skills before entering the workplace can only be beneficial to themselves and the UK, wages for themselves and taxes for the government to cover any costs an asylum seeker may cause whilst in the UK" (*Adam*, 2009). "They should be allowed to work as they then won't be living off benefits and will be contributing to the UK through tax and national insurance. It will then stop them working illegally and earning a wage that still getting benefits" (*Ella*, 2010). Asylum seekers should be able to work in the UK "because if they came to our country they should be able to give something back to the country" (*Dominic*, 2010).
As long as it does not negatively affect broader society	"They need to be able to earn money so that they can independently support themselves. However, letting all asylum seekers work may attract a large amount of asylum seekers that could lead to many asylum seekers and not enough jobs" (*Beth*, 2010). "I do think that asylum seekers should be allowed to work in the UK, providing there isn't a shortage of jobs for people within the UK. They deserve to be given the right to work and contribute to the UK's economy if they have be given no option but to move here" (*Sadie*, 2009).
Against as it will affect broader society	"We already have a high unemployment rate in the UK at the moment, I think it's important to get citizens working. When granted asylum they should be able to work" (*Rebecca*, 2010). "It would be bad for the economy if every asylum seeker was allowed to work in the country" (*Carter*, 2010).

they could make their point before the opposing team would be able to challenge their position.

Post-Debate Reflection

Immediately after the debate the students were given feedback from the tutor on the points each team made. The points made were to the entire team rather than to individuals within the team. This discussion emphasized the quality of the argument, alongside the use of evidence to support points. As part of the learning process it is necessary for the students themselves to reflect on what happened in the debate (Crone, 1997; Toohey, 1999). Through this reflection the students were able to engage with the strongest arguments from each side—those which had been evidenced and argued well by each team. Light & Cox (2001, p. 80) argue that "evaluations should not ... simply reflect what participants liked and disliked but how they felt it affected them". Consequently, after the debate the students were asked to reflect upon and make notes in relation to how they now felt about the question—should asylum seekers be allowed to work in the UK? And what they thought of the debate method of learning about the issue?[1]

The students' second reflective pieces illustrated how learning about the topic in this way has given some of the students a different perspective. The opportunity to engage with one issue affecting a particularly marginalized group in the UK offered the opportunity for students to recognize more general points about asylum seekers' daily lives.

> [The debate] opened my eyes more to the lose-lose situation which asylum seekers are currently in due to the stigma they have. (*David*, 2009)

> British citizens require a good education [about] asylum seekers and the difficulty they are under. (*Adam*, 2009)

David and *Adam* both illustrate how a lack of understanding—their own, or the general public—is a factor which influences the experiences of asylum seekers. In contrast to her original perspective (see Table 2), at the time of reflection, *Rebecca* believed that asylum seekers should be able to work on moral grounds.

> I think they should be given the option of working, and it depends on the person and their background as to whether they want to work, it's a human right. (*Rebecca*, 2010)

This student moved from being against asylum seekers working in the UK, to believing that on moral grounds it is a 'human right' that individuals be allowed to work. This was the most extreme change in viewpoint, for the majority critical engagement with the issue led to more nuanced opinions.

> Definitely should be able to work, but maybe not automatically. (*Lisa*, 2009)

> In the current system asylum seekers shouldn't work as it is not set up for them to work. However, I think in a new system, and with regulations they should be able to work. (*Ella*, 2010)

Both of these students were pro-asylum seekers working initially. Whilst their viewpoints have not changed entirely, they continue to believe that asylum seekers should work in the UK—their critical insight into the issue has led them to believe that the current system is inadequate and it would be inappropriate for asylum seekers to be working in the UK under such conditions. Although, in general, the students who wrote their reflections had similar perspectives when it came to asylum seekers working, their reasons for this had developed through the course of the debate. It is their justifications and critical understanding which have been challenged through debate and discussion.

Barnett & Coate (2005, p. 124) argue that it is important for students to "develop powers of self-reflection and so heighten their sense of self as emotional as well as intellectual beings". After the debates the team opposing asylum seekers working in the UK were asked how they really felt about the issue, given that the majority of students had noted in their original views that they thought asylum seekers should be able to work in the UK. In the 2009 cohort, each of the six members said that they still believed asylum seekers should be allowed to work in the UK. It is important to note here that these six students were therefore arguing against their own beliefs. Whilst none of the students commented on this in their reflection notes after the debate, it is likely to have influenced their learning through the debate process by forcing them to challenge their own position and identify the most convincing arguments against their personal viewpoint. In the 2010 cohort, arguing against asylum seekers working in the UK pushed Jo's opinion, from one who was very certain that asylum seekers should be able to work to a much more critically reflective position:

> It's a very complex problem. Theoretically, yes [asylum seekers should be allowed to work], but the [current] technicalities [prevent this], i.e. ensuring support is automatic ... [support] needs to be targeted towards individuals. The system needs reform. (*Jo*, 2010)

Both groups on either side of the debate examined the arguments on both sides. However, the detailed learning required by the opposing group, where the position they were arguing went against their personal beliefs, may have led to a deeper level of learning. Through investigating arguments which are in opposition to their own, and then presenting that position as a stronger argument than their personal beliefs, led these students through three stages. Firstly, they had to be prepared to challenge their initial perceptions; secondly, they had to work through the opposing views and look for appropriate arguments; and thirdly, they needed to reflect upon and consolidate the arguments supporting their own perspectives in order for them to continue to believe that asylum seekers should be able to work in the UK. The critical insight and more nuanced responses from the students were in contrast to some of the debate literature which found that debates shaped learners' attitudes, often away from their original perspectives (Green & Klug, 1990; Omelicheva & Avdeyeva, 2008; Kennedy, 2009). Here the majority of students still maintained their original views, but were more tempered in their statements, and nuanced in their reasoning.

Student Views on Debate as a Teaching Method

Having highlighted the critical thinking potential of a particular debate, this section focuses upon the student responses to the debate method. This highlights the more general

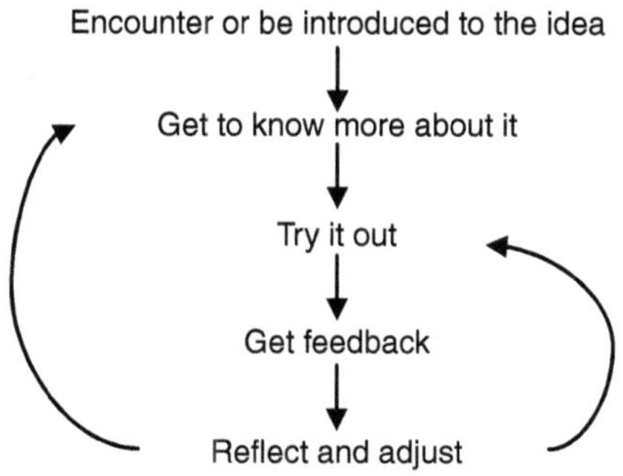

Figure 1. A simple model of the learning process. *Source*: Toohey (1999, p. 154). Reproduced with the kind permission of Open University Press. All rights reserved.

skills and knowledge, which the students highlighted as beneficial from the debate, and the views they had about the debate structure and format. These views are situated within Toohey's (1999) broader model of the learning process, illustrating how debate addresses different aspects of learning (Figure 1).

Skills and Knowledge

After the debate the students were asked to note down what they thought of the debate method of teaching. All of the students who participated in the debate found it to be a useful exercise:

> I enjoyed it—learnt more. (*Sophia*, 2010)

> The debate was really good as a different way of learning!! Really fun ☺ (*Larissa*, 2010)

A common theme in response to the debate was that students felt that they had learnt more than they would have done had the subject been taught as a lecture. Despite some of the students anticipating that the debate might be a nerve-racking experience, they were surprised to find it to be an enjoyable way of learning.

> Yes, was nervous at first but was different and fun, different way of learning and would like to do it again. (*Rebecca*, 2010)

> Very good, [and] not too scary. (*William*, 2010)

The initial fears about uncertainty, and managing to respond to potentially unexpected arguments, had led to some anxiety for the students, but for these students the process itself had been enjoyable.

Further comments on the debate link to features of critical thinking, defined earlier as reflective, analytical, interpretation, inference, explanation and evaluation skills (Facione, 2000). *Dillon* comments:

> I thought the debate format worked well. It is an interesting technique of learning—you end up learning things without really trying. (*Dillon*, 2009)

Table 3. Interpreting Toohey's model of the learning process in relation to the debate method

The process of learning though debate	Explanation
1. Engage with the student's current knowledge and awareness of an idea or issue.	In the introductory stage educators should engage with what students already know and check for misconceptions, if their previous learning contradicts or undermines what they are hearing in the classroom (such as media reports on asylum seekers), students "are more likely to cling to their previous conceptions unless they are pushed to confront inconsistencies" (Toohey, 1999, p. 154).
2. Students are given opportunities to explore new ideas, potentially challenging their original views.	Students are given the opportunity to get to know more about the topic. They are able to investigate the issue making the content of the topic meaningful and comprehensible to them. They begin to take ownership of the topic.
3. Students are given the opportunity to take ownership of the idea through an independent task allowing them to potentially discover the unexpected.	Students are given the opportunity to try out their knowledge. The student is given space in which to develop, "to come into herself and to feel a proper ownership of what she thinks and does" (Barnett & Coate, 2005, p. 125). In this case the class debate challenges each others' perspectives on the topic.
4. Through the debate itself and discussion with peers and the tutor, students gain feedback on their views and development of their understanding.	Students are given formative feedback to help develop their understanding of the topic.
5. Through reflection students may recognize some alteration in their original perspective as ideas become more embedded.	Reflecting on their experience takes account of the feedback and deciding how one's performance may be adapted on the next occasion. Reflection is a useful opportunity for students to clarify their focus and understanding of the topic.

For *Dillon*, the debate was a more effective form of learning than listening to a lecture. The debate offered him the opportunity to develop his interpretation and understanding of the topic without having to exert the same amount of effort which he would have in a lecture. Using Toohey's (1999, p. 154) model of the circular learning process, it is possible to illustrate the ways in which debate may support learning of a topic (Table 3). Further students illustrated the analytical and evaluative skills they had the opportunity to employ:

> The debate 'worked well for critically evaluating points'. (*Ella*, 2010)

> The "method of debate was useful to hear everybody's opinion yet have the opportunity to counteract and cross-examine what the opposing team said". (*Fiona*, 2009)

The dialogue between opposing sides offered students the chance to evaluate and analyse different arguments and forming counter points. Through this interchange students were able to reflect upon all the arguments on either side of the topic.

> The debate is a good method of getting a full viewpoint and the arguments surrounding them. (*Beth*, 2010).

> ...a debate is a good way to voice your opinion and to listen to that of others. (*Susan*, 2009)

Such reflection and inference enabled students to develop their own opinions through the course of the preparation and the debate itself.

Alongside this, the debate offered more practical skills in terms of constructing and orally presenting an argument in response to the point under discussion in the debate:

> Helps you to learn to think on your feet—good skills in life. (*Caroline*, 2010)

As *Caroline* points out the skills students are developing through debate are broader than just the academic topic under discussion. The decision to utilize a panel debate format also offers the opportunity to develop team work skills. However, not all students felt that this was entirely successful. As *Sadie* comments:

> It would of worked better if worked more together. (*Sadie*, 2009)

As an unassessed exercise, it is inevitable that not everyone put the same level of effort in, particularly when it came to working together and organizing themselves in advance of the debate. In the 2010 cohort, two students did not get involved in the preparation, or show up for the debate itself. In the previous weeks, these students' engagement had been variable, and consequently in organizing the groups, it was made sure that they were not in the same group. Overall, the response to the debate, however, was highly positive. Having had the opportunity to experience learning through debate, the student representatives for the module formally commented at both Staff Student Liaison Committees in 2009 and 2010 that they found the debate enjoyable and useful, and would like to have more opportunities to learn in this way.

Format

After the 2009 debate a couple of students requested that in future they had more time to cross-examine the opposing team, having found that they ran out of time in the structured debate format.

> I really liked the debate format, however, [I] would have liked more time to cross-examine. (*Lisa*, 2009)

The timings in this structure allowed double the amount of time for students to present their arguments (six minutes) in comparison to the opportunity for cross-examination (three minutes). For the 2010 cohort, the timings were changed to allow equal time for cross-examination as for presenting the argument (Appendix 1). Once again the students requested that they had more time for cross-examination of one another's ideas. This is indicative of the developing abilities of some of the students to critique the opposing team's arguments, and the desire to do a more thorough questioning of the arguments presented. Although not everyone was as comfortable with oral communication or with developing rebuttals spontaneously as part of the flow of the debate, as part of a panel debate method, the students could work collaboratively and include the strengths of different individuals in the group with the different tasks required in debate preparation and execution. The more reserved members of the class may participate to a lesser extent in the cross-examination of arguments; however, the preparation of arguments enabled students with different styles of learning to engage with the material in a variety of different ways.

Broader Applicability

The previous two sections have highlighted the potential for debate in terms of teaching for critical thinking, alongside the positive response from this group of students to learning in this way. This section considers broader applicability of such teaching by, firstly, briefly discussing how debate might be appropriate in different contexts and, secondly, the potential of debate for teaching for social transformations.

Debate in Different Contexts

The appropriateness of debate in different contexts predominantly relates to the format of the debate. Different variations of a debate relate to different topics and classes, and the purpose of the debate. As already mentioned, many controversial subjects characterize the present study of geography. Whether it is the most appropriate way to address climate change, what is a sustainable way of living or why particular people experience exclusion, there are many different opinions and arguments to be had. In terms of establishing a successful debate, it is necessary to focus the topic around a key question or issue that has distinct positions which the students may be able to draw upon. The preparation may then focus upon clear differences in perspectives, with more able students illustrating the more nuanced and complex points as part of their argument in the debate itself. If the different perspectives are too closely related, it may make it difficult to start off with a broad enough distinction between positions for the students to critically analyse the material focusing on the complexities of the discussion. The example illustrated here has been within a particular context and may not be an appropriate format for other classes. The way in which this debate was structured worked with a relatively small number of students, but would be more challenging in a larger class. It may be possible to divide a larger class into tutorial groups and run the debate with smaller numbers; where this is not possible Table 4 illustrates some different types of debates and how these might be used in different contexts. The table illustrates a range of different formats in which a debate may take place, and the potential points at which critical thinking may occur within them. Readers may use this table to stimulate ideas on the most appropriate way to adjust the use of this teaching method to local circumstances. These examples are not mutually exclusive categories, for example, it is possible to have a structured and graded debate or a role play, instructor debate. The different debates below offer variable levels of tutor control. For readers who are less comfortable with the unexpected, the options with greater structure may be less disconcerting.

Each of these different forms of debate has the potential for engaging students in critical thinking in the three stages of the debate:

1. *Preparing for the debate*: Critical thinking may occur through appraising, analysing and evaluating material in order to plan their arguments from different

Table 4. Examples of different types of debates

Type of debate	Example
Two students debating	The whole class is told to prepare for the debate, possibly asking them to focus on one side. The two participants are picked at random at the beginning of the debate class. The student audience could then be involved by, once again randomly, choosing people to ask questions of the two students debating. Although the students will all have been asked to prepare for the debate in advance, after randomly selecting the individuals it might be sensible to give them some time to reflect upon their notes before beginning the debate itself.
Panel debate	The students are organized into panel groups. When working in a team to prepare for a debate, students have the opportunity to learn effectively through comparative collaboration rather than working alone (Green & Klug, 1990). This can "foster a sense of camaraderie among students" (Dundes, 2001, p. 242). Depending on numbers this might be done so that there are several groups on each side of the argument who are then randomly selected to participate in the debate.
Instructor debate	The whole class is asked to prepare the questions for a debate around a particular topic. Two tutors are then asked the questions and offered the opportunity to explain their position. This could be run in a similar way to Presidential election debates. The tutors could be chosen on the basis of their differing perspectives, or they could role play a particular position.
Role play debate	Students are given defined roles in a debate. This works well when there are different arguments to be made from several different stakeholders. The students have to identify their position on the basis of the role they have been given (N. Tyrell, personal communication, 20 July 2010). In a large class it may be preferable not to run the debate in front of the whole class. Instead, students could be paired off, on the basis of the different positions they are role playing, and debate the issue with one another before feeding back the key ideas to the class. If the debate is run in front of the whole class people can again be put into pairs, with the audience asking questions of each person, or as a panel debate with people who are broadly for or against an issue putting forward their perspective.
Graded debate	A debate may be used as a teaching exercise or as an assessment. As a graded debate students may be given marks for when they 'win' a particular point (they present the strongest argument), the quality of their contributions overall, and who wins the debate overall. The features which are focused on will depend upon the class degree level and the aims of the assessment.
Structured debate	Like the one used in the example in this paper. A clear structure is in place where by each point to be presented in detail with the opportunity for follow up questions and comment from the opposing team (Tomlinson, 2009).
Unstructured debate	This is a less rigid structure than the model shown in the appendix, in which students are allowed to focus on discussion and in which overt competition is discouraged (Helenius *et al.*, 2006). It is likely that within this format the lines on either side may break down and there are no outright winners as the groups challenge and discuss more, rather than concentrate on clearly defined points and rebuttals. This more open structure still requires some guidance, but is underlain by greater flexibility.
Online debate	An online synchronous discussion offers the opportunity to engage people at a distance, and perhaps appeal to students who are not as confident speaking in front of the class (Dengler, 2008). This could be established for a debate.

perspectives. When working with others they may reflect on and discuss their arguments.

2. *During the debate*: When students are participating in the debate critical thinking may occur when they are thinking on their feet and considering how to phrase their arguments. When they are listening to the debate critical thinking may occur as students gain an understanding of the different positions being put forward and appraise, evaluate and reflect upon the strength of the different arguments.
3. *After the debate*: Critical thinking may occur as they formulate questions of panellists and when they reflect upon tutor feedback.

Potential of Debate for Teaching for Social Transformations

Debates have the potential to be a method for teaching for social transformations. Robinson (1988) illustrates how teaching about social transformation can deconstruct students' initial hostility, sympathy or paternalism towards the 'other', moving towards accepting another person as equal, an understanding of the context within which the other person lives, and an acceptance that the other person's value system is a valid alternative to their own. Educating people about social transformation may lead to social transformation (Ilkkaracan & Ercevik Amado, 2005). Yet, defining the specific form of social transformation itself is likely to be different depending upon the perspective of the teacher or teachers involved. What is more common is the desire to challenge in multiple ways in which the distinction between 'us' and 'them' is played out (Cook, 2000). This involves developing an understanding that "all identities are socially constructed ... we are all members of many overlapping social entities which help to make up ourselves" (Rogers, 2006, p. 134). Debate offers an opportunity for students to explore different perspectives on an issue rather than just focus on the viewpoint they already have. This may challenge the multiple ways in which the 'other' may be explored in the issues under discussion. One way which may be particularly effective at pushing students to engage with an alternative perspective is to have students who have clearly defined viewpoints on one side of a debate to prepare the arguments for the opposing view. The aim here is for them to develop greater awareness of the other perspectives. This may alter their initial beliefs or reconfirm them, either way they will be more informed opinions.

Conclusion

> "Students can come to look at the world differently and act in it differently" as a result of the things they learn. (Barnett & Coate, 2005, p. 145)

This paper has illustrated the potential for teaching controversial subjects in geography through debate. The skills developed from debate include the ability to construct and defend arguments in advance and in the moment of the debate; the ability to examine critically a variety of different sources of material; and enhanced communication skills. Alongside this, supporting other authors, this illustrative case found that students thought that it was a 'fun' way to learn. The students in both cohorts thoroughly enjoyed the debate; they were excited and engaged, alongside as seeing it as a valuable learning experience (Dundes, 2001; Helenius *et al.*, 2006). Significantly, and in contrast to some of the previous literature, this paper has found that these students rarely significantly altered their view from their original perspectives (Green & Klug, 1990; Omelicheva & Avdeyeva, 2008; Kennedy, 2009); rather the debate offered an opportunity to develop more nuanced and critical justifications in support of their original opinions. This may have been different if the students in both cohorts had not been positively predisposed towards the topic. Instead, the skills that the students developed through the activity offered them greater critical insight into the issue rather than leading them to contrasting perspectives. Given the potential power of debate for teaching a whole range of skills, this paper calls for geographers to further experiment with this method of teaching and to share their experiences with the wider geographic community. We need to help students to value the critical thinking skills that they take away from such work, not just in relation to understanding a topic for an assignment, but the significance of how such skills help them to interpret and understand other issues in the world. We need to show "that 'critical' thinking can be every bit as useful and world-changing as more technical, vocationally-centred knowledge" (Castree, 2000, p. 969).

Acknowledgements

I would like to thank the students who participated in this debate and Dr Chris Ribchester for reading and commenting on earlier drafts of this paper. I would also like to thank the three anonymous referees who offered valuable insights on how to enhance this paper.

Note

[1] Throughout the debate the tutor remained neutral about her own views on the subject (Hitchings (2011) explores the challenges and potential issues of this choice in detail). However, in both cohorts the students clearly wanted to know what the tutor thought on the issue, and so after the second reflections had been completed, the tutor expressed her own position and discussed why she believed this was the right course of action.

References

Angus, T., Cook, I. & Evans, J. (2001) A manifesto for cyborg pedagogy? *International Research in Geographical and Environmental Education*, 10(2), pp. 195–201.

Barnett, R. (2000) *Realizing the University in an Age of Supercomplexity* (Buckingham: The Society for Research into Higher Education/Open University Press).

Barnett, R. & Coate, K. (2005) *Engaging the Curriculum in Higher Education* (Berkshire: The Society for Research into Higher Education/Open University Press).

Bellon, J. (2000) A research based justification for debate across the curriculum, *Argumentation and Advocacy*, 36(3), pp. 161–173.

Castree, N. (2000) Professionalisation, activism, and the university: Whither critical geography? *Environment and Planning A*, 32(6), pp. 955–970.

Cook, I. (2000) Nothing can ever be the case of 'us' and 'them' again: Exploring the politics of difference through border pedagogy and student journal writing, *Journal of Geography in Higher Education*, 24(1), pp. 13–27.

Crawley, H. (2005) Evidence on Attitudes to Asylum and Immigration: What We Know, Don't Know and Need to Know, Centre on Migration, Policy and Society, Working Paper No. 23. Available at http://www.compas.ox.ac.uk/fileadmin/files/pdfs/Heaven Crawley WP0523.pdf (accessed March 2011).

Critical Geography Co. (2011) *What is Critical Thinking?* Available at http://www.criticalthinking.com/company/articles/critical-thinking-definition.jsp (accessed September 2011).

Crone, J. A. (1997) Using panel debates to increase student involvement in the introductory sociology class, *Teaching Sociology*, 25, pp. 214–218.

Dengler, M. (2008) Classroom active learning complemented by an online discussion forum to teach sustainability, *Journal of Geography in Higher Education*, 32(3), pp. 481–494.

Dundes, L. (2001) Small group debates: Fostering critical thinking in oral presentations with maximal class involvement, *Teaching Sociology*, 29(April), pp. 237–243.

Facione, P. A. (2000) The disposition toward critical thinking: Its character, measurement and relationship to critical thinking skill, *Informal Logic*, 20(1), pp. 61–84.

Fullan, M. (2003) *Change Forces with a Vengeance* (London: Routledge Falmer).

Green, C. S. & Klug, H. (1990) Teaching critical thinking and writing through debates: An experimental evaluation, *Teaching Sociology*, 18, pp. 462–471.

Hay, I. (2001) Editorial: Critical geography and activism in higher education, *Journal of Geography in Higher Education*, 25(2), pp. 141–146.

Helenius, I. M., Goldstein, C. E., Halm, E. A. & Korenstein, D. (2006) Using structured debate to teach a controversial topic: Mammography for women in their 40s, *Teaching and Learning in Medicine*, 18(4), pp. 292–296.

Heymann, R. (2000) Research, pedagogy, and instrumental geography, *Antipode*, 32(3), pp. 292–307.

Hitchings, R. (2011) Do you ever disagree with your students? Avoiding personal politics in human geography teaching, *Journal of Geography in Higher Education*, 35(1), pp. 85–101.

Ilkkaracan, P. & Ercevik Amado, L. (2005) Human rights education as a tool of grassroots organizing and social transformation: A case study from Turkey, *Intercultural Education*, 16(2), pp. 115–128.

Jackson, N. (2006) Imagining a different world, in: N. Jackson, M. Oliver, M. Shaw & J. Wisdom (Eds) *Developing Creativity in Higher Education: An Imaginative Curriculum*, pp. 1–9 (London: Routledge).

Kennedy, R. (2009) The power of in-class debates, *Active Learning in Higher Education*, 10, pp. 225–236.

Korkmaz, Ö. & Karakuş, U. (2009) The impact of blended learning model on student attitudes towards geography course and their critical thinking dispositions and levels, *Turkish Online Journal of Educational Technology*, 8(4), pp. 51–63.

Lewin, L. M. & Wakefield, J. A. Jr. (1983) Teaching psychology through an instructor-debate format, *Teaching Psychology*, 10(2), pp. 115–116.

Light, G. & Cox, R. (2001) Designing: Course and curriculum design, in: G. Light & R. Cox (Eds) *Learning and Teaching in Higher Education: The Reflective Professional*, 1st edn, pp. 69–96 (London: PCP).

Livingston, I. (1999) Role-playing planning public inquiries, *Journal of Geography in Higher Education*, 23(1), pp. 63–76.

Maddrell, A. M. C. (1994) A scheme for the effective us of role plays for an emancipatory geography, *Journal of Geography in Higher Education*, 18(2), pp. 155–162.

Moon, J. (2008) *Critical Thinking: An Exploration of Theory and Practice* (Abingdon: Routledge).

Omelicheva, M. Y. & Avdeyeva, O. (2008) Teaching with lecture or debate? Testing the effectiveness of traditional versus active learning methods of instruction, *PSOnline*, (July), pp. 603–607.

QAA (2007) *Subject Benchmark Statements: Geography* (Mansfield: QAA), Available at http://www.qaa.ac.uk/academicinfrastructure/benchmark/honours/geography.asp (accessed 26 January 2011).

Robinson, R. (1988) Development issues: Sympathy and paternalism, empathy and realism, in: R. Gerber & J. Lidstone (Eds) *Developing Skills in Geographical Education*, pp. 152–155 (Brisbane: International Geographical Union Commission on Geographical Education/Jacaranda Press).

Rogers, A. (2006) Escaping the slums or changing the slums? Lifelong learning and social transformation, *International Journal of Lifelong Education*, 25(2), pp. 125–137.

Schuster, L. (2003) Common sense or racism? The treatment of asylum-seekers in Europe, *Patterns of Prejudice*, 37(3), pp. 233–255.

Selwyn, N. (2007) Web 2.0 Applications as Alternative Environments for Informal Learning—A Critical Review, Background paper for the CERI-KERIS International Expert Meeting on ICT and Educational Performance, Cheju National University, South Korea. Available at http://www.oecd.org/dataoecd/32/3/39458556.pdf (accessed 26 January 2011).

Smith, B. (1918) The father of debate: Protagoras of Abdera, *Quarterly Journal of Speech*, 4(2), pp. 196–215.

Smith, D. M. (1995) Moral teaching in geography, *Journal of Geography in Higher Education*, 19(3), pp. 271–283.

Thomas, D. (Ed.) (1996) *The Concise Oxford Dictionary of Current English*, 9th edn (London: BCA).

Tomlinson, J. (2009) Argumentation, Available at http://facstaff.bloomu.edu/jtomlins/debate_formats.htm (accessed 26 January 2011).

Toohey, S. (1999) *Designing Courses for Higher Education* (Buckingham: The Society for Research into Higher Education/Open University Press).

United Nations (1951) Convention and Protocol Relating to the Status of Refugees, Available at http://www.unhcr.org/protect/PROTECTION/3b66c2aa10.pdf (accessed 26 January 2011).
Wellens, J., Beradi, A., Chalkley, B., Chambers, B., Healey, R. L., Monk, J. & Vender, J. (2006) Teaching geography for social transformation, *Journal of Geography in Higher Education*, 30(1), pp. 117–131.
Wolf, J., Stanton, M. & Gellott, L. (2010) Critical thinking in physical geography: Linking concepts of content and applicability, *Journal of Geography*, 109, pp. 43–53.

Appendix 1. Team Debate Format 2010

As part of the groups' preparation each student was allocated a consecutive number depending upon who was going to address which part of the debate. This model allows up to 12 students to participate.

Affirmative: For asylum seekers working (odd numbers before rebuttal, even numbers for rebuttal)

Negative: Against asylum seekers working (even numbers before rebuttal, odd numbers for rebuttal)

1. First affirmative constructive—five minutes

- A good introduction that attracts the audience's attention and interest in the topic
- Clearly state the resolution
- Clearly state each of your contentions and support these with reason and evidence
- Conclude effectively

2. Cross-examination of the affirmative by one of the negative—five minutes

- You ask questions—have a strategy or at the very least a direction to your questioning
- Be courteous

3. First negative constructive—five minutes

- A good introduction that attracts the audience's attention and interest in the topic
- Clearly state the negative's position on the topic
- Clearly state the negative's observations and support this with reason and evidence
- Attack and question the affirmative's contentions/evidence
- Conclude effectively

4. Cross-examination of the negative by one of the affirmative—five minutes

- You ask questions—have a strategy or at the very least a direction to your questioning

5. Second affirmative constructive—five minutes

- A good introduction that attracts the audience's attention and interest in the topic
- Clearly state each of your contentions—support with reason and evidence

- Respond to negative arguments/attacks
- Conclude effectively

6. Cross-examination of the affirmative by the other negative—five minutes
- You ask questions—have a strategy or at the very least a direction to your questioning
- Be courteous

7. Second negative constructive—five minutes
- A good introduction that attracts the audience's attention and interest in the topic
- Clearly state the negative's observations—the second negative can introduce additional observations—support with reason and evidence
- Attack and question the affirmative's contentions/evidence
- Conclude effectively

8. Cross-examination of the negative by the other affirmative—five minutes
- You ask questions—have a strategy or at the very least a direction to your questioning

9. First negative rebuttal—four minutes
- Rebuild the negative case
- Summarize how the negative position is superior, and that the affirmative has not carried the burden-of-proof
- Conclude effectively

10. First affirmative rebuttal—four minutes
- Respond to the negative arguments, rebuild the affirmative case and contentions—extend arguments and give additional support for them
- Conclude effectively

11. Second negative rebuttal—four minutes
- Respond to latest affirmative arguments
- Make your final case to the audience that the negative position is superior to the affirmative
- Try and convince the audience the affirmative has failed to carry the burden-of-proof
- Summarize the debate and conclude effectively and ask for the audience to agree with the negative position

12. Second affirmative rebuttal—four minutes
- Respond to final negative arguments
- Summarize the debate and show the audience how the affirmative position is superior—and the affirmative has carried the burden-of-proof
- Conclude effectively

Source: Adapted from Tomlinson (2009).

Construction of Student Groups Using Belbin: Supporting Group Work in Environmental Management

MARK SMITH*, GILES POLGLASE** & CAROLYN PARRY†

*Institute of Geography and Earth Sciences, Aberystwyth University, UK, **Centre for the Development of Staff and Academic Practice, Aberystwyth University, UK, †Careers Advisory Service, Aberystwyth University, UK

ABSTRACT *Belbin team role self and observer perceptions were applied to a large cohort (145) of Geography, Earth and Environmental Sciences undergraduates in a module assessed through two separate group projects. Students self-selected groups for the first project; for the second, groups were more 'balanced.' Results show slight improvement in group performance compared with that of previous years, with a significant increase in first-class grades. No evidence was found linking group balance to performance; however, students recognized the value of their Belbin report when entering the job market. Belbin usefully contextualizes regularly occurring group work issues and provides students with ways of managing these issues.*

Background

Geographers regularly encounter group work situations, be it on fieldwork or as teams addressing broad multi-disciplinary problems which form a central theme for environmental management. There is a longstanding recognition of the need to incorporate group work projects in higher education curricula (Springer *et al.*, 1999), reflecting the requirement of graduate recruiters for work-related or transferable skills. Both in the UK and internationally, there has been increased recent effort to develop employability statements, generic graduate attributes or personal development plans that clarify the role of education offered to students, beyond disciplinary content knowledge (Bowden *et al.*, 2000; Barrie, 2006, 2007; CBI/UUK, 2009). These core outcomes of higher education are often produced by individual institutions seeking to demonstrate the quality of their graduates, or they may be developed at a national or international scale to ensure comparability and compatibility of education systems (e.g. the European Higher Education Area). A common theme of graduate attributes worldwide is inclusion of the ability to work flexibly in a team, to communicate, to collaborate, to listen seriously to the insights of others and to compete in the future economic environment (e.g. Kuh, 2008; WAG, 2009). Wheatley (1992) reports the anticipation by managers in the UK that more of their work will be conducted in teams and that possession of good team-working skills is becoming increasingly necessary. Recent evidence suggests that team working is now the structural norm in high-performance organizations seeking to react and adapt to dynamic pressures of markets and shareholders (Partington & Harris, 1999).

The formulation of group work activities is driven by any of a wide range of motivating forces, including those directly aimed at addressing the employability agenda (Gedye, 2009), encouraging enterprise skills (Healey, 1992), implementing a problem-based learning exercise (Spronken-Smith, 2005; Pawson *et al.*, 2006), efficiently deploying teaching resources (Gibbs, 2010) or may be part of a broader tradition of group work, in fieldwork projects for example. Livingstone and Lynch (2002, p. 218) note that group projects "can be a method of increasing complexity in the learning experience, which thus strengthens students' preparedness for the complex environments into which they move after completing their degrees."

Yet, several observers have noted that group work at undergraduate level can be problematic (Gold *et al.*, 1991) and may reduce individual student motivation (Kerr & Bruun, 1983). Perceived difficulties include issues of high workload associated with group work exacerbated by unproductive time in group meetings (Healey *et al.*, 1996), freeloading group members and the 'sucker effect' where hardworking students reduce their effort in response to such freeloaders (Houldsworth & Matthews, 2000). Personality clashes are reported frequently suggesting that students lack group management and facilitation skills. Rather than reflecting on inherent weaknesses in the concept of group projects, such problems emphasize the importance of designing group projects carefully and supporting students throughout the process (Gibbs, 2010). Reported problems may be down to insufficient opportunities to develop or rehearse group work skills at university level; such unfamiliarity may manifest itself as anxiety and disorientation upon encountering a group work situation for the first time (Gibbs & Dunbar-Goddett, 2007).

Given the current competitiveness in graduate recruitment schemes, the focus on project-orientated employment and emergence of 'portfolio' careers (Henderson & Robertson, 1999), it seems appropriate that the benefits of existing group work projects should be maximized and some opportunity for student reflection on transferable skills be incorporated into project design. Reflection need not be particularly onerous or time consuming; it can be accomplished relatively easily. Bradshaw (1989) points to the use of team role classifications as a way of making more of group work in higher education and developing the inter-personal and team skills understood in employment. Moreover, a well-balanced team, which works well together, should experience an enhanced learning opportunity as the focus is on the task at hand and not on personalities.

It has long been recognized that the performance of a group, as a mix of individuals, is influenced by the combination of personality styles within that group. Attempts to design ideal teams through categorization of individuals into team roles date back over 60 years (Benne & Sheats, 1948; Bales, 1950). In recent decades, the team role categorization scheme of Belbin (1981, 1993a) has built up considerable momentum with management development professionals (Partington & Harris, 1999). Based on extensive observations of the behaviour of managers during training courses during the 1970s, Belbin (1981) hypothesized that team balance was more important for success than combined intellect, focusing on the emergence of informal, functional roles during training exercises. Rather than considering collective team behaviour, Belbin (1981) categorized individual behaviour within the team into eight types, later expanded to nine (Belbin, 1993a). These are described in Table 1. Since different people interact in different ways, successful teams are characterized by the compatibility of the preferred roles of their members.

An individual's natural team role preferences are rapidly identified through the Belbin self-perception index. There is a general acknowledgement that the Belbin scheme's intuitive appeal, ease of application, empirical support and widespread use in many organizations including government bodies, FTSE-100 companies and multinational agencies render it a useful tool for managers (Parkinson, 1995; Aritzeta *et al.*, 2007). The central claim of the Belbin team role theory is that a 'balanced' team, as judged by a spread of high-scoring individuals in each team role, has a greater propensity to perform highly. However, a variety of different group balance metrics (GBMs) have been reported previously (e.g. Senior, 1997; Partington & Harris, 1999). The Belbin theory also recognizes that behaviours are contextual and will change over time in response to new circumstances.

While some studies have questioned the psychometric properties and reliability of the Belbin team role self-perception inventory (Furnham *et al.*, 1993; Fisher *et al.*, 1996), Belbin (1993b) emphasizes that it is not a psychometric instrument. Team roles measure behaviour rather than personality (one of several factors that influence behaviour). A recent review by Aritzeta *et al.* (2007) identified mixed evidence on the convergent validity of the self-perception index. Indeed, more support exists for the Belbin team role model (e.g. Fisher *et al.*, 1994; Dulewicz, 1995; Fisher *et al.*, 2000), which assesses the potential for team role contribution of individuals based on behaviours and clusters of characteristics (Belbin, 1981). The main criticisms of Belbin question the identification of an individual's preferred team roles based on self-perception alone (Parkinson, 1995; Senior, 1996). The latest version of Belbin, administered through the *e-Interplace* computer program, offers the additional option to integrate observers' assessments into the analysis. When the observers are familiar with the individual, this offers a good opportunity to increase the robustness of the analysis. Broucek & Randell (1996) found significant correlations between self- and observer assessments; however, Senior & Swailes (1998) and van Dierendonck & Groen (2011) note that little research has been conducted using these observers' assessments.

This paper documents the inclusion of both the Belbin team role self-perception index and observer assessments into a large second-year undergraduate geography module. The aim of this project is to introduce team role analysis to support the student learning experience and provide students with a greater understanding of the roles of individuals within groups. It is hoped that this experience will encourage students to engage with employability issues and focus on the transferable skills they have acquired during their degree. Moreover, the implementation of the Belbin analysis on such a large student cohort permits empirical assessment of the central argument of the Belbin scheme: that more 'balanced' groups are more successful. This study is novel in that it documents a Belbin analysis using both self- and observer assessments on a large number of student groups (42) within the standardized assessment of an undergraduate module, thereby controlling for many variables confounding workplace empirical tests (e.g. differences in group tasks, environments and experience).

Two main research questions are addressed:

(1) Does the performance of student groups support the underlying assumption of the Belbin scheme that more differentiated groups function better?

Table 1. Individual roles in teams

Team role	Team contribution	Allowable weaknesses
Plant (PL)	Creative, imaginative, unorthodox. Solve difficult problems	Ignores incidentals. Too pre-occupied with own thoughts to communicate effectively
Resource investigator (RI)	Extrovert, enthusiastic, communicative. Explores opportunities. Develop contacts	Over-optimistic. Can lose interest once enthusiasm has passed
Co-ordinator (CO)	Mature, confident. Clarifies goals. Brings people together to promote team discussions	Can be seen as manipulative. Offloads personal work
Shaper (SH)	Challenging, dynamic, thrives on pressure. Has the drive and courage to overcome obstacles	Prone to provocation. Liable to offend others
Monitor-evaluator (MI)	Serious minded, strategic and discerning. Sees all options. Judges accurately	Can lack drive and ability to inspire others
Teamworker (TW)	Co-operative, mild, perceptive and diplomatic. Listens, builds, averts friction	Indecisive in crunch situations
Implementer (IMP)	Disciplined, reliable, conservative in habits. A capacity for taking practical steps and actions	Somewhat inflexible. Slow to respond to new possibilities
Completer finisher (CF)	Painstaking, conscientious, anxious. Searches out errors and omissions. Delivers on time	Inclined to worry unduly. Reluctant to let others into own job
Specialist (SP)	Single-minded, self-starting, dedicate. Provides knowledge and skills in rare supply	Contributes on only a limited front. Dwells on specialized personal interests

Source: Belbin (1981) and Belbin (1993a)

(2) Is Belbin team role analysis a valuable tool to support the implementation of effective undergraduate group work projects?

Method

Belbin team role analysis was integrated into a large group level 2 undergraduate module entitled *Environmental Management*. Student enrolment on the module was 145, an increase from 116 in 2010. Summative assessment entails two separate group work projects. The first of these requires student groups (of 5–6) to prepare a management report. Several formative assignments are set within the 5-week period of this project to encourage groups to work to internal deadlines including the preparation of a draft for anonymous peer review. The second project is less structured and more challenging. Larger student groups (of 8–9 members) are required to pitch for a grant at an end-of-module conference. They must prepare a 2-min video, 3-min oral presentation, A0 poster and five sides of project details for the conference handbook. Students must also defend their poster in a Q&A breakout session during the conference. Many of these are 'live' projects created in discussion with the local council, several of whom attended the conference to informally judge the submissions. Further details of this pitch assessment are outlined in Smith (in press).

The second project is more challenging in that it consists of four separate elements that need to clearly deliver the main selling points of the proposed projects to the conference attendees in an appropriate and professional manner. Many activities required in this second project are unfamiliar to most level 2 Geography, Earth and Environmental Science undergraduates. To work effectively, groups need to form sub-groups that have to be co-ordinated. The project is designed to be sufficiently large to require co-operation and cannot be completed or dominated by an individual. The formative steps and weekly internal deadlines of the first project were not enforced, although students were encouraged to arrange this internally.

Previous feedback suggested that the biggest challenge in the assessment was not the unfamiliarity of the tasks, but rather the difficulties of working in groups. In 2011, the Belbin team role analysis was introduced with the aim of encouraging greater reflection on general group work skills and to promote a greater understanding of individual strengths and weaknesses. The first project proceeded, as in previous years, with self-selected groups. In the final week of the first project, 143 students completed the nine-role version of the Belbin team role self-perception index (Belbin, 1993a). Students distributed 10 points between 10 statements according to the strength in which they felt the statement reflected their own behaviour. This was repeated for seven categories of statements. For each student, this generated a score between 0 and 100 for each team role presented in Table 1.

Given the concerns over the validity of the self-perception index discussed earlier, observer assessments were also conducted by the members of each project group. Observers must distribute points on a two-part checklist split between positive and negative traits; if a phrase describes the person's behaviour in the team, then it can be ticked or even double-ticked. In total, 557 observer assessments were made. Thus, the overall team role assessment of each student was based on a self-perception index and up to five observer assessments. Both self-perception and observer assessments were administered independently through the Belbin *e-Interplace* computer program.

Since one aim of this project was to familiarize students with the ideas underlying team role assessment, a lecture was given outlining the concept of Belbin, including how to recognize and work more effectively with individual Belbin types. The lecture also supported correct interpretation of individual reports. After the first project, students completed a short questionnaire on their perceived effectiveness of their group and their thoughts on the Belbin team roles.

For the second project, students were asked to use their preferred team roles to select a more 'balanced' team. They were initially divided based on choice of project and wrote their preferred team roles on a card. While groups remained self-selecting, they were required to state their Belbin team role preferences on a group submission form before their group could be approved. After the second project, students were asked to fill in the same short questionnaire on their group's effectiveness and their thoughts on the Belbin scheme.

To test statistically whether the balance of teams was reflected in their performance and address the first research question, objective criteria for both elements are necessary. Group performance was quantified through two different metrics: the mark awarded for the summative assessment and students' perceptions of their effectiveness as a group. Students graded their perceived group effectiveness on a scale of 1 (perfect) to 5 (disjointed). Group balance (based on Belbin team roles) must also be quantified through the calculation of objective metrics. Following Partington & Harris (1999), four GBMs were calculated to represent different elements of a balanced group. The four GBMs differentiate balanced and unbalanced groups based on different criteria:

GBM1: Distribution of strong examples of team roles (following Senior, 1997). This metric represents the presence of at least one very strong example of each role in a group on a scale of $0 - 100$ with a score of 0 representing a team with no strong examples of any role and a score of 100 representing a team with a strong example of each role. GBM1 is defined as

$$\text{GBM1} = 100\frac{r}{B},$$

where B represents number of Belbin team roles (in this study, following Belbin, 1993a, $B = 9$) and r denotes number of roles strongly represented by an individual in the group (i.e. a Belbin score of >80 per cent).

GBM2: Duplication of roles. Since Belbin (1993a) suggests that, with the exception of the team worker and implementer, team roles should not be duplicated, this metric represents duplication of strong team roles on a scale up to 100 with a score of 0 representing a team where each member is a strong example of the same role (and no others) and a score of 100 representing a team with no duplicated roles. Since an individual can be a strong example of more than one team role, a negative GBM2 can be observed in some circumstances. GBM2 is defined as

$$\text{GBM2} = 100\frac{(n-1)-d}{(n-1)},$$

where d represents the total number of duplicated strong examples in any role in a team (i.e. for a team with just three strong examples of 'Plants' and no other duplicated roles, $d = 2$ as the role is duplicated twice) and n represents the number of group members.

GBM3: Average Belbin score (out of 100) of the weakest two roles measured by the highest scoring individual in that role.

GBM4: Average Belbin score (out of 100) of the weakest two roles measured by the group average score

Thus, a group that contains strong examples of each Belbin team role with no duplicates will be more balanced according to Belbin team role theory and will score highly on each GBM. In addition, to examine the influence of each team role, the number of strong examples of each team role in a group was also compared with the group performance indicators.

Evaluation and Discussion

First, the assumption that more balanced groups perform more effectively was tested quantitatively. Overall group marks were compared against a baseline data-set from 2010 as both cohorts met identical entry requirements. Group performance was compared with GBMs to assess whether more balanced teams led to a better overall group performance. Second, to establish the usefulness of Belbin as a method to encourage student focus on group work, feedback from the first project was analysed to establish commonly occurring group work issues. The ways in which students planned to alter their approach to group work in the second project after being exposed to the Belbin team role theory were examined. Post-project questionnaires assessed students' perceptions of working in each group, which were examined alongside Belbin team reports to identify a broad typology of student groups and problems encountered. Student and instructor reflections on Belbin were also compiled.

Testing the Belbin Group Balance Assumption

Figure 1 displays the number of students with strengths in each Belbin team role. 'Plant' was the dominant role of a large number of students, although in the wider population the prevalence of this role is thought to be rarer (Belbin Associates, 2009). This may reflect the fact that the existing assessment structure of the undergraduate degree often requires students to work as individuals and encourages the development of this behaviour. Other common roles were 'Implementer,' 'Teamworker' and 'Co-ordinator.' Several roles were poorly represented in the student group: students preferring 'Resource Investigator,' 'Monitor-Evaluator,' 'Shaper' and 'Specialist' roles were rare. Kneale (1996) suggested that such behaviours may become more developed later in life as Belbin role preferences change with age. Inspection of Belbin team reports confirmed that the groups were indeed more balanced in the second project than in the first based on three of the four GBMs. The groups for the second project scored lower on the duplication of strong roles metric; however, this is to be expected as the group size was larger.

In previous years, as students became more practiced at group work, the average mark improved for the second project when compared with the first (Figure 2(a)). The average mark increase in 2010 was 2.7 per cent. After implementation of the Belbin team role analysis in 2011, the average increase improved slightly (3.6 per cent; Figure 2(b)) and proportionately fewer students recorded a mark decrease for the second project. More significantly, the increase in the number of first-class grades between the two projects in each year was 44 per cent in 2010 and 138 per cent in 2011 (Figure 2(c)) suggesting that

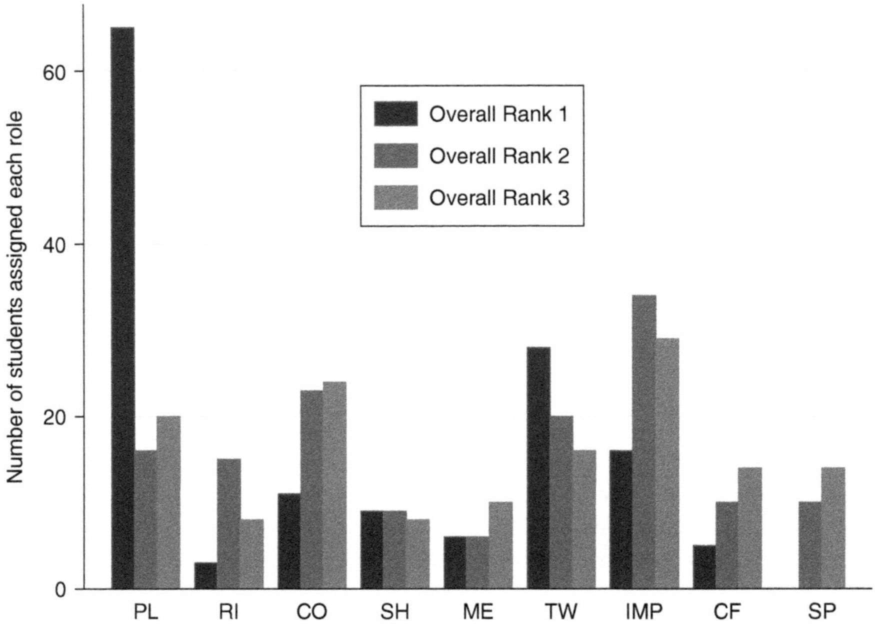

Figure 1. Top three overall Belbin team role ranks of 143 level two geography and environmental science undergraduates (based on both self-perception and observer assessments). For a description of the Belbin team roles, see Table 1.

students at the top end of the mark range were responding well to this implementation of the Belbin analysis.

As a direct test of the assumptions of the Belbin scheme, it was hypothesized that the more balanced groups would be the most successful and least prone to internal conflict. However, quantification of team role balance is problematic given the complex relationships between the team roles. Table 2 displays the correlations between the four GBMs and both the group mark for the summative assessment and the group's perception of their effectiveness. Each variable was found to be normally distributed by the skewness–kurtosis test for normality (D'Agostino *et al.*, 1990) except for GBM2 for which Spearman's rank correlation coefficients were calculated.

In common with Senior (1996) and Partington & Harris (1999), Table 2 shows that GBMs have little association with group performance. Only one significant relationship between group performance (measured through summative assessment) could be found: a negative relationship between GBM2 (the role duplication metric) and project mark suggesting that, contrary to Belbin (1981), more than one strong example of a team role can be advantageous to group performance. However, since GBM2 is strongly related to group size which increased for the second project (in which students scored highly; see Figure 2), this correlation may reflect the differences in group performance between projects. Additionally, the prevalence of the 'Teamworker' and 'Implementer' roles in the student cohort (Figure 1), which Belbin (1981) suggests can be duplicated in a successful team, may explain this result.

A significant correlation was found between the average Belbin score of the two weakest roles of any individual and group perception of effectiveness, though the direction

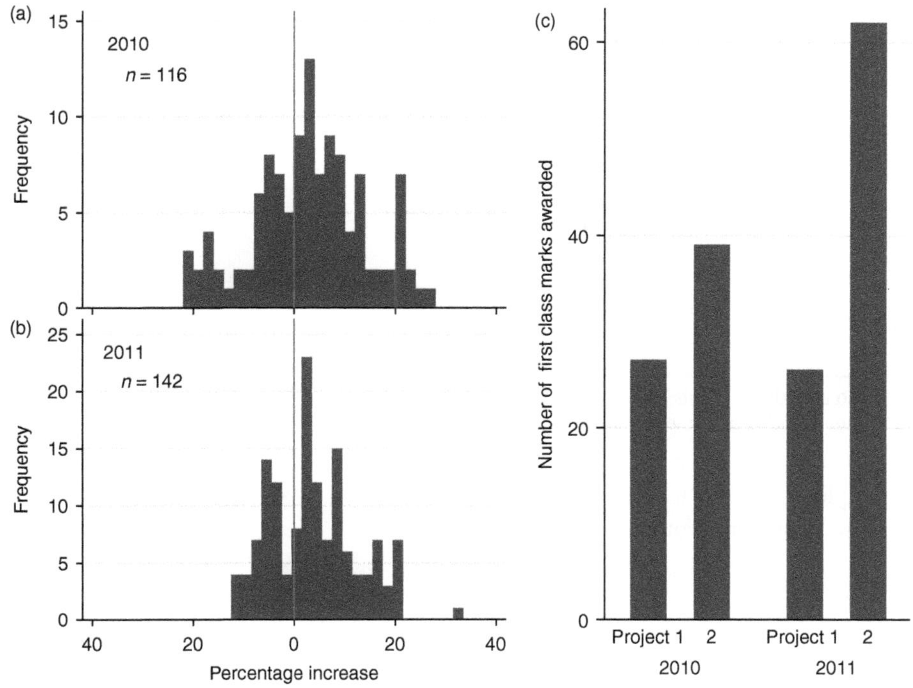

Figure 2. Summative assessment before and after Belbin implementation. Distribution of mark increases between the two projects (a) in 2010 and (b) in 2011. (c) Number of first-class marks awarded in each project over 2 years.

Table 2. Pearson correlations between group performance and GBMs (Spearman correlation coefficient is presented for GBM2)

$n = 42$	Group mark	Group perception
Distribution of strong examples (GBM1)	0.157	0.054
Duplication of roles (GBM2)[a]	−0.420[b]	−0.092
Lowest individual roles (GBM3)	−0.021	0.255
Lowest group roles (GBM4)	0.047	0.427[b]

[a] Spearman's rank correlation presented (the Pearson correlation value was slightly more negative).
[b] Correlation significant at the 1 per cent level.

of this relationship is such that a more balanced group is related to the student perception of a less effective group. This may represent a greater tendency for deeper discussion and more regular disagreement within balanced groups. To explore this idea further, the number of strong examples of each role was correlated against the two team performance indicators (Table 3). As each of these variables was found to be non-normally distributed by the skewness–kurtosis test, a Spearman's rank test was performed. Again, few significant relationships were observed. Visual inspection of the negative relationship between group mark and number of strong 'monitor-Evaluators' suggests that this significant relationship (at the 5 per cent significance level) is dominated by a single outlier with considerable leverage. However, the strong relationship between group perception of effectiveness and number of 'Shapers' in the group is quite apparent. Perhaps the tendency

Table 3. Spearman's rank correlations between group performance and number of examples of strong roles in each team

$n = 42$	Group mark	Group perception
Number of strong PL	0.062	−0.108
Number of strong RI	−0.013	0.122
Number of strong CO	0.189	−0.043
Number of strong SH	0.083	0.422[b]
Number of strong ME	−0.380[a]	0.201
Number of strong TW	0.223	−0.003
Number of strong IMP	0.085	−0.219
Number of strong CF	−0.004	0.067
Number of strong SP	0.258	−0.149

[a] Correlation significant at the 5 per cent level.
[b] Correlation significant at the 1 per cent level.

of multiple 'Shapers' to provoke and offend others has a detrimental effect on the coherence of student groups.

Using Belbin to Enhance Student Group Work Skills

The second research question analyses the usefulness of the Belbin scheme to encourage students to focus explicitly on group work skills and to encourage a greater understanding of individual strengths and weaknesses. Since the increase in high-performing groups appears unrelated to their balance of team roles, the use of Belbin as a teaching tool to expose students to the expected problems of group work may be a more appropriate focus of investigation. After the first group project, anonymous feedback highlighted several group work issues that were repeated across the cohort. Five commonly occurring group work problems were identified:

(1) General lack of communication. "Not all of us knew exactly what each other were doing and in the beginning there was quite a lot of overlap with research."
(2) Problematic individuals not completing work. "Limited work provided by individuals, with no possibilities of removing them (like what would happen in a 'real professional' environment)."
(3) Controlling individuals. "Unwillingness of some members to share work, took control over the majority of work and would not listen to suggestions."
(4) Lack of an identified leader. Laid back groups drifting along or procrastinating. "I feel that the biggest problem with our group was that towards the start we were all too friendly and nobody was taking up the role of leader."
(5) Clashes of individuals within the group. "There was a clash with some members of the group, and disagreements did occur."

A lecture on Belbin team role theory and completion of the self perception and observer assessments gave students the opportunity to identify their own group issues within this broad conceptual framework and develop ways of moving forward with their experience to improve teamworking in the second project. Typical responses included:

> Focus will be pushed towards the idea of working together and providing better understanding of individual roles within the group context.
>
> Take more control in the areas that I am more confident and competent with.
>
> I can place myself into roles in the team which suit my strengths, and with others who can offer different skills.
>
> I think that taking time to consider team roles would be a better approach to the "shotgun" style that jobs were claimed during our first project.

Each of the generic group work issues highlighted above can be conceptualized within the Belbin team role theory. Indeed, several students did just that. For example, a member of a group which struggled with keeping to topic on team meetings now thought of this problem in a new way: "Lack of a shaper—so we sometimes drifted off topic in meetings" (this perception agreed with the findings of the Belbin team report). Many students aimed to address the issue of communication through the Belbin scheme; the lecture made them understand why miscommunication can easily occur and how to communicate better with those of different team roles. Overall, it appeared to foster an attitude of tolerance within the groups; however, several individuals were critical of the Belbin scheme and raised some valid points about Belbin team roles changing according to the task at hand (as recognized by Belbin):

> I don't think that we should be placed into groups depending on the Belbin report as in real life I'm not sure that it necessarily works in all situations.

When asked to rate how well the group worked together on a scale of 1 (perfectly) to 5 (disjointed), the average mark improved from 2.41 on the first project to 2.20 on the second. In common with the first project, communication issues and controlling group members were also frequently cited group difficulties.

Based on the preferred Belbin roles of students seen in Figure 1, commonly occurring student group profiles and associated group work problems should be expected. Indeed, analysis of the typical composition of student groups alongside qualitative feedback permits a general categorization to be proposed tentatively. The Belbin scheme provides a framework for dealing with such group problems. Here, three 'typical' student groups are identified based on group average Belbin team roles. Additionally, the profile of a particularly ineffective team is analysed.

(a) Typical student group

This represents the most common student group composition found in this project. 'Plants' were the most common category, with 'Implementers,' 'Co-ordinators' and 'Teamworkers' being well represented (Figure 3(a)). 'Shapers,' 'Resource Investigators' and 'Specialists' were rare. Thus, feedback from this group was that although it took them a while to start, "we all were able to contribute ideas evenly with no overburdening leader." Overall, the groups found that they worked well together although several comments suggested that they would benefit from a 'Shaper' to improve the mark, which was otherwise

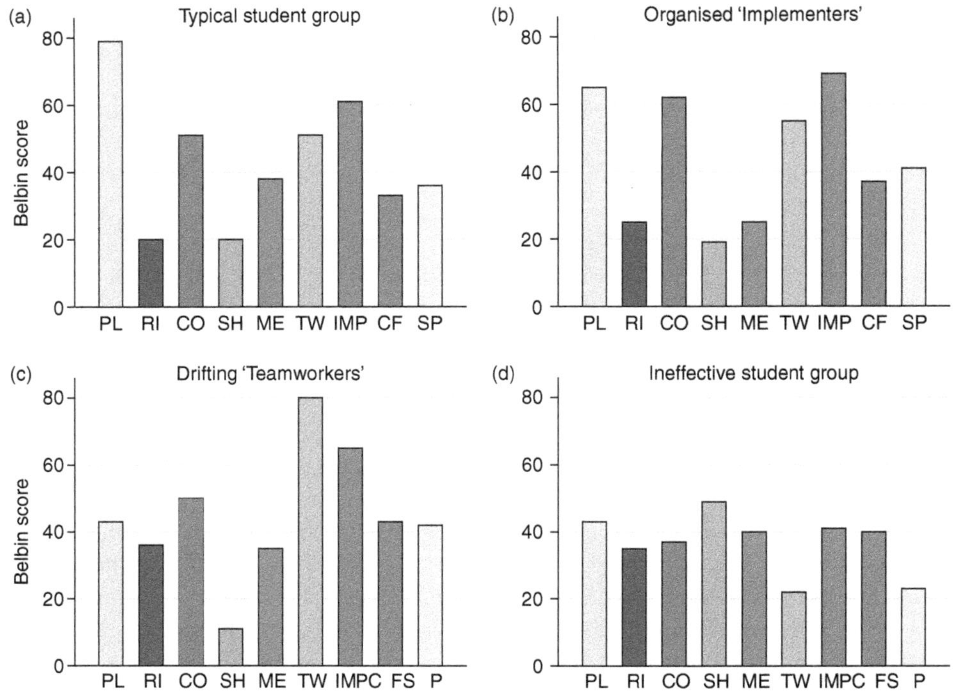

Figure 3. (a)–(c) Examples of the composition of Belbin team roles in three typical student groups referred to in the text; (d) composition of Belbin team roles in an ineffective student group.

quite average. Interestingly, one such group had just one example of a very strong 'Shaper.' This caused conflict in the group with feedback suggesting that "There was one individual in the team that wanted everything done their way . . . " (however, there is no firm evidence linking this to the actual 'Shaper'). Despite the seemingly negative effect of 'Shapers' on the students' perception of the group, this group achieved an above average mark. In another variation, groups with more under-represented 'Completer-Finishers' claimed to struggle with editing and finishing the final product and scored a below-average mark.

(b) Organized 'Implementers'

Several groups were dominated by 'Implementers' (e.g. Figure 3(b)). On the whole, these groups did substantially better than average and appeared to achieve this relatively harmoniously. These groups were quick to produce early versions of their work which they then improved: "We produced the first draft very quickly and to a good quality, this gave us longer to make small adjustments to it." These groups kept to time, setting and hitting internal deadlines along the way: "Everyone did exactly what they were assigned to do efficiently and always delivered on time. No one didn't want to do work, everyone was happy with what they did."

(c) Drifting 'Teamworkers'

Despite repeated requests otherwise, a few groups did keep close to their friends, either for comfort or for convenience. Based on student feedback, this may explain the category of teams that had more 'Teamworkers' than proportionately represented in the student cohort (Figure 3(c)). These groups typically scored below average. When combined with a deficit of 'Shapers,' these groups were extremely harmonious at the expense of quality control. As students in this example group (Figure 3(c)) suggested: "We were all great teamworkers but sometimes were too relaxed. We had to get motivation to get going with the work while we were distracted by things." This matches the team role type weakness characteristics identified by Belbin (1981, 1993a).

(d) The ineffective student group

During the first project (i.e. before the Belbin analysis), one group almost fell apart through disagreements and internal conflicts. Although they provided no qualitative feedback on their group dynamics, the group achieved one of the lowest marks and suffered heated internal disputes that frequently spilled over into the inbox of the module co-ordinator. The composition of this group was markedly different from the other student groups (Figure 3(d)). While 'Shaper' was one of the least common student roles (Figure 1), it was the *most* common role in this group; 'Teamworker' was the least common. This supports the findings of the statistical analysis above and agrees with Prichard & Stanton (1999) who also suggested that teams mainly composed of 'Shapers' underperform relative to more balanced teams. The statistical relationship between 'Shapers' and perceived ineffectiveness was not unduly influenced by this group and remained significant at the 1 per cent level when this team was removed from the analysis. This group's Belbin team report was remarkably insightful and is worth quoting directly: "The risk is that [the team's] energy can result in internal conflict which is not easily resolved. The problem about this team could be the unwillingness of its members to adjust to each other. There may be difficulty in developing team spirit." Thus, using Belbin profiles to avoid such problematic group structures may ease the administrative burden of group work modules.

In the post-project questionnaire (58 respondents), 35 students felt that their Belbin role matched their expectations, while 17 did not and 6 were undecided. Final student comments were predominantly positive, although many students raised concerns about the cost of the Belbin scheme. Typical responses are presented in Table 4.

As module co-ordinator, I found the introduction of the Belbin scheme to be a positive development to the module. The use of Belbin in graduate recruitment helped to highlight links between undergraduate group work and reality: a central theme of this module and broader personal development plans. Moreover, the acknowledgement that disagreements and clashes between group members are perfectly normal and not necessarily the 'fault' of any individual helped to de-personalize disagreements and bred an atmosphere of tolerance within the cohort. The description of team roles helped students to reflect on the activities they were best at and place themselves into that role within their group. As such, the implementation of the Belbin scheme was extremely useful for the students' personal development plans. Moreover, they have a professional report to keep for future job interviews. As an unsolicited student email suggested: "...it's a very snazzy piece of software and I think it will be very useful for when I apply for masters/jobs."

Table 4. Qualitative feedback on implementing the Belbin scheme in *Environmental Management*

"One of the most enjoyable modules undertaken in University so far. This module seems to be something that may actually be applied in a job once university life is over, unlike most other modules."
"The Belbin thing may have helped with the selection of the teams but a random grouping of people would probably have worked just as well."
"Overall the experience has opened my eyes and I have enjoyed it; however, I wish I had been able to do the pitch with my first group."
"I am grateful to have undergone this module. As suggested, I see it more as "work experience" rather than any other module. Thank you!"
"Working with people you don't know and not friends made it a lot better when trying to put the report together."
"The module is a great initiative and certainly breaks the monotony of academic style responses to module questions. Although initially I was concerned about the grade outcome from the start of this module due to the group submission and group grade, in hindsight as an individual, attained far higher success than expected. Group work is important in the real world; having worked for 3 years for an national limited company, I can see the merit of undertaking this style of study so that those with little exposure to the working environment, they may experience the highs and lows of employment. Thanks."

My initial concern surrounding Belbin was that strict adherence to assigned 'labels' would encourage students to defer their group responsibilities (e.g. "Of course I missed the deadline—I'm a 'Plant' *not* a 'Completer Finisher'"). Macrosson & Hemphill (2001) suggest that although Table 1 displays 'allowable' weaknesses for each team role, the conduct of some team members is often far from allowable. This was flagged in the initial Belbin lecture and fortunately, no evidence of this type of behaviour arose. There is some evidence to suggest that the Belbin implementation eased the administration of the module by reducing frequent and serious group conflicts as exemplified by the 'ineffective' student group of Figure 3(d). I question whether such a large group implementation was the best use of relatively expensive Belbin licenses, and share student concerns over the cost of these (around £24 per student). Perhaps the scheme would be better introduced at Masters level; however, a large group coverage helps address the employability agenda at the departmental level. Incorporating the Belbin analysis across the degree scheme, beyond this single module, would help spread the cost of the project.

I found the team role analysis and most elements of the Belbin report to be an extremely useful analysis tool; however, parts of the report that seemed rather speculative. Specifically, one part of the individual report, a counselling report of personal attribute descriptions based on self-perceptions and several observers' comments read quite like a horoscope, echoing the personal validation experiments of Forer (1949). Indeed, this element was especially unpopular with the students and partially undermined the Belbin scheme in their eyes. In future, I would suppress this section of the report, of which Geography, Earth and Environmental Science students seem naturally critical. However, I should stress that, for the most part, I found the Belbin scheme to be insightful and relevant to undergraduate group projects and a useful tool to direct students to focus on their future employability.

Conclusions

The Belbin team role analysis was introduced with the aim of improving group performance, encouraging a greater understanding of individual strengths and weaknesses. This was successful in that large mark increases were observed at the top end of the mark scale. The large group two-project study (with a total of 42 different student groups) offered the potential for an empirical test of the central argument of the Belbin scheme: that more balanced groups should be more successful. However, no robust statistical relationships were found to support this assumption. This result requires further testing since the complexity of the team roles and their relationships does not facilitate quantitative comparison. More general observations of the Belbin analysis matched students' perceptions of their groups. In particular, a surfeit of 'Shapers' within student groups (a rather underrepresented role in the wider student cohort) led to increased conflict and, in extreme circumstances (Figure 3(d)), a lower than average group mark on the summative assessment. Overall, students responded well to the Belbin scheme, were able to identify links between group work and employability and recognized the usefulness of such team role models to graduate recruiters. This in itself may explain the increased number of first-class grades awarded. Knowledge of the team role theory appeared to de-personalize internal conflicts and foster a tolerant attitude in the student group as students recast individual differences in attitudes and behaviours onto a broader conceptual framework.

Acknowledgements

This research was supported by the Aberystwyth University Learning and Teaching Implementation Fund. Thanks are offered to Joseph Wheaton who originally designed the workplace format for the *Environmental Management* module and to Carina Fearnley for help running the course in 2011.

References

Aritzeta, A., Swailes, S. & Senior, B. (2007) Belbin's team role model: Development, validity and applications for team building, *Journal of Management Studies*, 44, pp. 96–118.

Bales, R. F. (1950) A set of categories for the analysis of small group interaction, *American Sociological Review*, 15, pp. 257–263.

Barrie, S. (2006) Understanding what we mean by the generic attributes of graduates, *Higher Education*, 51, pp. 215–241.

Barrie, S. (2007) A conceptual framework for the teaching and learning of generic graduate attributes, *Studies in Higher Education*, 32(4), pp. 439–458.

Belbin, R. M. (1981) *Management Teams: Why They Succeed or Fail* (London: Heinemann).

Belbin, R. M. (1993a) *Team Roles at Work* (London: Butterworth-Heinemann).

Belbin, R. M. (1993b) A reply to the Belbin team-role self-perception inventory by Furnham, Steele and Pendleton, *Journal of Occupational and Organizational Psychology*, 66, pp. 259–260.

Belbin Associates (2009) Method, reliability and validity, statistics and research: A comprehensive review of Belbin team roles, Available at http://www.belbin.com/content/page/4279/BELBIN-MRVSR-AComprehensiveReview-2009.pdf (accessed December 2011).

Benne, K. D. & Sheats, P. (1948) Functional roles of group members, *Journal of Social Issues*, 4, pp. 41–49.

Bowden, J., Hart, G., King, B., Trigwell, K. & Watts, O. (2000) Generic capabilities of ATN university graduates, Available at http:/www.clt.uts.edu.au/ATN.grad.cap.project.index.html (accessed December 2011).

Bradshaw, D. (1989) Higher education, personal qualities and employment: Teamwork, *Oxford Review of Education*, 15, pp. 55–71.

Broucek, W. G. & Randell, G. (1996) An assessment of the construct validity of the Belbin self-perception inventory and observer's assessment from the perspective of the five-factor model, *Journal of Occupational and Organizational Psychology*, 69, pp. 389–405.

CBI/UUK (2009) Future fit: Preparing graduates for the world of work, Available at http://www.cbi.org.uk/media/1121435/cbi_uuk_future_fit.pdf (accessed December 2011).

D'Agostino, R. B., Balanger, A. & D'Agostino, R. B. Jr. (1990) A suggestion for using powerful and informative tests of normality, *American Statistician*, 44, pp. 316–321.

Dulewicz, V. (1995) A validation of Belbin's team roles from 16PF and OPQ using bosses' ratings of competence, *Journal of Occupational and Organizational Psychology*, 68, pp. 81–99.

Fisher, S. G., Macrosson, W. D. K. & Walker, C. A. (1994) The structure of new product teams, *Selection and Development Review*, 10(5), pp. 1–3.

Fisher, S. G., Macrosson, W. D. K. & Sharp, G. (1996) Further evidence concerning the Belbin team role self-perception inventory, *Personnel Review*, 25(2), pp. 61–67.

Fisher, S. G., Hunter, T. A. & Macrosson, W. D. K. (2000) The distribution of Belbin team roles among UK managers, *Personnel Review*, 29(2), pp. 124–140.

Forer, B. R. (1949) The fallacy of personal validation: A classroom demonstration of gullibility, *Journal of Abnormal Psychology*, 44, pp. 118–121.

Furnham, A., Steele, H. & Pendleton, D. (1993) A psychometric assessment of the Belbin team role self-perception inventory, *Journal of Occupational and Organizational Psychology*, 66, pp. 245–257.

Gedye, S. (2009) GEES, employability and the credit crunch, *Planet*, 22, pp. 2–5.

Gibbs, G. & Dunbar-Goddet, H. (2007) The effects of programme assessment environments on student learning, Available at http://www.heacademy.ac.uk/assets/York/documents/ourwork/research/gibbs_0506.pdf (accessed May 2011).

Gibbs, G. (2010) *The Assessment of Group Work: Lessons from the Literature. Assessment Standards Knowledge Exchange* (Centre for Excellence in Teaching and Learning in Higher Education, Oxford Brookes University), Available at http://www.york.ac.uk/media/staffhome/learningandteaching/documents/keyfactors/The_assessment_of_group_work_lessons_from_the_literature.pdf (accessed May 2011).

Gold, J. R., Jenkins, A., Lee, R., Monk, J., Riley, J., Shepherd, I. & Unwin, D. (1991) *Teaching Geography in Higher Education: A Manual of Good Practice*, 24 (Oxford: Blackwell), Institute of British Geographers Special Publications Series.

Healey, M. (1992) Curriculum development and 'enterprise': Group work, resource-based learning and the incorporation of transferable skills into a first year practical course, *Journal of Geography in Higher Education*, 16(1), pp. 7–19.

Healey, M., Matthews, H., Livingstone, I. & Foster, I. (1996) Learning in small groups in university geography courses: Designing a core module around group projects, *Journal of Geography in Higher Education*, 20, pp. 167–180.

Henderson, R. & Robertson, M. (1999) Who wants to be an entrepreneur? Young adult attitudes to entrepreneurship as a career, *Education and Training*, 41(5), pp. 236–245.

Houldsworth, C. & Matthews, P. B. (2000) Group composition, performance and educational attainment, *Education and Training*, 42(1), pp. 40–53.

Kerr, N. L. & Bruun, S. E. (1983) Dispensability of member effort and group motivation losses: Freerider effects, *Journal of Personality and Social Psychology*, 44(1), pp. 78–94.

Kneale, P. (1996) Organising student-centred group fieldwork and presentations, *Journal of Geography in Higher Education*, 20(1), pp. 65–74.

Kuh, G. D. (2008) *High-Impact Educational Practices: What They are, Who has Access to Them, and why They Matter* (Washington DC: AAC&U Publications).

Livingstone, D. & Lynch, K. (2002) Group project work and student-centred active learning: Two different experiences, *Journal of Geography in Higher Education*, 26(2), pp. 217–237.

Macrosson, W. D. K. & Hemphill, D. J. (2001) Machiavellianism in Belbin team roles, *Journal of Managerial Psychology*, 16(5), pp. 355–363.

Parkinson, R. (1995) Belbin's team role model: A silk purse out of a sow's ear? *Organizations and People*, 2(1), pp. 22–25.

Partington, D. & Harris, H. (1999) Team role balance and team performance: An empirical study, *Journal of Management Development*, 18(8), pp. 694–705.

Pawson, E., Fournier, E., Haigh, M., Muniz, O., Trafford, J. & Vajoczki, S. (2006) Problem-based learning in geography: Towards a critical assessment of its purposes, benefits and risks, *Journal of Geography in Higher Education*, 30(1), pp. 103–116.

Prichard, J. S. & Stanton, N. A. (1999) Testing Belbin's team role theory of effective groups, *Journal of Management Development*, 18, pp. 652–665.

Senior, B. (1996) Team roles and team building: Self-perception is no basis on which to build a team, *Iconoclastic Papers*, 1(1), December. Available at http://web.archive.org/web/20020203063823/http://www.solent.ac.uk/sbs/iconoclastic/Senior/senior.html (accessed August 2011).

Senior, B. (1997) Team roles and team performance: Is there "really" a link? *Journal of Occupational and Organizational Psychology*, 70, pp. 241–258.

Senior, B. & Swailes, S. (1998) A comparison of the Belbin self perception inventory and observer's assessment sheet as measures of an individual's team roles, *International Journal of Selection and Assessment*, 6, pp. 1–8.

Smith, M. W. (in press) Improving student engagement with employability: The project pitch assessment, *Planet*, 26.

Springer, L., Stanne, M. E. & Donovan, S. S. (1999) Effects of small group learning on undergraduate science, mathematics, engineering and technology: A meta-analysis, *Review of Educational Research*, 69(1), pp. 21–51.

Spronken-Smith, R. (2005) Implementing a problem-based learning approach for teaching research methods in geography, *Journal of Geography in Higher Education*, 29(2), pp. 203–221.

van Dierendonck, D. & Groen, R. (2011) Belbin revisited: A multitrait-multimethod in investigation of a team role instrument, *European Journal of Work and Organizational Psychology*, 20(3), pp. 345–366.

WAG (2009) *For our Future: The 21st Century Higher Education Strategy and Plan for Wales* (Department for Children, Education, Lifelong Learning and Skills, Welsh assembly Government), Available at http://wales.gov.uk/docs/dcells/publications/091214hestrategyen.pdf (accessed December 2011).

Wheatley, M. (1992) *The Future of Middle Management* (Corby: British Institute of Management).

RESOURCES

Reflective Journals as a Tool for Auto-Ethnographic Learning: A Case Study of Student Experiences With Individualized Sustainability

JON ANDERSON
School of City & Regional Planning, Cardiff University, UK

ABSTRACT *This paper critically evaluates the use of journals as a pedagogic tool to encourage reflection, critique and self-analysis by students. Based within a postgraduate teaching module that has operated annually since 2008 and was awarded the Royal Town Planning Institute's Award for Teaching Excellence in 2009, reflexive journals were employed as a method of assessment which, supplemented by conventional lectures and student focus groups, sought to explore the relations, opportunities and obstacles for sustainable development at the individual level.*

Introduction

This paper critically evaluates the use of journals as a pedagogic tool to encourage reflection, critique and self-analysis by students. Based within a postgraduate course on environment and sustainability, teaching and learning on this module were oriented around short lectures and longer focus group discussions, based on readings and questions. Reflexive journals encouraged students to apply these questions and concepts to the everyday realities of their own lives to critique the 'do-ability' of sustainability at the individual level (more information on the module can be found at Higher Education Academy funded website, www.citizensustainability.com[1]). This paper critically evaluates the use of such journals to enhance self-understanding and critical praxis amongst students. Due to limits of space, this paper will not explore in depth the lessons learned on sustainability itself through the process of journal composition, rather the paper focuses on the generic opportunities offered by reflexion through journals. It will demonstrate that, if undertaken in a supportive and respectful scenario, journals and auto-ethnography can enable critical self-reflection, engage students both intellectually and emotionally in their learning process, and by doing so help realize the higher order pedagogic goals of critical, deep and transformative learning.

Reflexive diaries have been used in academia predominantly in the ethnographic tradition. As Cook and Crang (1995, p. 29) suggest, such journals and diaries have been adopted as

> some kind of record to how the research progresses, day by day, and to chart how the researcher comes to certain (mis)understandings. Diaries... represent the doubts, fears, concerns, feelings, and so on that the researcher has at all stages of her/his work.

In this vein, diaries and journals have a conventional role, oriented around keeping a record of events soon after the activities have occurred, for posterity and future reflection. They position the researcher as the conduit through which experiences of others are understood and explained. [Webblogs may be considered a virtual version of the 'traditional' reflexive diary (see Cobanoglu & Berezina, 2011; and for a critique, see Gleaves *et al.*, 2008).] Reflexive diaries can, however, have an alternative emphasis. Journals and diaries can remain a record of events, but the entries themselves become not only the record but also the opportunity for reflexion; they may thus involve a synthesis of real-time recording, immediate reactions and more considered responses that occur at a later date. Reflexive diaries in this sense are a composite rather than a one-off register; they are an assemblage of records, reflections and considered arguments oriented around a theme or issue, rather than being specifically around chronological dates or events.

As a specific pedagogic tool, reflexive diaries have been predominantly used in this latter sense (see Langer, 2002; Park, 2003; Hubbs & Brand, 2005). As Gleaves *et al.* (2007, p. 632) put it,

> Diaries and journals have been used for several years as an effective means of monitoring higher education students' reflective capacity... as part of a reflexive meta-cognitive strategy to assist students in... their ability to transform their ... critical thinking, argumentative writing, [and] inquiry-based skills.

In this sense, diaries can be used as 'praxis analysers' (Nogueira & Canelhas, 2009, p. 471), offering a potentially, 'innovative and flexible approach to teaching, learning and assessment that encourages deep learning' (Dummer *et al.*, 2008, p. 459). Deep learning refers to the analysis of praxis that involves the acquisition and application of higher order skills such as analysing, interpreting and evaluating information. It encourages students to actively engage with complex materials, evaluate and critique them, and become integral assessors to their appropriateness or resonance within a specific context (Anderson, 2002). Such involvement contrasts markedly with passive, rote or surface learning which reduces the scholar to a simple database, retaining information which can be called up and tested. Reflexion is a key process to achieve deep(er) learning, as Dummer *et al.* (2008, p. 460) state:

> Reflexive learning is essentially the process of considering what we know or have learnt in order to generate new knowledge.... It involves the 'reprocessing' of ideas or knowledge in order to make sense of complicated or unstructured situations.

> In reflexion we describe, analyse and evaluate our practice in order to learn from that practice.

The reflexive diary can, therefore, fulfil an ideal function in enabling students to, as Gleaves *et al.* (2007, p. 632) put it, 'realize for the first time that their progress as successful learners is not merely a function of acquiring knowledge, being also predicated upon attitude, self-esteem and the mutability of their own ideas.' Assessment is a vital component in facilitating this 'self education,' and the role that diaries can play in this arena is increasingly acknowledged; as Gleaves *et al.* (2007, p. 632) confirm:

> When used as an assessment method ... diaries and journals have been shown to contribute to changes in critical awareness, growth in autonomy, and transformations from strategic approaches to learning to deeper and authentic engagement with subject matter.

This project was wholly assessed through reflexive journals, requiring students to realize the potential of this process in order to reflect successfully on issues concerning sustainability at the individual level. From these experiences, this paper seeks to review the ways in which reflexive diaries can expand the space in which students can encounter, (re)process and (re)consider ideas and actions, and by doing so contribute meaningfully to their critical awareness and enhance their 'self-education' repertoire (after Nogueira & Canelhas, 2009, p. 471).

'Individuated' Sustainability and the Ethnographic Eye/I

Reflexive diaries were used in this project as a tool for students to reflect on the obstacles and opportunities in realizing sustainability at the individual level. This objective was motivated by the broader societal context in which environmental problems have become 'individuated' (see Anderson, 2012); in other words, functionally, theoretically and politically the individual has become one of the key actors deemed both the cause of and solution to sustainability issues. As Spaargaren & Mol (2008, p. 354) put it:

> The individual is no longer seen as 'the small polluter' whose contribution can be dismissed in the light of the huge impacts of big industrial polluters.... the relative contribution of [individuals has now] come to the fore as the main obstacle for realizing ambitious environmental goals.

Within a structure of conventional lectures and student-based discussion groups, this module required students to write the story of their own approach to environmentalism and sustainable development. In terms of journals in particular, it encouraged students to consider their own positioning and philosophy in relation to environmental issues, compose entries that critiqued the academic, policy and media-related material they had exposure to and reflect on the (in)actions these combined to stimulate in their everyday life. In short, it encouraged students to adopt an 'ethnographic eye' to their own activities.

Geography has a long history of adopting ethnography to study far-away cultures and spaces (see Clifford, 1997). Yet, despite turns into the 'expanded field' (see Katz, 1994; Amit, 2000), auto-ethnography has played a minor role as a research method through which to gain understanding of key geographical issues and processes (but see Kideckel, 1997; Ellis, 2004; Cook, 2001). Auto-ethnography as a method connects the 'autobiographical and personal to the cultural, social, and political' (Ellis, 2004, p. xix). Through introspection, reflexion and action, auto-ethnography switches our critical gaze from others to ourselves, as Moss (2001, p. 3) tells us with respect to her own self-reflexion:

> I decided to use myself as a source of information. . . . I wanted to use my experience the way [I would] use theirs [my research respondents]—to elaborate empirical links with concepts . . .

As such, auto-ethnography is not a straightforward exercise. Ellis, in her well-balanced examination of the subject, outlines how this method requires insightful observation, considered introspection and good narrative skills; as a consequence, the tasks it poses students are not easy:

> [Auto-ethnography] is amazingly difficult. It's certainly not something that most people can do well. Social scientists usually don't write well enough. Or they are not sufficiently introspective about their feelings or motives, or the contradictions they experience. Ironically, many are not observant enough of the world around them. The self questioning auto-ethnography demands is extremely difficult. Often you confront things about yourself that are less than flattering (Ellis, 2004, p. xviii).

As Ellis outlines, turning the auto-ethnographic eye on ourselves can be a difficult process. Through the process of self-reflexion it is possible to discover elements of our identity that we may generally choose to ignore. Auto-ethnography, therefore, risks exposing inconsistency in our values, incoherence in our arguments or even hypocrisy in our practices (see Anderson, 2010). As both a cause and consequence, this process involves not simply rational, considered and cognitive responses, but also provokes embodied affect, relational sensibilities and emotional reactions (which are not always pleasant). This is perhaps especially true in the case in the arena of sustainability and environmentalism. As Thomashow (1995, p. 161) suggests, there are so many 'tension, uncertainties, and ambiguities' in attempting to become sustainable that all, 'environmentalists lead lives of contradiction.' Due to the potential for 'less than flattering' consequences produced through auto-ethnography, a sensitive, non-judgemental form of respect (see Markell, 2003) was consciously nurtured in all teaching delivery and focus group discussion, and encouraged in diary reflexion. It was made clear that individuals were not being tested in terms of their environmentalism, rather their reflexions upon it. Through this respect, the auto-ethnographic journal offered students the opportunity to insightfully interrogate the complex phenomena associated with sustainability and environmentalism.

Reflecting on and Through Diaries: in Trepidation of the Naked Scholar

Despite—but perhaps because—of the opportunities for self-reflexion and -critique offered by diaries, students were initially apprehensive about the prospect of composing them. As the following journal entry outlines:

> My initial thoughts about this auto-ethnography is that it doesn't really apply to *me* . . . it applies to those 'others' who are not doing anything for the environment. I've even chosen to study sustainability, surely that has to count for something!?

The composition of diaries thus gave students the opportunity to talk not only in the first person—attaching themselves and their reflexions to the world through which they are formed and constituted—but also to react personally and honestly to the requirements of doing so. This student is honest enough to admit that they would have preferred not to have to analyse their own behaviour; they felt that their commitment to enrolling on an environmentally related MSc course meant they had already gone further than many in the environmental cause—they had 'done their bit.' One can suggest a number of reasons why students felt the urge to sidestep sustained reflexion. Firstly, many felt daunted and disquieted by the task, in many cases it was a strange and alien experience for them.[2] With this in mind, it appears that the process of self-reflexion and critique, although an aspirant objective for many modules in the move towards deep learning, is rarely attempted and achieved by students, in particular with relation to sustainability. Secondly, the process of diary-composition threatened the possibility of revealing something (or some aspect of themselves) the students had been unaware of (as Ellis outlines above). As such, this introspection would really bring 'home' to them (literally so in the case of energy use or recycling for example) their contribution to (un)sustainability. Revealing their selves to themselves[3] in this way—becoming the naked scholar if you like—is something they felt understandably wary of. The scientific default position of standing outside and beyond the academic social/environmental critique was a far safer and more familiar position for them to be located in. As a consequence, some students used the diary as an opportunity to resist the task, as the following respondent outlines:

> Admittedly, I was very willing to rubbish the idea of a reflexive journal detailing my experiences and was determined not to let this piece of work change me or my life.[4]

In an attempt to mitigate the strangeness and perceived risk threatened by this project, the initial module introduction outlined to students suggestions of what good diaries might involve (e.g. emplacement, critique, self questioning, examples from everyday experience, connections drawn to and from theory, policy and the media). Students did not have to cover the range of topics or ideas raised in lectures and focus groups; they were free to engage with whatever topics that were most relevant and interesting to them.[5] Students were also given the opportunity to hand in two draft diary entries for comment (of the six they had to do by the semester's conclusion). This opportunity was welcomed by those particularly challenged by the process, as the following statement from a student module evaluation outlines:

> At first, it was confusing, but the lecturer gave us some feedback of two entries so that I understood better the nature of the assessed work.

Through the support of focus groups that ran throughout the module and encouraged peer discussion, self-reflexion and critique, initial trepidation evolved into considered acceptance of the utility of auto-ethnography, as the following diarist insightfully writes:

> I have a theory that the most difficult thing about auto-ethnography is most likely how to begin it, so it seems an apt place to start (although as I am reaching the end of this sentence that theory is starting to lose its footing already). The whole premise of auto-ethnography is a difficult one to grasp at first and I found myself wondering what the point was. It is easy to understand that ethnography allows valid ways of analysing an individual or perhaps a group in that it allows the researcher to discover what an individual does as well as says, and thus be able to analyse any discrepancies (Herbert, 2000, p. 4). The possibilities of this seem endless when I began to wonder how fantastically 'juicy' a collection of ethnographic research papers on various historical figures of power, both bad and good, would be. This idea soon dissipated however when I realized that even in the case of ethnography there's no reason why it would be immune to propaganda. This thought in turn led me to doubt auto-ethnography: how can it be reliable if an individual knows someone else is going to read it? After a quick read of a paper by Sarah Wall (2006), I came to the conclusion that this attitude defeats the object entirely. Sarah mentions (p. 43) that she used 200 words of her critique of the paper's subject defensively and apologetically, which defeated the personalized point of the work. This kind of self reflexive work is not being written for the sake of anyone else, but rather as a chance to be honest, think, open up and look at what oneself is doing which I have come to realize has already started taking place. So far I have changed my mind about the first sentence and decided that the hardest part is all of it; especially considering the fact I thought I would find it easy to criticize and analyse my thoughts and my actions and although this seems to be the case so far, I am interested to see how open I manage to be and how honest I am in reflecting upon that.

Initial trepidation thus evolved into considered acceptance for this new form of assessment, despite the fact that continual reflexion remained a challenge and involved an affective dimension rarely explored through academic analysis (although see Widdowfield, 2000; Askins, 2009). Reflexion ignited the embodied scholar within students (Haraway, 1988), and the space offered within the diary to express this emotion led to affective language being adopted to communicate these sensibilities (see Keith, 1992). Emotion and frustration were turned outwards to broader society but also inwards to the diarist themselves:

> These reflexive journals were useful in uncovering my own part and feelings about environmentalism. My previous entry displays a lot of my frustration towards implementing environmentalism ... and push[ing] humanity into a new way of living.
>
> I read through some lecture notes today that had some provocative suggestions as to what I could write about in these auto-ethnographies and it actually got me quite riled up, not because of the slides or the work but because of the answers I gave to the questions, it really irritated me that these were my answers because I don't think it's my fault, because I don't feel I have any other option. I tried to think about my apathy and why it think it's here. This question provoked me: When do you feel most politically empowered? Answer: Seldom, but mostly when spewing rhetoric in small groups.

> The biggest hitting point is that the changes I want are not evident in my own behaviour or desires, it feels a bit like I'm torn between my current identity and the values it holds, and the value I see in an environmental identity and change.

This affective reflexion was thus successful in moving students away from a detached, abstracted cognition concerning the world and their relation to it, and towards a broader, embodied understanding that combined rationality with emotional sensibility. Students engaged with a passionate geography (drawing on Game & Metcalfe, 1996), and certainly an embodied environmentalism, with all its concomitant dilemmas (see Thomashow, 1995).

A Space of Self-Education

Due to their embodied engagement with the process, the composition of reflexive journals often led students to critically reflect on a range of sustainability-related issues that were meaningful to them. The potential for diaries to become a space for self-education was enabled in practice, as the following diary extracts go some way to illustrate:

> I sit in lectures everyday listening to problems that effect, me, my community, my family, and will affect my children. The only way I can see to make changes practically is to accept, utilize and manipulate the current system. I don't think bright ideas can really change society holistically. I think there are ways I can be more environmental (recycle, reduce travel) I might even be able to change a few people's minds along the way, but this is limited to very few people. It's not going to have an effect on a grand scale at all. This says something about *real* democracy, liberty and freedom, all tenets of sustainable development.

> Many people assert that the UK is a democracy and that we have liberty and freedom, but those who suggest this do not understand the true meaning of freedom. That is: the ability to think in abstract ways outside the mould of society, the ability to be what you want and change and live how you want. I think that there are very few arenas, perhaps sometimes they are temporal, where people can truly act outside the realms influenced by dominant societal discourse.

> I have been questioning my thoughts and beliefs a lot recently. Has this course enabled me to become a better environmental citizen? Have I become more of an environmental dictator to friends? Or am I just the same as before? I believe to a certain extent 'yes' to the first question, a definitive 'yes' to the second question and a 'no' to the last. I feel that I have made great personal strides during this year, I understand many of the concepts that underpin sustainability and the environment and I actively seek to apply them to my everyday life.

> I am happy that I use my car less, I eat less meat and ready meals than previously and also I have embraced recycling. But I realize that this is because I put myself in the privileged position to study this MSc and I cannot help feeling my life would have continued under a 'business as usual' approach to the environment without the education I received. I understand my position better within the role of the

> environment, I can be a destructor or I can be a facilitator of positive environmental change but economics still provides a dominant counter balance to what actions I perform. I am happy to sacrifice some comforts since I believe that I have a responsibility towards the environment but I am not willing to sacrifice all my actions to the significant detriment of my lifestyle. I still enjoy travelling too much.
>
> . . . I feel a closer connection to the environment over the past 8 months than I have ever before; I hope that I can continue to grow this bond for my own benefit, if not the benefit of others.

The process of reflexive diary composition thus became a source of critique, but also emancipation and empowerment for many students. Over the course of the module, and due to the interest sparked by their reflexions, students attempted to reduce their carbon footprint, they researched local sources of food, altered their diet and politicized their decision making. Making the connections between abstract academic ideas and their everyday lived experiences became crucial for them, as the following diary entry confirms:

> what the process of reflexion has taught me is that having knowledge in some areas means nothing—I need to actively engage with my beliefs, actions, views, and inactions before this knowledge can have any impact on myself and the world around me. In short—I need to move from storing my 'environmental identity' within my practical consciousness and put it into my discursive consciousness. I need to be more reflective and reflexive when it comes to what I see as sustainable, and not rest upon ideas formed previously—such as in relation to my resource consumption, and try not to compartmentalize my environmentally friendly consumption practices . . .

By the conclusion of the module, the process of reflexion through discussion and diary composition appeared to have been successful in the now auto-ethnographic eyes of students:

> On a personal level I have learned that environmentally friendly practices need reflexivity and discursive consciousness in order to keep up with the changing environment, and this reflexive journal has been the perfect basis for such activity.
>
> The four months I have spent questioning myself about my green identity have raised some interesting questions and led me down many paths of thought and realizations that I wouldn't normally have entertained. For as long as I can remember I have cared about the environment and actually felt that I was one of the 'good guys'. Maybe not so much proactively doing everything but at least doing what is possible when feasible. It seems this is entirely not the case. Upon reflection the green aspects of my life such as recycling, switching off lights and so on are the results of nothing more than my situation and general common sense. Likewise, growing vegetables started out as something to do as fun and not a conscious effort to help the environment. Did I take on green aspects of my lifestyle because I wanted to, or simply because it suited my lifestyle? The fact that I simply throw away my recycling in normal trash when I run out of green bags suggests the latter.

> As uncomfortable as it has sometimes been to actually think about why I do what I do, and even though I have come out of it feeling slightly beaten, I am at least happy that my eyes are now open to the reality of how I lead my life. Instead of plodding along as I have been doing, I want to make a conscious effort to be more pro-active and to fit the role that for years I thought I already filled.

Conclusion

The combination of auto-ethnography and reflexive diaries can be seen to offer a useful opportunity not only to enhance student learning in relation to sustainability, but also to secure generic advances in critical analysis and deep involvement in learning. Through the creation of a mutually supportive and respectful learning environment (enabled through focus group discussion, the absence of pejorative judgement and advice on diary writing), the alliance of auto-ethnography and reflexive journal entries was successful in broadening the academic space in which students could engage with and begin to understand complex structures and geographies of the world around them. Due to the auto-ethnographic nature of the task, the reflexive diary allowed students the space to engage with their everyday lived experiences and make political connections between these and more abstract policies and concepts. These important connections led students to form a resonant understanding of the obstacles and opportunities for sustainability at an individual level, but also their positioning as citizens, voters, consumers and activists. This embodied learning prompted reflexive diaries to be a source of inspiration and transformation for students, as well as securing for them positive grades in the module assessment.[6] The attachment and engagement generated by students to the topics under discussion led individuals to want to change their lives and contribute to developments in their actions and practices. As Mezirow has outlined, this process contributed to deep and transformative learning, where "the learner is ultimately changed, or 'transformed' through the process, the learner's prior patterns of thinking ultimately growing and changing or 'transforming' over time" (cited in Gleaves *et al.*, 2008, p. 220). Through sensitive introduction, supported discussion and engaged lecturing, in this case reflexive diaries successfully facilitated deep learning and empowered students to move towards the synthesis of theory and practice, and the critical (self) analysis of broader geographic processes and activities.

Notes

1 In more detail, the module is taken by approximately 20 students every year, with 75 per cent of these being full-time students and 25 per cent part-time students. The gender balance is approximately 50:50, and the home:overseas student ratio is also approximately 50:50. Attendance is excellent at both lectures and focus groups, although a register is not taken, and attendance is not compulsory. In general terms, there are no significant patterns identifiable in grades achieved when analysed in terms of gender, age or point of origin.

2 This was particularly true for students from overseas, as the following Chinese student maintains:

> In order to be an auto-ethnographer, it is necessary to start writing my own life diary about how I live, which kinds of energy utilization I have everyday and what benefits I can get if I change my life styles. To reflect upon my thoughts and actions with reference to specific texts has been challenging for me.

However, many 'home' students experienced similar apprehension at the beginning of the module:

> This exercise has made me think about my life differently and pause for thought over the reasons why I make the choices in life that I do. I was initially daunted by the idea of the journal but it has been an interesting exercise in reflection.

[3] And of course to the marker. With this in mind, it was also made clear to students at the beginning of the module that these diaries would be anonymised for marking, and would be archived anonymously for research use such as these. All students signed consent forms to allow this latter use of their material.

[4] Over the course of the module, however, this student did modify their opinion:

> However, having been given allocated time to think about sustainability and environmental issues, I can conclude that there are many problems that modern society has created and I have been given some self-empowerment to deal with some of these issues.... Without this reflective journal and auto-ethnography, my thought would not have been provoked.

[5] In practice, a range of subjects were considered by students; however, the vast majority were connected in some way to those raised in lectures and focus groups (for more information on these subjects, see www.citizensustainability.com).

[6] In academic year 2010–2011, the average (mean) student grade on this module was 64 per cent. This figure also represents the average mean grade taken across all modules within the MSc degree of which this module is a part.

References

Amit, V. (Ed.) (2000) *Constructing the Field* (London & New York: Routledge).

Anderson, J. (2002) Researching environmental resistance: Working through Secondspace and Thirdspace approaches, *Qualitative Research*, 2(3), pp. 301–322.

Anderson, J. (2010) From 'zombies' to 'coyotes': Environmentalism where we are, *Environmental Politics*, 19(6), pp. 973–992.

Anderson, J. (2012) Managing trade-offs in 'ecotopia': Becoming green at the Centre for Alternative Technology, *Transactions of the Institute of British Geographers*, 37(2), pp. 212–225.

Askins, K. (2009) 'That's just what I do': Placing emotion in academic activism, *Emotion, Space and Society*, 2(1), pp. 4–13.

Clifford, J. (1997) *Routes: Travel and Translation in the Late Twentieth Century* (Cambridge: Harvard UP).

Cobanoglu, C. & Berezina, K. (2011) The impact of the use of blogs on students' assignment engagement, *Journal of Hospitality, Leisure, Sport & Tourism Education*, 10(1), pp. 99–105.

Cook, I. (2001) You want to be careful you don't end up like Ian. He's all over the place', in: P. Moss (Ed.) *Placing Autobiography in Geography*, pp. 99–120 (Syracuse: Syracuse University Press).

Cook, I. & Crang, M. (1995) *Concepts & Techniques in Modern Geography: Doing Ethnographies* (Durham: CATMOGS).

Dummer, T., Cook, I., Parker, S., Barrett, G. & Hull, A. (2008) Promoting and assessing 'deep learning' in geography fieldwork: An evaluation of reflective field diaries, *Journal of Geography in Higher Education*, 32(3), pp. 459–479.

Ellis, C. (2004) *The Ethnographic I. A Methodological Novel About Ethnography* (Walnut Creek, CA: Alta Mira).

Game, A. & Metcalfe, A. (1996) *Passionate Sociology* (London: Sage).

Gleaves, A., Walker, C. & Grey, J. (2007) Using digital and paper diaries for learning and assessment purposes in higher education: A comparative study of feasibility and reliability, *Assessment & Evaluation in Higher Education*, 32(6), pp. 631–643.

Gleaves, A., Walker, C. & Grey, J. (2008) Using digital and paper diaries for assessment and learning purposes in higher education: A case of critical reflection or constrained compliance? *Assessment & Evaluation in Higher Education*, 33(3), pp. 219–231.

Haraway, D. (1988) Situated knowledges: The science question in feminism and the privilege of partial perspective, *Feminist Studies*, 14(3), pp. 575–599.

Herbert, S. (2000) For ethnography, *Progress in Human Geography*, 24(4), pp. 550–568.

Hubbs, D. & Brand, C. (2005) The paper mirror: Understanding reflective journaling, *Journal of Experiential Education*, 28(1), pp. 60–71.

Katz, C. (1994) Playing the field. Questions of fieldwork in Geography, *Professional Geographer*, 46(1), pp. 67–72.

Keith, M. (1992) Angry writing: (Re)presenting the unethical world of the ethnographer, *Environment & Planning D: Society & Space*, 10, pp. 551–568.

Kideckel, D. (1997) Auto-ethnography as political resistance, in: D. E. Reed-Danahay (Ed.) *Auto/Ethnography. Rewriting the Self and the Social*, pp. 47–70 (Oxford & New York: Berg).

Langer, A. M. (2002) Reflecting on practice: Using learning journals in higher and continuing education, *Teaching in Higher Education*, 7(3), pp. 337–351.

Markell, P. (2003) Tragic recognition. Action & identity in Antigone and Aristotle, *Political Theory*, 31(1), pp. 6–38.

Moss, P. (Ed.) (2001) *Placing Autobiography in Geography* (Syracuse, NY: Syracuse University Press).

Nogueira, P. V. & Canelhas, S. (2009) Learning experience through diary writing: A case study, *International Journal of Learning*, 16(7), pp. 471–486.

Park, C. (2003) Engaging students in the learning process: The learning journal, *Journal of Geography in Higher Education*, 27(2), pp. 183–201.

Spaargaren, G. & Mol, A. (2008) Greening global consumption: Redefining politics and authority, *Global Environmental Change*, 18, pp. 350–359.

Thomashow, M. (1995) *Ecological Identity. Becoming a Reflective Environmentalist* (Cambridge MA, London: MIT Press).

Wall, S. (2006) An auto-ethnography on learning about auto-ethnography, *International Journal of Qualitative Methods*, 5(2), Article 9. Available at http://www.ualberta.ca/~iiqm/backissues/5_2/html/wall.htm (accessed December 2011).

Widdowfield, R. (2000) The place of emotions in academic research, *Area*, 32(2), pp. 199–208.

SYMPOSIUM

Mapping the Journey Toward Self-Authorship in Geography

NIAMH MOORE*, ERIC J. FOURNIER**, SUSAN W. HARDWICK[†], MICK HEALEY[‡], JOHN MACLACHLAN^ & JÖRN SEEMANN[§,||]

*School of Geography, Planning and Environmental Policy, University College Dublin, Ireland, **Department of Geography, Samford University, USA, [†]Department of Geography, University of Oregon, USA, [‡]Healey HE Consultants, UK, ^School of Geography and Earth Sciences, McMaster University, Canada, [§]Department of Geography and Anthropology, Louisiana State University, USA, [||]Departamento de Geociências, Universidade Regional do Cariri, Brazil

ABSTRACT *Learning is a developmental journey, and geography curriculum plays a key role in supporting student progression. In this article, we argue that the concept of 'self-authorship' is a useful guiding principle in supporting curriculum revision and reform. A series of international case studies illustrate how self-authorship can be enacted in different ways within geography curricula in a range of contexts. The role of a range of collaborators and the co-curriculum in supporting the student journey are highlighted. The article concludes by suggesting that the key strength of the concept is its non-prescriptive nature.*

Introduction, Purpose, Goals and Research Question

> I want you to imagine that you have been asked to form a new department of geography. Given the rare opportunity to write without constraint, would your curricula bear much resemblance to most of the formal courses of study to be found today? With any luck, your answer will be something like, good grief no! If your answer is something else... there is not much hope for the future! (Gould, 1973, p. 253).

The quote above serves as a reminder that the development and dissemination of innovative curricula remain a challenge for geography educators for almost four decades after Peter Gould made his observation. Despite the efforts to re-vision, re-create and rethink the curriculum in geographic education since the early 1970s, a great deal remains to be accomplished. Our primary concern in this article is to present a vision that deeply embeds opportunities within the geography curriculum for students to engage in self-directed learning, to think independently, to make more informed choices and achieve their academic and personal goals.

This article provides specific examples of how 'self-authorship' can be successfully fostered in geography curricula. Our overarching goal is to encourage geographers collectively to think about ways through which authentic student learning in geography classes and programs can lead students toward this goal. Self-authorship is based on a student's "capacity to internally define a coherent belief system and identity that coordinates engagement in mutual relations with the larger world" (Baxter Magolda, 2004, p. xxii). It encourages students to balance external influences with their internally held beliefs and to engage in intense reflection of their own learning process. Typical skills associated with this stage of intellectual development include the ability to critically analyze and evaluate information, to view 'known truths' from multiple perspectives, to embrace diversity and ambiguity and to develop an independent learning capacity and their own voice. Although many of these ideas and aspirations are very much in line with current educational thought on the importance of greater levels of student autonomy and more experiential learning, self-authorship as a concept was first discussed in the 1960s. In the past decade, it has gained increasing attention as the concept has been expanded, tested and refined, and there is now an extensive body of research supporting the concept (Baxter Magolda, 1992, 2001, 2006; Kegan, 1994; Hodge *et al.*, 2007, 2008a, 2008b, 2009). Convincing evidence to support self-authorship as a goal for intellectual development has recently been provided by Bekken and Marie (2007). In their discussion of a four-semester earth sustainability pilot project, they report that "a remarkable transformation in student development as evidenced by four snapshots of written reflective work over time that capture how the students conceptualize complexity and authority, whether they recognize fundamental assumptions and arguments, and how they apply or transfer disciplinary knowledge beyond disciplinary boundaries" (Bekken & Marie, 2007, p. 65) became evident.

We offer 'self-authorship' as a guiding principle for curriculum reform and revision efforts in geography. In this article, we examine what self-authorship could look like in geographic education and provide some suggestions for implementing this approach in geography classrooms and programs in the sections that follow. Our conceptual discussion of self-authorship is followed by a series of international case studies that provide examples of the approach embedded within higher education geography curricula. Each of these case studies provides a concrete example of an effective way to encourage students to understand, apply and create their own meanings through mastery of specific content and skills in geography. In addition, they illustrate the use of a holistic approach that addresses the affective as well as the cognitive dimensions of student development. In the final section of this article, we offer some preliminary conclusions and reflections that may point the way toward a greater role for self-authorship in geography education.

Rethinking and Designing the Geography Curriculum

Curriculum is an ambiguous concept and may be defined in a variety of ways depending on the cultural context, educational system and level of education. A curriculum is, perhaps, most often understood in terms of the specific content to be covered in a course or set of courses.

Thus, a geography curriculum usually refers to a prescribed package of materials designed to help students master a specific set of goals and objectives. The most useful curriculum has the potential to be used by a novice teacher, but can also be expanded and changed over time in a fluid, flexible and organized way and created, assessed and then (re)created in the hands of each individual educator. In this view, curricula can perhaps best be considered as a 'map' that is designed according to an established scale, projection and language.

Curriculum is also heavily influenced by the socio-cultural context within which it is initially designed and then reformed. In the USA, the higher education curriculum in most disciplines is controlled at the department level. There are some state college and university systems that require common course titles, descriptions and numbering to facilitate what are called 'articulation agreements'. That means students can transfer from one college or university to another with a guarantee that their course work will also transfer from one institution to another without constraints. However, there is a great deal of freedom usually governed by some sort of university-level approval or oversight committee. Increasingly, there is a call for assessment of programs where departments are required to develop goals and standards and then to evaluate how their students are meeting those standards, but this varies from one college or university to another (Southern Association of Colleges and Schools, 2010). In summary, curriculum in US geography is guided by disciplinary tradition, inertia and the standards of individual departments. In Ireland and the UK, curriculum is broadly conceived as key topics or content that must or should be covered and the learning outcomes that should be achieved on completion. Designing the curriculum is a key strategic issue for geography departments, but it has not received the attention it deserves (Healey *et al.*, 2010). This may partly be related to the difficult decision making required from all stakeholders and the potentially contentious nature of curriculum design and reform. Key questions arising during this process of revision include (adapted from Jenkins, 1998, p. 3):

- What curriculum is most essential for students to master and experience?
- What geography content and skills are to be mastered by students before they enroll in a course?
- How can staff most effectively present learning materials and content to students?
- How can students be taught to go beyond mastering only the basic content included in a certain curriculum so that it becomes a part of himself/herself and remembered and used many years later?

Some of these questions are much easier to answer than others. As a discipline, geography is particularly well positioned to help facilitate student engagement and learning due to its wide-ranging and diffuse subject matter and the range of teaching practices adopted. In their general study on student engagement, Nystrand and Gamoran (1991) identify a range of teaching practices with a highly positive impact on substantive engagement. They concluded that learning was most effectively achieved when collaborative and peer learning approaches, as well as group work and fieldwork, were utilized. Geographers typically use many of these teaching approaches, creating opportunities for students to experience meaningful content-based learning.

In recent years, the discipline has also rediscovered its role as an interdisciplinary bridge. Public debate and concern for issues broadly geographical from global interdependence to climate change have provided a context within which the learning environment should be easily connected by both teachers and learners to 'real world' events. Encouraging students to connect ideas from both academic and personal experiences to their current learning is a highly effective teaching and learning method (Dummer *et al.*, 2008) and one that should be comparatively easier to achieve in geography than in other disciplines. This kind of active engagement prepares students to think and inquire spatially (Spronken-Smith & Kingham, 2009; Spronken-Smith *et al.*, 2008) and 'do geography' in practice (Boyle *et al.*, 2007). The opportunity, therefore, to define a relevant and engaging curriculum is significant in the disciplinary context, but the overarching framework for doing so remains challenging.

At the current time, one of the biggest challenges for curriculum reform is how best to encourage students to go beyond mastering subject content so that it becomes a part of himself/herself and remembered and used many years later? Although we may believe that the curricula we design are relevant and engaging, how do we encourage students to recognize this relevance and make meaning from it. In this article, we hypothesize that the concept of self-authorship provides a useful method for facilitating better engagement and deeper learning as well as for developing more critical and informed world citizens.

Conceptualizing the Development of Self-authorship

During the last four decades, many educators have shifted their focus from a teaching to a learning paradigm (Barr & Tagg, 1995) and toward a more student-oriented design of curricula that aim to empower students as critical thinkers. The aim is to move away from the scholar's "traditional role of the expert or the tendency to seek students' approval and instead push students to gain intellectual, relational and personal maturity through continuous feedback and high expectations" (Hodge *et al.*, 2009, p. 19). Recently, a number of writers have suggested the need for a further shift toward a 'discovery paradigm' in which students become the creators and constructors of knowledge not just the consumers (Table 1, Hodge *et al.*, 2008b; Healey & Jenkins, 2009).

In these debates, the metaphor of a journey is frequently used to describe the learning processes of a student during college education. The educator's role is to accompany the students on their voyage:

> We offer maps (e.g. the curriculum and co-curriculum) and highlight possible pathways (e.g. majors, leadership, opportunities). We point out important stops along the way and activities to pursue at those stops that will make the journey more productive. We advise students when to slow down, when to let traffic pass and when to use their brakes. We encourage students to explore on their journey, but at the same time to be mindful of others along the way. Educators also monitor travel

Table 1. Changing educational paradigms (Hodge *et al.*, 2007, p. 3)

Paradigm	Approach
Teaching	Telling students what they need to know
Learning	Engaging students in learning how to learn; emphasis on learning what they need to know
Discovery	Encouraging students to seek and discover new knowledge

> difficulties. We help students who get stuck along the way and seem to have trouble getting back on the road. We deal with instances of road rage—travelers who violate others' rights in their exploration, or who lose control of themselves to their own detriment. Being good company for literally thousands of students, all of whom are on different journeys, is a complex challenge (Baxter Magolda, 2001, p. xvi).

Building on the work of Perry (1970) and Kegan (1994), Baxter Magolda has been conducting a longitudinal study on the ways of knowing among students who started in college almost 25 years ago (Baxter Magolda, 1992, 2001, 2004). On the basis of interviews in intervals of one year, she has gained insights into their learning processes and experiences, and developed a framework for analyzing the intellectual growth of students through the lens of self-authorship and the Learning Partnerships Model (LPM) that aims to *transform* rather than *redesign* educational practice (Baxter Magolda, 1992, p. xiv). In her writings, she suggests that students' construction of meaning (or ways of knowing) progresses over time from absolute knowing, through transitional and independent knowing, to contextual knowing or self-authorship (Table 2). The LPM also incorporates elements of the co-curriculum—activities that involve students outside the traditional bounds of the classroom—such as debating societies, Model UN, environmental organizations or sports clubs (DeZure, Babb & Waldman 2005).

Such transformative learning (Mezirow, 2000) adopts a holistic approach to student intellectual development:

> Moving away from uncritical acceptance of knowledge to critically constructing one's own perspective, however, is more complex than learning a skill set. It is a transformation of how we think—a change in our assumptions about the certainty, source and limits of knowledge (Baxter Magolda, 2006, p. 50).

The LPM provides guidance for the journey toward self-authorship by bringing together the curricula that students experience in the classroom with its emphasis on knowledge, with what they concurrently experience outside the classroom, the co-curricula and with its emphasis on affect:

> Learning partnerships support self-authorship via *three principles*: validating learner's capacity as knowledge constructors, situated learning in learners' experiences and defining learning as mutually constructing meaning. Validating learners' capacity to learn and construct knowledge is necessary for them to realize that they can go back to the potter's wheel. Situating learning in their experience instead of the experience of authority gives them a context from which to bring their identity to learning. Defining learning as a mutual process of exchanging perspectives to arrive at knowledge claims supports their participation in the social construction of knowledge (Baxter Magolda, 2004, p. xix).

Hodge *et al.* (2008b) have drawn on Baxter Magolda's ideas to map out a developmental journey for students (Table 3). However, as students go through the stages at different rates and many may not reach the self-authorship stage by the end of their undergraduate course, it would be a mistake to map the three levels neatly to a three-or four-year curriculum. One of our biggest challenges in designing learning activities and assessments is to facilitate

Table 2. Development of students' construction of meaning (Baxter Magolda, 1992, p. 30)

Domains	Absolute knowing	Traditional knowing	Independent knowing	Contextual knowing
Role of learner	• Obtains knowledge from instructor	• Understands knowledge	• Thinks for self • Shares views with others • Creates own perspectives	• Exchanges and compares perspectives • Thinks through problems • Integrates and applies knowledge
Role of peers	• Share material • Explain what they have learned to each other	• Provide active exchanges	• Share views • Serve as a source of knowledge	• Enhance learning via quality contributions
Role of instructor	• Communicates knowledge appropriately • Ensures that students understand knowledge	• Uses methods aimed at understanding • Employs methods that help apply knowledge	• Promotes independent thinking • Promotes exchange of opinions	• Promotes application of knowledge in context • Promotes evaluative discussion of perspectives • Student and teacher critique each other
Evaluation	• Provides vehicle to show instructor what was learned	• Measures students understanding of the material	• Rewards independent thinking	• Accurately measures competence • Student and teacher work toward goal and measure progress
Nature of knowledge	• Is certain or absolute	• Is partially certain and partially uncertain	• Is uncertain—everyone has own beliefs	• Is contextual; judge on basis of evidence in context

Table 3. The developmental journey of the student (Hodge *et al.*, 2008b)

Developmental level	Student traits
Reliance on external references [*foundations*]	Knowledge viewed as certain Reliance on authorities (e.g. professors and parents) as source of knowledge Externally defined value system and identity Act in relationships to acquire approval
At the crossroads [*intermediate learning*]	Evolving awareness of multiple perspectives and uncertainty Evolving awareness of own values and identity and of limitations of dependent relationships
Self-authorship [*contextual knowing*]	Awareness of knowledge as contextual Development of internal belief system and sense of self Capacity to engage in authentic, interdependent relationships

students at different stages in their intellectual journey. A schema similar to the one above has been developed by Taylor and Haynes (2008, p. 5), who have devised outcomes for a three-tiered curriculum and co-curriculum "that steadily lead students toward the capacity to engage in sophisticated intellectual and creative activities, build mature relationships, and make decisions based on an internal belief system."

Enacting Self-authorship in the Geography Curriculum: International Exemplars

Students do not all progress along the path to self-authorship at similar rates so it would be disingenuous to attempt to map particular milestones in the journey to specific points in the undergraduate curriculum. Having said that the journey to self-authorship does need to be carefully managed to ensure that early experiences are perceived as worthwhile and challenging rather than overwhelming. With this in mind, different kinds of activities to support the process of self-authorship may be considered more or less appropriate depending on the stage at which they are introduced. In terms of educational paradigms (Table 1), it may be that for successful self-authorship to occur elements of the three paradigms need to be followed at different stages. Some basic information may need to be taught initially, to facilitate better learning based on the discovery principle. This section of the paper introduces a series of five case studies, all of which provide specific examples of how self-authorship has been fostered in a variety of educational settings through a learning partnership approach. All of the case studies come from a Western and predominantly Anglophone cultural perspective. We believe that the process of self-authorship is broadly transferable cross-culturally, but there may be certain cultural considerations, particularly in an East Asian context, that would require the concept to be introduced at a different pace and stage in the student career.

The first example is from Miami University, Ohio, USA, and represents a very specific approach given that it has been introduced as part of a university top-down policy to foster better student engagement. Figure 1 illustrates how one particular course, *Global Forces/Local Diversity*, has challenged students to engage with conceptual material and to begin to explore their role as knowledge constructors within the geography curriculum. Increasingly, inquiry-based learning approaches such as this are being effectively used as a

Encouraging self-authorship in introductory human geography via student-created, collaborative, place-based case studies

The Department of Geography at Miami recently redesigned its introductory human geography course, *Global Forces/Local Diversity*. The primary goal was to provide support for undergraduate geography students along the journey to self-authorship by involving them in their own learning. This is accomplished by moving away from traditional teaching modes (e.g. lectures, textbook reading assignments, exams) toward an approach based on weaving student-created case study inquiries into everyday teaching and learning. Students are asked to do conceptual and thematic research in collaborative groups on real issues in particular places. They then share their multi-media projects with other students enrolled in the class (digitally and in person) and these student-created geographical portraits are ultimately incorporated into class discussions.

Students learn about their own place in a changing and complex world by exploring globalization as it unfolds in local places and current events. Thematic case studies forming the center point of this course can easily be adapted, expanded, and/or revised by new faculty or graduate student instructors over time. This not only allows the class to be taught by a variety of diverse faculty, it also keeps course content up-to-date.

Self-authorship is also encouraged in the course by (1) conveying to students that knowledge is complex and socially constructed; (2) validating learners' 'capacity to know' by trusting in their ability to conduct collaborative research and sharing in mutually constructed meaning; and (3) showing students how to develop a framework for authoring and analyzing multiple perspectives.

Geography is only one of several introductory courses which have been converted to inquiry-based learning as part of the 'Top 25' project at Miami (Hodge *et al.*, 2008b).

Sources and further information:

http://www.units.muohio.edu/celt/engaged_learning/top25/

Figure 1. Miami University, OH, USA.

mechanism to enthuse students about their role as learners and researchers in higher education institutions even from their very first encounter at the University.

Figure 2 illustrates how this is enacted at the University of Gloucestershire, UK, where such an approach forms an important component of the university induction program. Responses to evaluations of the program by staff and students highlight the importance of developing social and academic networks to promote higher-order thinking and engagement:

> 'It gave staff and students an opportunity to begin professional relationships. The project worked very well because it engaged students in bonding and active learning' (Staff Induction Evaluation Questionnaire, 2007 in Centre for Active Learning, 2009).
>
> 'I was apprehensive about the trip ... I loved it! Great way to bond with fellow students' (Student Focus Group, FCH, 2008 in Centre for Active Learning, 2009).

The advent of new technologies has also facilitated this new emphasis as exemplified at University College Dublin in Ireland, where an introductory Human Geography module has recently been transformed from a more traditional lecture style course to one that uses a blended-learning approach extending well beyond the media used to incorporate all aspects of, and inputs into, the learning process. Drawing on work by Garrison and

Undergraduate Research at the University of Gloucestershire, UK begins at Induction/Orientation

In 2007, over 650 students in the Faculty of Education, Humanities and Science undertook discipline-based inquiry projects during induction week. This involved them working in small groups to collect information from the library and in the field, analyse it, present it to tutors in novel ways and receive formative feedback. For example, the human geographers and the sociologists researched the experience of Gloucester residents of 'the Great Flood of 2007'. Physical geographers investigated a scenario of establishing a wolf park in the Forest of Dean. Other faculties in the University are developing their own versions of undergraduate research as part of induction. It has also proved a significant staff development activity both for the many academic tutors involved in designing inquiry-led activities and for the library staff who changed their approach to library induction to support the specific student research projects.

This approach helps to start the students on the journey to self-authorship by giving them an enjoyable inquiry-led learning experience before they start their first formal class. The challenge is to build on this approach to learning in the way the curriculum is designed in the first and subsequent years of the students' programmes.

Sources and further information:

Centre for Active Learning (2009); Healey and Jenkins (2009)

Figure 2. University of Gloucestershire, Cheltenham, UK.

Vaughan (2008, p. 30) who suggest that "the fusion of real and virtual experiences [through blended learning] creates unique communities of inquiry that are accessible regardless of time and location", the course leaders viewed the VLE as a tool for facilitating the mutual construction of meaning through collaborative work.

Inquiry is also at the heart of the approach at McMaster University in Ontario to improve student writing from the first year (Figure 3). The structure of this course ensures that not only do students get the opportunity to work collaboratively on small research projects but they also must reflect on their, and others, learning processes. Empowering students as critical thinkers and providing a range of types of feedback, from both instructors and peers, is critical to the self-authorship approach.

As each of our case studies shows, the theme of collaboration is one that is critical in effectively supporting the journey toward self-authorship but collaboration need not necessarily be confined to peer or learner–instructor relationships. Baxter Magolda (2004) argues that a holistic approach to learning which draws on extra-curricular as well as classroom-type experiences is critical to student progression as it facilitates the construction of meaning over time and the grounding of learning experiences. This is particularly evident in one approach taken within the second-year undergraduate curriculum at the Australian National University (Figure 4). The Vietnam Field School is an intensive learning experience where classroom and conceptual knowledge is enhanced, challenged and refined through interactions in the field with other students, instructors and residents of the various locations visited. Through the production of an individual learning goal statement, students have the opportunity to produce their own learning objectives. Feedback from fieldcourse participants suggests that the learning outcomes achieved went far beyond that was initially expected. In the words of one student:

Improving Student Writing through Small, Inquiry-based Classes

Students struggle with the act of writing while instructors struggle with quality of student writing, reporting in some cases that they feel writing quality is declining at the same time as undergraduate student class size is increasing making providing individualized writing feedback a greater challenge. McMaster University, a mid-sized, research-intensive, Canadian university has attempted to address this challenge in the Faculty of Social Sciences through the creation of inquiry-based small-sized (<30 students) in first year.

Students are provided the opportunity on four occasions in the course to engage in the entire writing process during a three-hour class. At the start of the class students arrange themselves in group of three to five students and are provided a short, usually editorial piece on a topic related to the course. Working in a combination of individual, small group and plenary sessions the students develop a research question in each group, identify the resources that they would require to answer the question, identify the possible library search strategies they would employ, develop a draft response which includes a potential layout for the paper, develop a time management strategy about how they would complete the paper and finally reflect on each group members (including their own) strengths and weaknesses. Students report an increased confidence in their ability to write and to locate library sources as a result of this exercise. In addition students report they were unaware that different students take different approaches to their writing.

Research on this approach has shown that students who have taken the course perform better in later courses than other students; and more go on to take honours (Justice *et al.*, 2007a, 2007b).

Sources and further information:

http://cll.mcmaster.ca/resources/inquiry.html

Figure 3. McMaster University, Ontario, Canada.

> 'It [the fieldcourse] has allowed me to test out the 'Development Studies' waters, and prompted me to question where I fit into things. I have recognized that my experiences and reflections are valuable; my confidence in my ability and capacity to share knowledge has been reinvigorated. I've come to realize that it's not so important to have definitive answers. It is important to try to reach a more holistic understanding of a perspective, and to question who's perspectives are being reflected'.

Another student immediately recognized the wider learning implications of the fieldcourse:

> 'I have no doubt the full impacts will not be evident till well after the submission of this piece … and will have much longer lasting influences on my thought process in relation to future learning'.

Reflection is a critical component of the course enabling students to monitor how their understandings and ideas mature through their learning experiences in the field and to

Vietnam Field School – Giving Australian National University Undergraduates Real Research Opportunities

This course is open to second year undergraduates from a range of disciplines. An 18-day field course focuses on the interaction between development and environmental issues in Vietnam. During a weekend of preparatory workshops students devise a learning goal statement, which is subsequently used to define their particular research interests. Small research groups are formed and proposals developed collaboratively, using online tools such as wikis. Students are also required to undertake a major research project prior to travelling to Vietnam and to share their work with students orally and through a learning management system.

In Vietnam, students spend the first week in the city of Da Nang, four days in the world heritage town of Hoi An and three nights in a rural village where they stay with local families. On their return to Da Nang, students share their preliminary research findings with each other and reflect on their personal and group experiences. A reflective learning journal is used to record the interconnections between different parts of the course, what students learn from peer presentations and their changing understandings of research throughout the course.

At the end of the course, all students are required to give a seminar on their interpersonal, intrapersonal and epistemological development. Each year the course has received very good evaluations and students have reported their participation as being a transformative learning experience. Such experiences foster the intellectual development of the students and speed their journey towards self-authorship.

Sources and further information:

http://fennerschool-people.anu.edu.au/richard_baker/ENVS2017/students%20work/index.html

http://fennerschool-people.anu.edu.au/richard_baker/ENVS2017/resources/groupwork.html

http://insight.glos.ac.uk/tli/resources/toolkit/resources/ugcs/Pages/default.aspx

Figure 4. Australian National University, Canberra, Australia.

chart their own intellectual and personal progression. A core component of the ANU approach is to make the implicit explicit through a final seminar where students are asked to self-identify their own development. The deliberate incorporation of these kinds of extra-curricular experiences and encounters into the curriculum in order to facilitate the intellectual, relational and personal development of students is becoming an increasingly important component of curriculum revision and reform.

At Samford University, Alabama, this is given expression through the *Geography In Practice* component of the curriculum (Figure 5). Following their progression through a number of core and elective courses, students have the opportunity to engage with external 'clients' or with academic supervisors to develop their research skills through specific projects. A similar approach is undertaken in Brazil, where this type of beyond-the-curriculum learning is practiced through the 'Program for Tutorial Education'. Faculty members tutor and train small groups of undergraduate students to develop projects in

A Curriculum Designed to Facilitate a Student's Journey Toward Self-Authorship

The geography department at Samford University recently redesigned their department's curriculum guided by goals of increasing student engagement with the discipline, improving their practical skills, and enhancing their ability to solve complex problems and engage in critical thinking. Core modules provide basic instruction, but these introductory modules incorporate case studies, problem-solving, and active engagement with the subject matter. Students then proceed through a series of elective courses and finally to a series of courses called "Geography in Practice". Here students have the option of doing a supervised externship, acting as a teaching assistant for an introductory class, or doing an independent research project.

These experiences provide students with an opportunity to link their prior coursework with practical workforce skills. Finally all students complete a capstone experience where they may either undertake a client-based project, or may elect to do a traditional research paper. With the client-based projects, students work in teams with an outside client to define a problem, devise a work plan and create some distinct output. As examples, students have produced a series of maps for a local bicycle club, worked with the university's disability services on an accessibility map of campus, and collaborated with an environmental agency to study sedimentation in a river.

All capstone students are assessed on a range of skills, as well as informational and quantitative literacy. As students progress through the curriculum they are expected to take increased responsibility for their own learning and to develop the intellectual skills necessary to move beyond the campus and into society.

Sources and further information:

http://howard.samford.edu/geography/

Figure 5. Samford University, AL, USA.

research, education and off-campus activities, foster their participation in academic events and publish their findings (Ribeiro & Marafon, 2005). As with many of the other examples mentioned, the overall goal is to ensure that students are facilitated in developing as independent, critical thinkers and responsible active citizens.

Conclusion

The move toward adopting more student-oriented approaches to teaching and learning and developing students as critical thinkers provides significant challenges to those engaged in curriculum development and reform. In this article, we have argued that the concept of self-authorship may provide a useful tool for guiding curriculum change and in promoting more holistic approaches to learning which support collaboration, reflection and more autonomous student learning. Learning is a partnership but one that can effectively develop and extend beyond the bounds of the classroom as exemplified particularly through the ANU and Samford case studies. If we are to be successful in guiding our students on their journey toward self-authorship we should also pay attention to supporting their development through the co-curriculum. Through our specific case studies, we have illustrated how a variety of geography departments internationally are facilitating students on their journeys toward self-authorship. We have shown that this need not necessarily depend on major capstone projects or the development of entirely new courses but can effectively be achieved through alternative approaches to teaching and learning in the existing curriculum. Geography departments can also take a more outward-looking approach to curriculum reform, and at least two of our case studies have demonstrated how fieldwork or internships can facilitate engagements with employers, community groups and others that support students on their intellectual and personal journeys. Although a useful tool for framing departmental or university discussions, the value of the self-authorship concept is its flexible and non-prescriptive nature. As each student will move along the path toward self-authorship at different rates, the challenge for educators is to facilitate the journey without attempting to impose particular developmental stages on to existing curriculum structures. Used effectively, the concept of self-authorship provides a useful framing concept for responding to Professor Gould's call for a more relevant and engaging curriculum for the students of today.

References

Barr, B. & Tagg, J. (1995) From teaching to learning: A new paradigm for undergraduate education, *Change*, 27(6), pp. 13–25.

Baxter Magolda, M. B. (1992) *Knowing and Reasoning in College: Gender Related Patterns in Students' Intellectual Development* (San Francisco, CA: Jossey-Bass).

Baxter Magolda, M. B. (2001) *Making their Own Way: Narratives for Transforming Higher Education to Promote Self-Development* (Sterling, VA: Stylus Publishing).

Baxter Magolda, M. B. (2004) Preface, in: M. B. Baxter Magolda & P. M. King (Eds) *Learning Partnerships: Theory and Models of Practice to Educate for Self-Authorship*, pp. xvii–xxvi (Sterling, VA: Stylus).

Baxter Magolda, M. B. (2006) Intellectual development in the college years, *Change*, 38(3), pp. 50–54.

Bekken, B. & Marie, J. (2007) Making self-authorship a goal of core curricula: The earth sustainability pilot project, in: P. S. Meszaros (Ed.) *Self-Authorship: Advancing Students' Intellectual Growth New Directions for Teaching and Learning*, No. 109, pp. 53–67, San Francisco, CA: Jossey-Bass.

Boyle, A., Maguire, S., Martin, A., Milsom, C., Nash, R., Rawlinson, S., Turner, A., Wurthmann, S. & Conchie, S. (2007) Fieldwork is good: The student perception and the affective domain, *Journal of Geography in Higher Education*, 31(2), pp. 299–317.

Centre for Active Learning (2009) *The Introduction of an Active Learning Induction* (Cheltenham: University of Gloucestershire). Available at http://insight.glos.ac.uk/tli/resources/toolkit/wal/sustainable/pages/activelearninginduction.aspx.

DeZure, D., Babb, M. & Waldman, S. (2005) Integrative learning Nationwide: Emerging themes and practices, *Peer Review: Emerging Trends and Key Debates in Undergraduate Education*, 7(4), pp. 24–28.

Dummer, T., Cook, I., Parker, S., Barrett, G. & Hull, A. (2008) Promoting and assessing 'Deep Learning' in geography fieldwork: An evaluation of reflective field diaries, *Journal of Geography in Higher Education*, 32(3), pp. 459–479.

Garrison, D. R. & Vaughan, N. D. (2008) *Blended Learning in Higher Education: Framework, Principles and Guidelines* (San Francisco, CA: Jossey Bass).

Gould, P. (1973) The open geography curriculum, in: R. J. Chorley (Ed.) *Directions in Geography*, pp. 253–284 (London: Methuen).

Healey, M. & Jenkins, A. (2009) *Developing Undergraduate Research and Inquiry* (New York: Higher Education Academy). Available at www.heacademy.ac.uk/assets/York/documents/resources/publications/Developing Undergraduate_Final (accessed June 2010).

Healey, M., Bradford, M., Roberts, C. & Knight, Y. (2010) *Bringing about Change in Teaching and Learning at Department Level* (Plymouth: Higher Eduaction Academy Subject Centre for Geography, Earth and Environmental Sciences). Available at www.gees.ac.uk/events/2009/deptchg09/deptchg09.htm and http://insight.glos.ac.uk/TLI/RESOURCES/TOOLKIT/EAL/Pages/default.aspx (accessed June 2010).

Hodge, D., Pasquesi, K. & Hirsh, M. (2007) From convocation to capstone: Developing the student as scholar. Keynote address at the Association of American Colleges and Universities Network for Academic Renewal

Conference, April 19–21, Long Beach, CA. Available at www.aacu.org/meetings/undergraduate_research/documents/Keynote.pdf (accessed June 2010).

Hodge, D., LePore, P., Pasquesi, K. & Hirsh, M. (2008a) It takes a curriculum. Preparing students for research and creative work, *Liberal Education*, 94(3), pp. 6–15.

Hodge, D., Haynes, C., LePore, P., Pasquesi, K. & Hirsh, M. (2008b) From inquiry to discovery: Developing the student as scholar in a networked world, Presented at Learning Through Enquiry Alliance Inquiry in a Networked World Conference, June 25–27, University of Sheffield. Available at www.ucm.muohio.edu/president/reports_and_speeches/pdfs/From_Inquiry_ to_Discovery.pdf (accessed June 2010).

Hodge, D., Baxter Magolda, M. B. & Haynes, C. (2009) Engaged learning: Enabling self-authorship and effective practice, *Liberal Education*, 95(4), pp. 16–23.

Jenkins, A. (1998) *Curriculum Design in Geography* (Cheltenham: Geography Discipline Network).

Justice, C., Rice, J., Warry, W., Inglis, S., Miller, S. & Sammon, S. (2007a) Inquiry in higher education: Reflections and directions on course design and teaching methods, *Innovative Higher Education*, 31(4), pp. 201–214.

Justice, C., Rice, J., Warry, W. & Laurie, I. (2007b) Taking inquiry makes a difference: A comparative analysis of student learning, *Journal of Excellence in College Teaching*, 18(1), pp. 57–77.

Kegan, R. (1994) *In Over Our Heads: The Mental Demands of Modern Life* (Cambridge: Harvard University Press).

Mezirow, J. (Ed.) (2000) *Learning as Transformation: Critical Perspectives on a Theory in Progress* (San Francisco, CA: Jossey-Bass).

Nystrand, M. & Gamoran, A. (1991) Instructional discourse, student engagement, and literature achievement, *Research in the Teaching of English*, 25(3), pp. 261–290.

Perry, W. (1970) *Forms of Intellectual and Ethical Development in the College Years: A Scheme* (Troy, MO: Holt, Rhinehart & Winston).

Ribeiro, M. A. C. & Marafon, G. J. (Eds) (2005) *Novos Caminhos Para Velhos Problemas: A Geografia no Programa de Educação Tutorial* (Rio de Janeiro: Gramma).

Southern Association of Colleges and Schools Commission on Colleges (2010) *The Principles of Accreditation: Foundations for Quality Enhancement* (Decatur, GA: Southern Association of Colleges and Schools). Available at http://sacscoc.org/pdf/2010principlesofacreditation.pdf (accessed November 2010).

Spronken-Smith, R. A. & Kingham, S. (2009) Strengthening teaching and research links: The case of a pollution exposure inquiry project, *Journal of Geography in Higher Education*, 33(2), pp. 241–253.

Spronken-Smith, R. A., Bullard, J., Ray, W., Roberts, C. & Keiffer, A. (2008) Where might sand dunes be on Mars? Engaging students through inquiry-based learning in geography, *Journal of Geography in Higher Education*, 32(1), pp. 71–86.

Taylor, K. & Haynes, C. (2008) A framework for intentionally fostering student learning, *About Campus*, 13(5), pp. 2–11.

Gaia: "thinking like a planet" as transformative learning

Martin Haigh

Department of Social Sciences, Oxford Brookes University, Oxford, UK

Transformative learning may involve gentle perspective widening or something more traumatic. This paper explores the impact of a transformative pedagogy in a course that challenges learners to "think like a planet". Among six sources of intellectual anxiety, learners worry about: why Gaia Theory is neglected by their other courses; the removal of humans from "crown of creation" status; Gaia's unfamiliar multibillion-year timescales, comprehending the world system as a whole; the array of weaker and stronger Gaia concepts; and the diversity of viewpoints among classmates. Analysis finds that most learners coped with this exercise although some achieved this by compartmentalization.

Introduction

For Jack Mezirow, transformative learning is about changing perspectives (Cranton, 2002; Taylor, 2007). Mezirow (2009) places his discussion in the context of critical reflection as a "rational process ... a metacognitive application of critical thinking that transforms an acquired frame of reference – a mind-set or worldview of orienting assumptions and expectations involving values, beliefs, and concepts – by assessing its epistemic assumptions" (Mezirow, p. 124, in Dirkx, Mezirow, & Cranton, 2006). Of course, there are many alternative conceptions of transformative learning and this variety has important implications for the development of Transformative Learning Theory (Taylor, 2008). Taylor (2008) also notes: "... much is not known about the practice of transformative learning in the classroom ... Furthermore, there is little known about the impact of fostering transformative learning on learner outcomes" (p. 13). This paper addresses these topics. It explores the consequences of posing Gaia Theory as a transformative challenge to the conventional worldview of Physical Geography.

Gaia Theory is a new "Theory of the Earth", which suggests that the global environment, including its atmosphere, temperature, water, air, soil and much of its geology, is produced/preserved and/or regulated by the combined activities of the planet's living organisms (Haigh, 2010). The theory was inspired by the remarkable stability of terrestrial conditions across long periods of time (Lovelock, 1979). Despite the chemical instability of the Earth's environment (compared to that of the neighbouring terrestrial planets), during the 4 billion years that life has been upon Earth, there has been an unusually small range of temperature variation. Despite a steady increase of solar radiation, it has never been so hot that all the ocean's boiled away, nor in the last 0.5 billion years it has been so cold that all the oceans froze. Biological processes, especially microbiological processes, tightly coupled with physical processes, seem to regulate the global environment in ways that (usually) favour their perpetuation (Lovelock, 1988a, 2005). So, Gaia Theory deploys the core notion of ecological succession at the planetary scale, which is that life creates a better environment for life itself.

In class, Gaia is presented only as a "good hypothesis", one that is worthwhile if it encourages new and interesting questions for research (Haigh, 2001). However, accepting this hypothesis would also suggests that Physical Geography should be a bioscience, and hence almost all of its textbooks are either misdirected or fundamentally wrong. Inevitably, this possibility is a major challenge to learners who, after several years of study, believe that they understand the fundamentals of Physical Geography: a body of knowledge that does not normally centre on the workings of micro-organisms.

Overview of Transformative Learning Theory

Transformative learning is about learning in adults, it is "andragogy" rather than pedagogy, "a uniquely adult form of metacognitive reasoning" (Mezirow, 2003). The spiral path of transformative learning is also "a generator of revolution for a society and revitalization for organizations [and] an instrument of education in almost every recorded culture" (McWhinney & Markos, 2003, pp. 16–17).

Transformative learning is designed for learners who have already self-created a coherent body of experience, associations, concepts, values and beliefs, from which they have constructed one or more sets of structured assumptions (or "frames of reference") that they use to interpret and evaluate experience. Transformative learning is the process by which these frames of reference are changed, ideally, towards something more inclusive and integrative (Mezirow, 1997).

In practice, this requires six conditions (Beck & Cowan, 1996). First, the learner must be open to change, not locked into a particular worldview. Second, they must be aware of dissonance, unresolved problems in their current frame. Third, they must feel free to explore the possibility of resolving these problems through the construction of a new mental framework. Fourth, they must have the opportunity space needed to explore alternatives and develop their own personal insights. Fifth, they must confront and overcome the barriers – psychological, emotional and socio-cultural – that lock them into their present mental frameworks. Finally, they must be given support during the time when they are building and consolidating any new understanding. The task of the teacher is to sow the seeds that make this perspectival shift possible and then help the seeds of change to grow. Appropriately, Beck and Cowan (1996) qualify their notes with the Biblical "Parable of the Sower".

Mezirow (1997), following Bateson's (1972) typology, explores this growth in terms of four learning styles, where only the last two are truly transformative. The first involves simply elaborating an existing perspective. The second recognizes other points of view without necessarily accepting them. The third expands the personal perspective to accept others perspectives, a process of accretional perspectival shifting. Esbjörn-Hargens and Zimmerman (2009) distinguish "perspective practices" (p. 39), which expand the capacity to address multiple perspectives simultaneously from "depth practices", which develop deeper levels of comprehension. Hence, the fourth, most deeply affective, involves building new, overarching, mental frameworks that subsume their predecessors.

Mezirow (1997) notes: "Such epochal transformations are less common and more difficult. We do not make transformative changes in the way we learn as long as what we learn fits comfortably in our existing frames of reference" (p. 7). Equally, Freire (1998,

p. 33) calls such learning, where participants engage authentically with "continuous transformation" in the "construction and reconstruction" of "the essence" of their subject of study, a *Pedagogy of Freedom.* Ideally, transformative learning is more than just a mode of education, it is a process that affects the ways learners interact with their world (Lange, 2004).

Of course, the approach is not without problems. It is constructivist, with all this implies, and hence it demonstrates "a concept of rationality that is essentially ahistorical and decontextualized" (Clark & Wilson, 1991, p. 90). After reviewing empirical studies of transformative learning, Taylor (1997) concludes that transformative learning needs exploration at greater depth, with more emphasis on the varying nature of the disorienting dilemmas that are the catalysts of learning and greater clarification what actually constitutes a perspective transformation. These are the goals of this paper. A decade later, Taylor (2007) found that research has shifted from identifying transformative experiences to fostering transformative learning and noted the increasing sophistication of, still mainly, qualitative, research designs. This paper is also about fostering transformative learning, and it also suggests an approach to quantification.

Thresholds and transformative learning

Transformative learning is an expansion of consciousness involving a critical reappraisal leading to fundamental changes in self-concept and worldview (Elias, 1997; McEwen, Strachan, & Lynch, 2011). As already indicated, there are two distinct views about how this may occur. In one case, change is a gradual process akin to that contained by Kolb's (1984) learning spiral. Here, transformational learning is a spiral where new perspectives evolve through reflection on experience, abstract conceptualization, active experimentation and the construction of a new perspective at a higher level. A spiral also features in the invitational education helix (Purkey & Novak, 1993), and Beck and Cowan's (1996) "Spiral Dynamics", which deconstructs social evolution into a series of mental frameworks that rise from an emphasis on material to an emphasis on spiritual concerns via alternations between worldviews that emphasize the individual and the collective.

However, equally embedded within Transformational Learning Theory is the concept of the transformative threshold. A threshold is a doorway and this style of thinking connects with the more abrupt mental restructurings, powered by the eureka Moment, envisaged by Bateson's (1972) fourth mode of learning. Meyer, Land, and Baillie (2010) describe the crossing of such a mental threshold in three stages – a *preliminal* stage, where there is an encounter with troublesome knowledge, in this case the Gaia Theory; a *liminal* phase, where the threshold is crossed through an ontological and epistemic shift; and a *postliminal* stage, where previous understandings are "transcended and included" within a more general system (Wilber, 1998). These stages are linked to three learning processes: critical reflection, reflective discourse and action (Merriam & Caffarella, 1999).

Kuhn (1962) is a major developmental influence on Transformative Learning Theory (Haigh, 2001; Mezirow, 2006). Kuhn's work discusses paradigm shifts in Science, pointing out that such thresholds are crossed suddenly to resolve "cognitive dissonance", an earlier formulation of the concept of "troublesome knowledge", and in response to a build-up of stress that forces the learner away from his or her own previous certainties and towards a new conceptual framework. Deep Ecological thinking contains a complementary notion. Its foundation is a wilderness epiphany, the Tvargastein experience of Naess (1987, 1989) when, alone in Nature, he gained the life-changing flash of intuition that he and Nature were an interconnected One.

Equally, Mezirow (2009) suggests that such thresholds of awareness may be crossed in one of the following two ways. The first is by way of a gentle personal transformation of consciousness. The second is through acute psychological shock, which causes the individual to re-evaluate their presumptions and the meaning of their existence. An equally troubling third option is observed by Meyer et al. (2010), who recognize that learners may become stranded in transit in a liminal state of confusion. The recognition of both "trauma" and "confusion" scenarios raise cautions about the ethics and applicability of transformative learning approaches.

In fact, the same problem is recognized by the Systems Theory that underpins Gaia (Goerner, 1994). Of course, Systems Theory enjoyed a vogue in Physical Geography, which spanned several decades from the mid-1960s. However, the Physical Geography version of Systems Theory was fundamentally different from that of the System concept's founder, Ludwig von Bertalanffy (1968). Typically, Geographical Systems thinking was mechanistic, its mental frameworks were dominated by thoughts of computer systems (Strahler, 1980; cf. Newson, 1992). By contrast, General System Theory was launched by identifying living organisms as Open Systems, its mental frameworks were organicist, dominated by the idea that the best models for all systems were living organisms – hence, ultimately, the Gaia worldview (von Bertalanffy, 1950a). Gaia is the notion that the Earth is best explored as if it were a single living system (Haigh, 2010). Also, relevant here is that General System Theory was constructed on recognition of the Isomorphic Laws of Science (von Bertalanffy, 1950b). It sought to show that different sciences had constructed different laws to explain what are essentially the same phenomena and that science could be streamlined by recognizing that these laws concerned particular classes of interaction such as growth, queues, hierarchical restructurings and what happens when complex systems reach and cross thresholds, as in Transformative Learning Theory. Homologous descriptions of transformative change are found in many different disciplines.

In modern Systems Science, dynamic systems, including the Earth, are shown to evolve by revolutionary changes, which may have external or internal origins (Lenton & Watson, 2011). A system, like Gaia, may be disrupted from the outside by something like a meteor impact or from the impact of an internal change, such as the oxygen crisis that followed the evolution of photosynthesis or the avalanche of mass extinctions that followed the expansion of human society. In either case, the system is destabilized, driven from its previous equilibrium state (system attractor) and forced to find a new one. This may be attained by collapse, system death is stability of a kind, and it may involve stabilization at a new level of complexity and organization that is either below (devolution) or above (evolution) the pre-existing state. In education, writers tend to emphasize the positive and neglect the negative alternatives.

However, in Geography, Chappell (2006) used transformative grieving theory to explore student learning journal accounts of coping with the demands of problem-based learning, while in New Age contexts, transformative learning is heavily linked to the remediation of negative perspective transformations. Typical is the work of Casto (2006), whose model is the transformation of caterpillar into butterfly. Stage 1 is *The Cocoon* where the barriers are emotional inhibitions, which have to be broken by the will to change. Stage 2, *Growing and Stretching* is where the fear of change and the unknown is confronted and overcome; this stage of questioning involves breaking the "hipnosis of social conditioning" (Casto, 2006, p. 36). Stage 3 is called *Changing Form*, which involves introspection, self-appraisal and a shift of energy guided by intuition. Stage 4 is

Flight, where the transformed takes to the air and explores new horizons "with a new movie in their projector" (Casto, 2006, p. 46), while Stage 5, *Conscious*, is a phase of self-realized peace where worries are resolved and the transformed becomes a confident co-creator of their own future. Even so, no one claims that these processes are easy or comfortable. The path is taken only by those who find their present position intolerable or unsustainable.

So, the question arises: is Mezirow's notion of a gentle evolutionary transformative shift reliably achievable or, in fact, dangerous (McEwen et al., 2011)? How effective is the Mezirovian perspective transformation compared to the more traumatic and dangerous revolutionary varieties? Does real perspective transformation involve altogether deeper soul searching (Boyd & Myers, 1988; Dirkx, 2000)? Tara Mata (1924 [1999]) talks of feeling "a kind of electrical drill boring out new thought channels" (p. 243), which may be too much to wish upon a Geography undergraduate.

Of course, Furedi (2004) leads those who object to the idea that learners are fragile and that educational processes may threaten their well-being. However, Furedi (2004) also points out: "One survey stated that 53 per cent of UK University students had 'anxiety at a pathological level' and the British Association for Counselling and Psychotherapy (BACP) claims that one in ten students seeking university counselling is already suicidal" (p. 8). Ecclestone (2007) also worries that such presuppositions of "emotional vulnerability" should not "distract professionals and students from educational experiences that encourage risk, challenge and discomfort as part of striving for autonomy" (p. 467).[1] However, no one wants to produce confused, unhappy or damaged, learners, so it is wise to consider the tipping point. Of course, at the heart of these concerns lies the ancient problem of how, given the diversity of undergraduate learners, to apply stress sufficient to challenge learners to learn but not so much as to affect negatively either their well-being or their enthusiasm for further exploration and learning (Selye, 1956).

Gaia as a learning invitation

"A learning invitation is a call to action. It is a powerful summons to reinvent oneself, to tackle seemingly impossible obstacles, to reconstruct the world, and to find challenges" (Purkey, 2011, p. 17). The possibility of Gaia challenges learners, intellectually. It is a "wicked problem" that affects previous understanding of the world, the sense of Being-in-the-World and, ultimately, the learners' value systems (Martin & Murray, 2011). It creates concerns that invite resolution and learning. More importantly, it invites responses at both scientific and personal levels (Slater, 1993). Gaia is more than just "a Big Idea" that is intriguing and invites objective investigation (Midgley, 2001). It is inherently transformative because its challenge to conventional wisdom is so large and direct that, if accepted, it affects some core understandings of one's place and purpose in the world (Lenton & Watson, 2011, p. 4).

Gaia's starting point is the notion that the Earth System is governed by biological forces rather than the mechanistic physical processes emphasized by the Earth Sciences (Lovelock, 1988a). Gaia Theory drives scientific understanding of the global environment towards a holistic, biocybernetic perspective (Midgley, 2001). It also portrays humans as a subordinate aspect of the Earth's living systems, a recently evolved nuisance more than the towering influence suggested by much Environmental Geography (Lovelock, 2005).

Gaia Theory is also something that is hard to dismiss. While the term "Gaia" is used in a variety of contexts that range from the scientific to the poetic, New Age and religious (Haigh, 2010), and while there are a spectrum of "Gaia Hypotheses" that range from "weak" to "strong", depending on how much control and autopoiesis is allowed to the Gaian system (Kirchner, 1989, 2002), Gaia is at heart a scientific theory with a robust mathematical base (Watson & Lovelock, 1983). It is a logical extension of General System Theory, which suggests that all open systems are best understood in organic terms (Haigh, 2010; Harding, 2006; von Bertalanffy, 1968). Its arguments are couched in the complexities of biogeochemical recycling, planetary science and microbiology. However, its core assertion is that the long-term stability of the Earth's environmental system has a biological control system (Lovelock, 1988a). Evidence includes the singularity and constancy of the Earth's atmosphere, which is kept very far from chemical equilibrium by the activities of its biota, especially the bacteria, which input gases such as oxygen and methane, and remove gases such as carbon dioxide to storage as coal, oil, limestone and all living tissues (Lovelock, 1979). Viewed from the Gaia perspective, almost every key process operating in the realm of Physical Geography from weathering through biogeochemical cycles, cloudiness and climate change to volcanicity and plate tectonics is biologically moderated (Kump, Kasting, & Crane, 2010; Lenton & Watson, 2011). Hence, Gaia seems an appropriate theoretical base for a transformed and modernized Physical Geography.

Hearing this can be quite exhilarating for learners but problems can begin when learners start asking the fundamental questions, "what does this mean for us humans – why are we here?" The "scientific" answer, that humans are recently evolved bacterial super-colonies of many billion quasi-independent cells, is not felt to be satisfactory. A more personal, emotional response tends to be elicited, which often searches for some positive purpose in human existence (Slater, 1993). For many learners, the possibility of there being a larger, more important and overarchingly more powerful living thing on this planet than humans themselves is disturbing and causes them to question their place in the world. The apocalyptic writings of Lovelock's later years, which suggest that the main effect of human activity shall be to drive the species to extinction and, in the process, push the Gaia System to a new, hot, desert-like state, are not entirely welcome. No more so are Lovelock's earlier portrayals of humans as some kind of planetary disease (Lovelock, 2005). Science has little truck with teleology, at least at scales beyond the gene. However, people are more comfortable when they can understand why things are the way they are and see some purpose. Equally, this may matter greatly to people who are actively questioning the *status quo* and defining purpose in their lives.

Classroom context

This study concerns a course that centres Gaia Theory in a re-envisioning of a traditional systems approach to Physical Geography and invites learners to partake in this process. The course has evolved from a conventional systems approach to Physical Geography, which featured the Gaia Hypothesis, to a course where Gaia Theory takes centre stage (Haigh, 1986, 2001). The current course is an elective pitched to a second-year undergraduate audience in the UK system, which equates to a 300–400 level course in the US system. This Oxford Brookes University course enrols around 45 learners biennially, mainly from Geography and the Environmental Sciences. It is taught across 11 weeks in a single semester through 9 sessions of up to 3 hours. Sessions mix together short lectures, films, discussion sessions, student presentations and, commonly, a short class quiz (Haigh, 2007).

The course curriculum invites learners to construct a new perspective on the Earth, one that is based neither in the piecemeal visions of the separate Earth Sciences, which grant priority to physical, geological and chemical processes, nor in the anthropocentric viewpoint of Environmental Geography, but one based on biocybernetics and systems thinking (Margulis, 1998). The curriculum is not prescriptive. Learners are made aware from the start that there is no single "right answer" and that whether they accept or reject Gaia Theory or adopt some particular point between on the strong–weak Gaia spectrum is not important in assessment terms (Haigh, 2010). Assessment is based on how much they engage with the subject, how much reading and research they undertake and how systematically they construct their own understanding.

The engine for this construction is a reflective learning journal exercise: "Awakenings", which has been described in detail by Haigh (2001). In class, this is supplemented by a system of quizzes that are designed to ensure that learners undertake revision, prepare for each class and remain engaged with the developing argument of the course. The merits and problems of the quiz system have been formally evaluated (Haigh, 2007). The Gaia course also offers a wide range of alternative coursework options, which are designed to allow learners with different aptitudes to engage effectively with the curriculum. It is accepted that not everyone is comfortable writing a learning journal or has the kind of mind that embraces quizzes, although every student is expected to attempt these exercises. So, a series of alternative options is offered, which include computer modelling, the critical deconstruction of a commercial Gaia computer simulation and spoken presentations based on readings of the Gaia critics or designed to help fellow learners with difficult systems concepts, typically: holons, hierarchical jumps and the role of (mathematical) chaos. The final assessment is based on "the best 100%" of work submitted.

However, the pedagogic instrument that provides the best guide to learning is the learning journal. This provides an effective way of monitoring the development of a learner's ideas (Haigh, 2001) as well as being a good way of involving learners in their own learning (Park, 2003; Wagner, 1999). Here, the "Awakenings" exercise requires learners to keep, week-by-week, in a hardbound notebook, a record of their thoughts, insights and discoveries from readings. The journal comes to class each week with the learner and, from time to time, they are invited to read from their latest entry, either to a classmate or to the entire class. Small penalties are levied against those unable to respond when so requested.

"Awakenings" begins with the prompt: *As Gaia sees the World.* It invites learners to use the substance of the course to explore two perspectives on the global environment: a human focussed, "anthropocentric", perspective and a Gaia focussed, "ecocentric", perspective. Instead of just thinking like a human, they are invited also to try and "think like a planet" and evaluate environmental change from this perspective and its very different (time)scales, after the fashion of Seed's "Council of all Beings" (Seed, Macy, Flemming, & Naess, 1988) and using the example of Schumm and Lichty (1965). Seed's "Think like a Mountain" risks anthropomorphism but it encourages learners to step outside their human framework and construct a different viewpoint. Schumm and Lichty (1965) demonstrate how the different timescales used by different schools of Geomorphology affects the significance of different variables considered by landform systems research. In each of their Cyclic, Graded and Steady State time spans of geomorphological research, different sets of variables are counted as Independent, Dependent or Not Relevant. Similarly, the "Awakenings" exercise asks learners to differentiate which environmental variables are relevant, independent or dependent at a human timescale measured in decades or centuries and a Gaian timescale measured in 100 millions or billions of years.

The exercise also asks learners to reflect upon their own learning processes and map the ways they self-create conclusions. The "Awakenings" journal is handwritten in a fixed-leaf volume precisely so that the learner can see how their thinking evolves through the course. The effectiveness of this use of learning journals was explored previously using data from the 1990s (Haigh, 2001). This paper uses post-2002 data that explore the goal of transformative learning. In this course, the designer's model was, appropriately enough, the three-step transformative pedagogy of Deep Ecology. This envisions the maturation of a learner in two transformative steps. The learner begins with prioritization of the personal self. At some point, one usually reached by college-age, learners subsume this individual self within the priorities of a larger social Self, commonly a family or peer grouping, sometimes a larger unit such as a Nation or, idealistically, the Family of all Humanity. The further step is taken when the learner realizes that the Human species means nothing on its own and cannot exist outside the matrix of Nature, and so recognize the priority of the still larger ecological Self (Naess, 1987).

Results

Gaia: six challenges to learners

The first step in transformative learning is the creation of a "troublesome" awareness in the learner of a possible need for change (Meyer et al., 2010). In Kuhnian terms, this process involves the accumulation of cognitive dissonance, which translates as that nagging feeling that something is not quite right despite all the arguments that can be mustered to the contrary (Kuhn, 1962).

Since 2000, the author has reviewed around 200 (additional) "Awakenings" learning journals (cf. Haigh, 2001). These suggest that Gaia Theory creates cognitive dissonance in undergraduate learners in six major ways. Table 1 presents a selection of quotations from the "Awakenings" journals of the year's 2003 to 2011, which illustrate these six causes.

The concern that emerges first, as a rule, is why a notion that seems so fundamental should have been ignored by their previous classes in Geography and the Environmental Sciences and is sometimes derided by fellow learners. Table 1, Row 1, provides some typical learner reactions. This concern is useful because it means that many learners enter their work with a healthy scepticism and apply unusual critical attention to the information they receive. Of course, since the evidence for Gaia is fairly persuasive, one knock-on effect is that many continue to question knowledge that they have obtained from other sources. It is an ambition of British Geography to consider scientific knowledge not as an absolute, but something socially constructed and value laden (QAA, 2007, p. 10). Interrogating Gaia Theory produces such questioning as a near inevitable side effect.

The second concern, which also appears early in learner journals, arises from Gaia Theory's challenge to anthropocentric thinking. Gaia Theory portrays humans as a recent and possibly superfluous development. If, hitherto, the learner assumed that humanity was the crown and keeper of all creation then this possibility can be deeply disturbing. Many find it hard to remove the human ego from centre stage (Table 1, Row 2). Few like the ideas that humans may be a disease-like disturbance of Gaia or a short-lived irrelevance and many, like Lovelock himself, prefer to portray humans in apocalyptic terms (Allaby, 1989; Lovelock, 2006, 2009; Sagan, 1990). Consequently, some "Awakenings" journals revert to an extended search for some special role for the human species. One outcome of the course can be a learner that thinks about humans in planetary terms rather than one conceives the world from the Gaia perspective (Haigh, 2001). Of course,

Table 1. 'Six Challenges' Quotations from student journals 2003–2011 (cf. Haigh, 2001).

Challenge	"Awakenings" quotations. (initials and page numbers key the quotation to individual student journals)
1. Ignored by other classes.	"I did enter the module quite suspiciously". (HC2009, p. 10.1) "... it is nice to think that some people, (scientists included) are grasping the idea that there is a higher power, an organism or being that is controlling human life and the world around us. I have taken many science classes both in the US and in Britain and have always struggled with the concept". (KA2003, p. 1) "Thinking in a Gaian perspective is proof of intellectual courage (against science's conformity). It should help us to evolve philosophically". (AB2003, p. 75) "'hippy' ... style ... encourages criticism". (EW2003, p. 32)
2. Decentres humans	"Before, I would have just seen the human view ... now I'm starting to think on a different level". (DT2011, p. 100) "Because I have been brought up in a world 'dominated' by humans ... my experience of life so far has been dominated by human concerns. My idea of history struggles to conceive of a world without humans in it". (DS2003, p. 2.8) "... It suggests that we have the key to ruin this planet and on the other hand have the solution to fix it when necessary. It puts us on a pedestal looking down on and being in charge of everything else. This is, I suppose, just a reflection of humankind's big-headed selfish nature...". (SK2003, pp. 1–2) "Egos are diminished here! Sorry, you'll have to leave, your ego is taking up too much space!". (JR2003, p. 16) "I now see myself as less significant within the World ...". (KE2011, p. 72)
3. Inhuman scales	"The idea that life began more than 3.5 billion years ago is hard to grasp". (DS2003, p. 2.8) "... we need to concentrate on the planet's timescales rather than that of human life". (SH2003, p. 25) "Is Gaia aware that humans exist? Probably not – we are just another life-form. ... except we haven't been around as long as others". (JR2003, p. 1) "The main thing that I find fascinating is the huge role that microbes ... have for this theory and that it is likely that they play a bigger role than humans" ... "We as individuals are miniscule compared to the enormity of Gaia". (KM2003, p. 8–9) "I have read an article by Lyn Margulis ... we are all walking communities, 10% or more of our body weight is bacteria". (SP2003, p. 2.7)
4. Holism	"... for the first time, I am asked not to learn, digest and re-use knowledge, but to enquire into what I think about the whole thing". (AB2003, p. 5) "Gaia exploits the terms of inter-relatedness. Although we are good at thinking about this socially and philosophically, we are not very good at looking at this at an environmental level" (KA2011, p. 75). "molecules do not think, they react to conditions ... organelles in cells too do not think, they act on the conditions they are subjected to. Cells, as combinations of organelles do not think in the same way that humans do". (NT2003, p. 33)

Table 1 – *continued*

Challenge	"Awakenings" quotations. (initials and page numbers key the quotation to individual student journals)
	"... you can tell the planet is alive because it dynamically processes energy, preserves itself in a changing environment, behaves differently to its environment ... has its own agenda". (SP2003, p. 5) "It struck me that Gaia does not have a conscious mind. Our bodies are in some ways the same. If we get warm, we do not choose that our blood vessels come to the surface to cool us down, it is a regulatory system in us". (LB2003, p. 6)
5. Too many Gaias	" Gaia can't be proved or disproved; it is in effect a circular argument with conclusions about its own assumptions". (AG2003, p. 78) "What can be used as evidence against Gaia is there anything that isn't Gaia ... the hypothesis is untestable!" ... is Gaia 'Lovelock's pension plan'? (JC2009, p. K8) "Gaia theory is very open to interpretation and that's is how I have interpreted it". (EP2011, p. 79) "Constant [analogical reasoning] makes me realise that I don't have to believe everything I read, I'm not being ignorant of Gaia but I would like an explanation with some scientific evidence and data". (EW2003, p. 41) "the more we get into scientific explanations of Gaia Theory, the odder and more metaphysical it appears to be ...". (AB2003, p. 21) "The name Gaia is misleading ... gives the impression of a personality, which must imply control over actions or thought". (NT2003, p. 27)
6. New "turquoise" worldview	"It has opened my eyes to seeing something completely outside of the box ... a new way of thinking ...". (JH2011, p. 45) "Gaia is not a theory but a development of the human mind". (JC2011, p. 59) "I have learnt to think in a Gaian way where everything is interconnected and related and to do this critically". (MJ2011, p. 59) "This has been a huge learning process for me, not just academically but also for self-aware thinking ...". (MJ2011, p. 47) "Gaia Theory represents ... a leap of faith to a new paradigm ... what was difficult was to engage with the science on a level where the rational sophistication matched the intuitive feeling about the interconnectedness of the systems. I found it a challenge not to consistently slip into a 'belief is enough' level ...". (AC2011, p. 26) "I do think that these problems arise as a result of the mythic themes that surround Gaia. Its motherly, alive concepts are difficult to understand or take seriously". (CH2003, p. 5)

such considerations, especially the last that connects with contemporary environmentalism, provide a safety net for those that find the core theory or its implications too difficult.

The third problem is scale. Humans have been a planet-influencing species for $2 \times 10^{2-3}$ years, which is dwarfed by the epic scales of Gaia beginning with its nearly 4×10^9 years of notional existence as a planet regulator (Lovelock, 1988b). In physical terms, the scale of humanity is dwarfed by at size of the planet (10^{+6} m) and the vastness and immortality of the microcosmos beneath (typical plant cell: 10^{-6} m) (Margulis & Sagan, 1986). Gaia also acts at two unfamiliar timescales – that of stellar evolution and that of bacterial reproduction. Human space and time rests disturbingly between these two extremes. Human *scale-ism*, the prejudice that "real" or "living" things have to be of a human perceptible scale, is introduced in class as a "political incorrectness". However, the problem of scale remains an assault on anthropocentrism (Table 1, Row 3). It makes the whole Gaia conception so overwhelmingly alien that many students feel obliged to reconcile it with their personal belief systems, a task that causes more problems for those from Abrahamic than Dharmic or New Age traditions. It is unusual, in these secular times, for British students to volunteer discussion of their belief systems in a classroom context. Religion has long held near taboo status in British Geography courses. However, between 2003 and 2011, more than 20% explored religion in their learning journal.

Concern four arises from the need for holism. Gaia thinking involves the Geographical skill of seeing, as an interconnected functioning whole, the processes that other subjects study separately (Conway-Gómez et al., 2011; QAA, 2007, p. 7.1). Table 1, Row 4, displays some of the learner reactions to the novelty of trying to see things together as dynamic systems. Sadly, at the outset, few of the learners know much about systems thinking or the system sciences (Goerner, 1994; Skyttner, 2001). Few find it easy to absorb this new approach to the world, with its associated metalanguage, within the few weeks of the course (Wolf, Stanton, & Gellott, 2010). Equally, at some point, almost every learner finds it necessary to revisit subject areas that he or she had previously abandoned, such as mathematics, chemistry and philosophy (cf. Wilson, 1998).

The fifth worry emerges from Gaia Theory's weaker points. Scientific Gaia is criticized for containing circular argument and wishful thinking and, especially, for being slippery to the point of 'untestability' (Kirchner, 2002; cf. Haigh, 1997). Kirchner (1989) recognizes a spectrum of scientific Gaia Hypotheses that run from a "Weak Gaia" where life influences conditions on the Earth to "Strong Gaias" that control the global environment for their own benefit. Gaia thinking ranges from the most mechanical Physical Science through to spiritual depths that lie beyond the ambitions of Deep Ecology (Haigh, 2010). This causes frustration in some learners (Table 1, Row 5). Many come to the view that, because they are taking the course, they should 'play along' and 'believe' in Gaia. However, this is confounded by the absence of any single "correct view". Their concept of Gaia has to be personally constructed, if only because there are just so many 'Gaias' to choose between. Some find this situation very challenging.

The sixth challenge arises from the fundamental differences of worldview contained within each student group. Integral Spiral Dynamics has not yet much impinged into Geographical thinking; it remains largely in the domain of the Business Schools (Beck & Cowan, 1996; Wilber, 2006). However, it provides a viable, if imperfect, map of human consciousness and its evolution. This is described in terms of an evolutionary spiral of eight worldviews. These are ordered in two tiers, a lower sequence dominated by survival-oriented perspectives and a higher tier oriented to "being"-oriented perspectives. The problem, which the theorists recognize, is that those holding particular worldviews regard those holding others as either primitive, if looking down, or ludicrously idealistic and incomprehensible, if looking up (Beck & Cowan, 1996). A mapping of the "Awakenings" journals (cf. Haigh, 2011) finds most learners entering with lower tier survivalist worldviews (mainly the so-called 'Orange' and 'Green' ᵛMEMES) while Gaia is definitively a higher tier ('Turquoise' ᵛMEME) concept (Graves, 1974).

Comparing the anthropocentric and Gaia viewpoints

The final part of the "Awakenings" reflective learning diary includes an exercise designed to determine how far learners can differentiate between the Gaia and human viewpoints. In this context especially, the key to transformative learning is an ability to transcend and include an earlier viewpoint (Wilber, 1998). Most learners begin with a firmly anthropocentric view of the world, whilst those who have absorbed the core concepts of the class are also capable of seeing the world in ecocentric, Gaia-centric, terms. In this exercise, learners are asked to compare in two columns the differences between the Gaia and human viewpoints on one from a range of environmental issues. One of the most popular selections is "global warming".

Table 2 is based on the reports produced by these learners. The author has abstracted the number of anthropocentric and Gaian concerns identified by each learner and the number of these concerns assigned correctly or incorrectly, according to Gaia Theory, to each column (Lovelock, 2005). Results show that the average student identified 3.18 (SD: 1.21) anthropocentric issues correctly and only 0.34 (SD: 0.64) incorrectly, as well as 2.93 (SD: 1.04) Gaia issues correctly and, rather more, 0.96 (SD: 1.81) incorrectly. Fisher's exact test suggests that the probability that these two response patterns come from the same population is less than $p = 0.0001$, which simply confirms that the learners find using the anthropocentric viewpoint much easier than the acquired Gaia viewpoint.

Further statistical analysis finds a significant Spearman's ρ correlations linking the number of Gaia issues identified correctly and the final mark for the course, which is a marker of overall achievement (ρ: 0.371; $p = 0.026$) (Haigh, 2001). Human issues connected with the global warming debate could have been obtained in other parts of the Geography programme, so it is useful to note that the number of these issues identified correctly does not correlate with the overall mark for the course. However, Human issues do correlate significantly with the numbers of Gaia issues identified correctly (ρ: 0.291; $p = 0.015$). There is also a weak correlation between the number of human issues correctly assigned and the number of Gaia issues identified as human issues (ρ: 0.225; $p = 0.048$), which is caused by some learners listing unselectively.

Table 3 displays the individual issues most commonly identified correctly in reports from 58 students who attempted this task between 2003 and 2011. It also displays the strength and significance of the correlation between recognition of each individual concern and its association with the correct identification column.

This exercise demonstrates that while the learners retain a stronger grasp on the more familiar ideas of human impacts on the environment, many of them have grasped the character of the Gaia perspective, which transcends and includes it after the fashion of Schumm and Lichty (1965) and Wilber (1998). Relatively few learners displayed signs of confusion associated with the misassignment of issues to the wrong column, so the study shows that learners had achieved the intellectual capacity to distinguish between processes that are significant to Gaia and those that matter to humans. The study also reassures that the grading for the course as a whole, only 10% of which is associated with the exercise described, is rewarding those with the better grasp of Gaia.

Table 2. Spearman's correlations between the overall grade for the course and numbers of issues identified in comparing the human and Gaia perspectives assignment.

Spearman's ρ	Overall grade	Human issues correctly assigned	Gaia issues correctly assigned	Gaia issues in human issues column
Overall grade			0.371 ($p = 0$*.026)	
Human issues correctly assigned			0.291 ($p = 0.015$)	0.225 ($p = 0.048$)
Gaia issues correctly assigned	0.371 ($p = 0$*.026)	0.291 ($p = 0.015$)		
Gaia issues in human issues column		0.225 ($p = 0.048$)		
Human issues in Gaia issues column				
Issues correctly assigned – Total	0.364 ($p = 0.029$)			

Table 3. Comparing the human and Gaia perspectives: topics correlated with correct perspective assignment (Spearman's rank correlation coefficients and their significance).

Anthropocentric perspective issue	Gaian perspective issue
• Human carbon emissions: 0.50 ($p < 0.0005$); • Ozone layer depletion: 0.44 ($p = 0.002$); • Pollution: 0.41 ($p = 0.001$); • Greenhouse effect: 0.39 ($p = 0.003$); • Technological solutions: 0.354 ($p = 0.004$); • Impact of sea-level rise: 0.31 ($p = 0.011$); • Climate change – desertification: 0.28 ($p = 0.019$)	• Microcosmos' capacity to survive environmental change: 0.42 ($p = 0.001$) • Capacity to adapt through speciation: 0.41 ($p = 0.001$) • Geological regulation of the "greenhouse" effect: 0.36 ($p = 0.005$) • "Fevered" condition described by Lovelock (2005); 0.34 ($p = 0.005$) • Capacity for long-term self-regulation: 0.32 ($p = 0.008$).

Discussion

Learning journals provide an unequalled window on the thinking and learning processes of learners, which those learners tend to describe with surprising candour. The technique was the focus of an earlier paper, but Table 4 provides further illustration of how learning journals foster the metacognitive self-awareness of learners (Haigh, 2001). The key point here is that exploring learning journals is a good way of trawling the learners psyche for signs of stress, confusion and resistance to learning.

Recently, Wolf et al. (2010) have commented on some avoidance strategies that learners use to avoid critical thinking; the same were in evidence among less committed learners here. In this Gaia case, the dominant form of rejection or avoidance took the form of an exaggerated anthropocentrism. This manifested as a focus on "planetary engineering", which is a subtext of some Gaia thought. As such, this provided a creative

Table 4. Learners on writing learning journals (initials and page numbers key the quotation to individual student journals (cf. Haigh, 2001).

"I have learnt to think in a different way – a more reflective way". (BP2011, p. 66)
"I wish that all my thoughts and deliberations with friends could appear in this journal .... My journal jumps from place to place with many of the questions not being answered. This is not because of me being lazy but usually the result of bigger questions appearing as the result of my initial questions ...". (AB2003, p. 30–31)
"Today, we read over our past 2 journal entries and I was surprised to see what I had previously written". (KA2003, p. 31)
"I am hoping to obtain through this course, the science behind my childhood dreams and beliefs". (AC2003, p. 2–3)
"I found it hard not to look at topics in a scientific way but to try and look at them as part of Gaia". (MH2011, p. 51)
"Thinking of Gaia makes me think again of things 'big' and 'small' ... how unpacking/rediscovering their processes and energies can help us to gain a more sensitive understanding of how they relate to the world and how the world relates to them (and me!)". (DS2003, p. 2)
"For me, Gaia is just a big photo album that is recording the history of time ... The meaning of life is the last photo in the album. I bet it has fallen out when I get there". (ER2003, p. 32–33)

and educationally valuable escape route for learners who chose not to face the challenge of "thinking like a planet" for whatever reasons. True, they may not have grasped the main point of the course but neither did they come away empty handed.

So, despite its novelty, in subject matter, learning journal mode of assessment and transformative goals, the Gaia course does not seem to have stressed its learners unduly. Student comments on their "Awakenings" learning journal tasks, as Table 3 demonstrates, show that the learners are aware of their developing understanding (metacognitive development). They also show that those who skipped classes and the reading sometimes end up "confused", but this may be inevitable for learners whose preoccupations are other than learning. However, there is little sign of the struggling with cognitive dissonance that is implied by the Kuhnian model of transformative learning and, if there are changes of perspective as anticipated in Mezirovian theory, these tend to be indicated more by assertion than actual evidence (Table 2, Row 6). The main evidence of intellectual tension is the aforementioned recourse to religious discourse. For the most part, the emotional range of the journals varies narrowly between the seriously academic and the light and "chirpy". There is no hint of any kind of intellectual crisis. This contrasts with the outcome of another experiment, which sought to introduce an "internationalized" curriculum containing non-Western ideas, where segments of the class reacted against or, more usually, cold-shouldered the "other-cultural" elements (Haigh, 2009).

In this course, however, another avoidance process seems to be at work, one that may be indicative of a kind of "modular" mentality. Learners have a well-known, if unfortunate, tendency to compartmentalize their learning into its course-sized packages. In this case, they seemed happy to adopt the lore of the Gaia course within its confines and for its duration. In the early years of this course, it was common to receive complaints from colleagues about learners introducing Gaia-based ideas into their own courses. However, whether due to learner compartmentalization or familiarity or changed perceptions on the part of colleagues, no such comment has been received in this century. The evidence from the learning journals is that there is relatively limited cross-fertilization of ideas from other courses studied.

A central ambition of the module is to help students "think like a planet" as well as a human and to overcome an unconscious anthropocentrism (an inability to conceive the world in other than human terms), which is very widespread and widely recognized as a major cause of our current environmental crisis (Berry, 1999). The course aims to replace such thinking but, sometimes, succeeds only in expanding the framework of anthropocentric thinking towards a planetary scale. Whether such a development is sufficient to be considered transformative, even in the Mezirovian sense, is doubtful.

Certainly, the existence and survival of this module is evidence of an educational transformation, which Goerner (1994) calls, the "ecological transformation of society ... away from [its worldview of] a Newtonian clockwork machine universe, towards a vision of a living evolving ecological universe" (p. ix); namely the Turquoise ^{V}MEME (Haigh, 2011; Wilber, 2006). However, for many, faith in the technological genius of humanity, with its capacity to resolve all of the problems of the world, remains intact (cf. Berry, 1988, pp. 30–31). These same learners, often self-styled "scientists", are often those most irritated with the romantic personification of Gaia that has popularized the notion (Joseph, 1990; Turney, 2003). Many of these same learners never fully come to grips with the significance of the differences in scale between a planet-sized superorganism composed of billions of interacting species and a lifespan measured in billions of years and a human, composed of billions of interacting cells and with a lifespan measured in a few decades. A few even complained that it was unreasonable to ask them to try to "think like a planet".

Possibly, this is correct. Ecological self-realization is presented by Naess (1987) as a process of personal maturation, not of learning. The human condition is such that childhood and adolescence are times when the person discovers who they are and where they belong in society. It may be asking the average 20-year-old a lot to move so quickly to the next step, which is to understand where they belong in the community of all life, especially where this is not the received wisdom of their social group.

In summary, the most this course achieves is a widening of the learner's perspective to include the Gaia point of view, possibly temporarily, but sufficiently for them to transcend and include their earlier understandings of Physical Geography and Environmental Science. However, there is another, widely echoed, positive outcome from this Gaia course. While, in many cases, it may not change the participants and, in some cases, it may do little to change the way they think, in most cases, as the journals witness, it certainly does make them think for themselves. The module may not have overthrown anthropocentrism as a dominant mindset, but perhaps, at least, it planted the seeds for a challenge.

So, if transformative learning is difficult to achieve in the context of a single course, how might that transformation be achieved? Recently, the author has worked on the biographies of two spiritual leaders who underwent major transformative change while in Higher Education. In the first case, Mahatma Gandhi who studied Law in London, 1888–1891, isolation in an alien culture began a desperate search for a socially acceptable mask, which expanded into a re-evaluation of his whole persona and, forged by experience, a new way of being. It also awoke him to the spiritual strengths of his own culture and introduced the arts of political campaigning (Gandhi, 1927–1929; Haigh, 2013; Hunt, 1993). In the second case, the American Guru Adi Da found his experience of study at Columbia University in the 1960s "completely devastating"; its approach was a "market place of doubt" that disturbed his search for Reality (Da, 1995, pp. 66–67). While both dismiss their classroom studies as totally worthless, there is no doubt that their contemporaneous non-formal, above all, experiential learning sparked their emergence as the change agents they ultimately became. If there is a message here, it may be at one with that of Gaia Theory, namely that transformative learning must be conducted at the level of the whole person, perhaps by immersion (Jorge, 2011). It is something formed by experiences that go beyond the formal classroom curriculum and something, more than all other forms of deep learning, that has to be brought out by the learner from within the learner's self.

Conclusion

The essence of transformative learning is that it challenges learners to cross a threshold of perception, usually by means of confronting them with troublesome knowledge, information that does not fit within their current worldview. This paper concerns such a challenge, which invites learners to revision Physical Geography from the perspective of Gaia Theory and to learn how to "think like a planet". The aim is to overcome anthropocentrism, which is an inability to conceive the world in other than human terms. The problem the course faces is one that is inherent in every transformative educational agenda, which is that the outcomes of transformative learning can be negative as well as positive and result in confusion and distress among learners. Transformative learning works through the deliberate application of intellectual stress, which provides a stimulus for the crossing of the transformative learning threshold to a new perspective. The pedagogic challenge is to make sure that those who accept the transformative challenge can make the adjustment smoothly while the learning of those who do not is not disadvantaged. The ethical problems include whether it is good or wise to try to break through the barriers that divide the learner from their learnt material and whether setting out to transform a learner's psyche, with all of its unpredictable and potential negative consequences, should be attempted at all?

Nevertheless, analysis of the "Awakenings" learning journals (Haigh, 2001), reassures, first, that few learners experienced any undue stress as a result of this course and second, that some managed to cross the learning threshold and gain the germ of an ability to "think like a planet" or at least grasp the "Gaia perspective". In some cases, it was clear that this perspectival shift was adopted purely for the duration of this course although a significant few vowed to take their new understanding forward. For others, there was no transformation, only some questioning of a deeply rooted anthropocentric worldview and, in some cases, simple incomprehension or disbelief. In several cases, the learning sparked more extravagant anthropocentric thinking, inspiring dreams of human intervention at the planetary scale. Some found reinforcement for beliefs already deeply held while, in others, the study inspired an energetic reassessment of their personal worldviews, present knowledge and belief systems. Of course, such questioning is also a symptom of intellectual stress, although, within the context of the "Awakenings" learning journal dialogues, there were no signs of that such stress reached levels that were harmful. In fact, almost all participants report that they enjoyed the course.

However, it also seemed that many learners had developed ways to protect themselves from deep engagement with "troublesome knowledge". This was achieved by means of a process of "modular thinking"; they accept the lore of a classroom course for the duration of that course and do not worry overmuch if this contradicts the rules of a different course or their wider beliefs. This emphasizes that real transformative learning must be a whole person, whole life experience and not something limited to a single course or timeframe. Meanwhile, while it may be that for most students, encountering Gaia did no more than initiate new questioning of these learners' previous worldviews. Possibly, it is the stimulation of this process of personal questioning and re-contextualization that remains the greatest benefit of studying Gaia.

Note

1. Interestingly, one anonymous peer reviewer of this paper also scoffs at this author's molly-coddling approach: "boring out new thought channels is exactly what we should be wishing upon geography undergraduates. … if we're not trying to get students over multiple tipping points and transforming how they think and view the world… what we are here for? … I find all this talk of 'well being' to be rather overstating the case. Can the author provide an example of where an undergraduate student's well being has been adversely affected because of the stress of being exposed to new intellectual material and perspectives?" Obviously, the answer is sadly yes, several examples, but then undergraduates run into trouble by many means and for many reasons. Of course, a major part of the discourse about the well-being of international students concerns exactly this problem. Haigh (2013) reviews this theory ahead of using the journal of the Mahatma Gandhi as a case study. Haigh (2009) also describes a case where an experimental attempt to internationalize a Geography curriculum by building it from non-Western foundations overstepped its mark with negative consequences.

References

Allaby, M. (1989). *Guide to Gaia*. London: MacDonald Optima.
Bateson, G. (1972). *Steps to an ecology of the mind.* New York: Ballantine.

Beck, D. E., & Cowan, C. C. (1996). *Spiral dynamics: Mastering values, leadership and change.* Oxford: Blackwell.

Berry, T. M. (1988). *The dream of the earth.* San Francisco, CA: Sierra Club.

Berry, T. M. (1999). *The great work: Our way into the future.* New York: Belltower.

Boyd, R. D., & Myers, J. G. (1988). Transformative education. *International Journal of Lifelong Education, 7*(4), 261–284.

Casto, M. L. (2006). *Life transformation: How to make smart life changes and let your soul be your guide.* Corpus Christi, TX: Get Smart!.

Chappell, A. (2006). Using the 'grieving' process and learning journals to evaluate students' responses to problem-based learning in an undergraduate Geography curriculum. *Journal of Geography in Higher Education, 30*(1), 15–31.

Clark, M. C., & Wilson, A. (1991). Context and rationality in Mezirow's theory of transformational learning. *Adult Education Quarterly, 41*(2), 75–91.

Conway-Gómez, K., Williams, N., Atkinson-Palombo, C., Ahlqvist, O., Kim, E., & Morgan, M. (2011). Tapping Geography's potential for synergy with creative instructional approaches. *Journal of Geography in Higher Education, 35*(3), 409–423.

Cranton, P. (2002). Teaching for transformation. *New Directions for Adult & Continuing Education, 93*, 63–72.

Da, Adi (1995). *The knee of listening (Unabridged, New Standard Edition).* Middletown, CA: Dawn Horse Press.

Dirkx, J. M. (2000). Transformative learning and the journey of individuation. *ERIC Digest, 223*, 1–2. Retrieved from http://www.calpro-online.org/eric/docs/dig223.pdf (accessed July 2011).

Dirkx, J. M., Mezirow, J., & Cranton, P. (2006). Musings and reflections on the meaning, context, and process of transformative learning: A dialogue between John M. Dirkx and Jack Mezirow. *Journal of Transformative Education, 4*(2), 123–139.

Ecclestone, K. (2007). Resisting images of the 'diminished self': The implications of emotional wellbeing and emotional engagement in education policy. *Journal of Education Policy, 22*(4), 455–470.

Elias, D. G. (1997). It's time to change our minds. *ReVision, A Journal of Consciousness & Transformation, 20*, 2–6.

Esbjorn-Hargens, S., & Zimmerman, M. (2009). *Integral Ecology: Uniting multiple perspectives on the natural World.* Boston: Integral Books.

Freire, P. (1998). *Pedagogy of freedom: Ethics, democracy and civic courage.* Lanham, MD: Rowan and Littlefield.

Furedi, F. (2004). *Therapy culture; cultivating vulnerability in an uncertain age.* London: Routledge.

Gandhi, M. K. (1927–1929). *An autobiography: The story of my experiments with truth.* Harmondsworth, UK: Penguin (1982). Retrieved from http://wikilivres.info/wiki/The_Story_of_My_Experiments_with_Truth/Part_I/In_London_at_Last (accessed March 2011).

Goerner, S. J. (1994). *Chaos and the evolving ecological universe.* Langhorne, PA: Gordon and Breach.

Graves, C. W. (1974, April). Human nature prepares for a momentous leap. *The Futurist, 8*, 72–87.

Haigh, M. (1986). The evaluation of an experiment in Physical Geography teaching. *Journal of Geography in Higher Education, 10*(1), 133–147.

Haigh, M. (1997). Testing hypotheses of biocybernetic regulation. *Biometeorology, 14*(2), 53–62.

Haigh, M. (2001). Constructing Gaia: Using journals to foster reflective learning. *Journal of Geography in Higher Education, 25*(2), 167–198.

Haigh, M. (2007). Sustaining learning through assessment: an evaluation of the value of a weekly class quiz. *Assessment & Evaluation in Higher Education, 32*(4), 457–474.

Haigh, M. (2009). Fostering cross cultural empathy with non-Western curricular structures. *Journal of Studies in International Education, 13*(2), 271–284.

Haigh, M. (2010). Gaia. In B. Warf, P. Jankowski, B. D. Solomon & M. Welford (Eds.), *Encyclopedia of Geography* (pp. 1175–1178). Thousand Oaks, CA: Sage. Retrieved from http://www.sage-ereference.com/geography/Article_n458.html (accessed January 2011).

Haigh, M. (2011). Transformative learning for global citizenship: Turning to turquoise. *Learning & Teaching in Higher Education (LATHE), 5*, 7–16, Retrieved from http://www2.glos.ac.uk/offload/tli/lets/lathe/issue5/Lathe_5_M%20Haigh.pdf (accessed August 2011).

Haigh, M. (2013). Toward the intercultural self: Exploring the effects of Mahatma Gandhi's International Education. In J. Ryan (Ed.), *Cross cultural teaching and learning for home and international students: Internationalization of pedagogy and curriculum in higher education* (pp. 191–210). London: Routledge.

Harding, S. (2006). *Animate earth: Science, intuition and Gaia*. Totnes: Green Books.

Hunt, J. D. (1993). *Gandhi in London* (2nd ed.). New Delhi: Promilla.

Jorge, E. (2011). Assessing the value of a community-based approach to language and cultural learning: A longitudinal study. *Journal of the Scholarship of Teaching and Learning*, *11*(1), 33–52.

Joseph, L. E. (1990). *Gaia: the growth of an idea*. London: Arkana.

Kirchner, J. W. (1989). The Gaia hypothesis: Can it be tested? *Reviews of Geophysics*, *27*, 223–235.

Kirchner, J. W. (2002). The Gaia hypothesis: Fact, theory and wishful thinking. *Climatic Change*, *52*, 391–408.

Kolb, D. A. (1984). *Experiential Learning: Experience as the source of learning and development*. Englewood Cliffs, NJ: Prentice Hall.

Kuhn, T. (1962). *The structure of scientific revolutions*. Chicago, IL: University of Chicago Press.

Kump, L. R., Kasting, J. F., & Crane, R. G. (2010). *The Earth system* (3rd ed.). Upper Saddle River, NJ: Pearson Educational – Prentice Hall.

Lange, E. A. (2004). Transformative and restorative learning: A vital dialectic for sustainable societies. *Adult Education Quarterly*, *54*(2), 121–139.

Lenton, T., & Watson, A. (2011). *Revolutions that made the Earth*. Oxford, UK: Oxford University Press.

Lovelock, J. E. (1979). *Gaia: A new look at life on Earth*. Oxford, UK: Oxford University Press.

Lovelock, J. E. (1988a). Self-regulation of the earth as a living organism: Mini review. *Microbiología Sem*, *4*, 141–148, Retrieved from http://www.ncbi.nlm.nih.gov/pubmed/3271062 (accessed January 28, 2013).

Lovelock, J. E. (1988b). *The ages of Gaia: A biography of our living Earth*. Oxford, UK: Oxford University Press.

Lovelock, J. E. (2005). *Gaia: Medicine for an ailing planet* (2nd ed.). London: Gaia Books.

Lovelock, J. E. (2006). *The revenge of Gaia: Why the Earth is fighting back – and how we can still save humanity*. London: Penguin – Allen Lane.

Lovelock, J. E. (2009). *The vanishing face of Gaia: A last warning*. London: Penguin – Allen Lane.

Margulis, L. (1998). *The symbiotic planet: A new look at evolution*. London: Orion-Phoenix.

Margulis, L., & Sagan, D. (1986). *Microcosmos: Four billion years of evolution from our microbial ancestors*. Berkeley: University of California Press.

Martin, S., & Murray, P. (2011). The role of wicked problems: Values in personal and organisational change. *Learning and Teaching in Higher Education (LATHE)*, *5*, 163–169, Retrieved from http://www2.glos.ac.uk/offload/tli/lets/lathe/issue5/Lathe_5_Martin_Murray.pdf (accessed August 2011).

McEwen, L., Strachan, G., & Lynch, K. (2011). 'Shock and awe' or 'reflection and change': Stakeholder perceptions of transformative learning in higher education. *Learning and Teaching in Higher Education (LATHE)*, *5*, 34–55, Retrieved from http://www2.glos.ac.uk/offload/tli/lets/lathe/issue5/Lathe_5_McEwen_Strachan_Lynch.pdf (accessed August 2011).

McWhinney, W., & Markos, L. (2003). Transformative education across the threshold. *Journal of Transformative Education*, *1*(1), 16–37.

Merriam, S. B., & Caffarella, R. S. (1999). *Learning in adulthood*. San Francisco, CA: Jossey-Bass.

Meyer, J. H. F., Land, R. & Baillie, C. (Eds.). (2010). Editors' preface. *Threshold concepts and transformational learning* (pp. ix–xlii). Rotterdam, NL: Sense.

Mezirow, J. (1997). Transformative learning: Theory to practice. *New Directions for Adult & Continuing Education*, *74*, 5–12.

Mezirow, J. (2003). Transformative learning as discourse. *Journal of Transformative Education*, *1*(1), 58–63.

Mezirow, J. (2006). An overview on transformative learning. In P. Sutherland & J. Crowther (Eds.), *Lifelong learning: Concepts and contexts* (pp. 24–38). London: Routledge.

Mezirow, J. (2009). Transformative Learning Theory. In J. Mezirow & E. W. Taylor (Eds.), *Transformative learning in practice: Insights from community, workplace, and higher education* (pp. 18–32). San Francisco, CA: Jossey-Bass.

Midgley, M. (2001). *Gaia: The next big idea*. London: DEMOS.

Naess, A. (1987). Self-realization: An ecological approach to being in the world. *Trumpeter*, *4*(3), 35–42, Retrieved from http://trumpeter.athabascau.ca (accessed July 2010); Reprinted in: A. Drengson, & Y. Inoue, (Eds.). (1995). In *The Deep Ecology movement, an introductory anthology* (pp. 13–30). Berkeley, CA: North Atlantic.
Naess, A. (1989). *Ecology, community and lifestyle*. Cambridge: Cambridge University Press.
Newson, M. (1992). Twenty years of systematic physical geography: Issues for a 'New Environmental Age'. *Progress in Physical Geography*, *16*(2), 209–221.
Park, C. (2003). Engaging students in the learning process: The learning journal. *Journal of Geography in Higher Education*, *27*(2), 183–199.
Purkey, W. W. (2011). Letters to the Editor. *Invitational Education Forum*, *32*(3), 17.
Purkey, W. W., & Novak, J. M. (1993). The invitational helix: A systemic guide for individual and organizational development. *Journal of Invitational Theory & Practice*, *2*(2), 77–86.
QAA (2007). *Subject benchmark statements: Geography*. Gloucester: QAA. Retrieved from http://www.qaa.ac.uk/Publications/InformationAndGuidance/Documents/geography.pdf (accessed January 2013).
Sagan, D. (1990). *Biospheres: Metamorphosis of planet Earth*. London: Arkana.
Schumm, S. A., & Lichty, R. W. (1965). Time, space and causality in geomorphology. *American Journal of Science*, *263*, 110–119.
Seed, J., Flemming, P., & Naess, A. (1988). *Thinking like a mountain: Towards a council of all beings*. Gabriola Island, BC: New Society.
Selye, H. (1956). *The stress of life*. New York: McGraw-Hill.
Slater, F. (1993). *Learning through Geography*. Indiana: National Council on Geographic Education. (Pathways in Geography 7).
Skyttner, L. (2001). *General systems theory: Ideas and applications*. Singapore: World Scientific.
Strahler, A. N. (1980). Systems theory in Physical geography. *Physical Geography*, *1*(1), 1–28.
Tara Mata (1924). A forerunner of the new race. In Self Realization Fellowship Council (Ed.), (1999). *A world in transition: Finding spiritual security in times of change* (pp. 237–246). Los Angeles, CA: Self Realization Fellowship.
Taylor, E. W. (1997). Building upon the theoretical debate: A critical review of the empirical studies of Mezirow's Transformative Learning Theory. *Adult Education Quarterly*, *48*(1), 34–60.
Taylor, E. W. (2007). An update of Transformative Learning Theory: A critical review of the empirical research (1999–2005). *International Journal of Lifelong Education*, *26*(2), 173–191.
Taylor, E. W. (2008). Transformative Learning Theory. *New Directions for Adult and Continuing Education*, *118*, 5–16.
Turney, J. (2003). *Lovelock and Gaia: Signs of life*. Cambridge, UK: Icon Books.
von Bertalanffy, L. (1950a). The theory of open systems in physics and biology. *Science*, *111*, 23–29, Reprinted in: Taschdjian, E. (ed.). (1975). In *Ludwig von Bertalanffy: Perspectives on General System Theory, scientific – philosophical studies* (pp. 127–136). New York: Braziller.
von Bertalanffy, L. (1950b). An outline of General System Theory. *British Journal for the Philosophy of Science*, *1*, 134–165.
von Bertalanffy, L. (1968). *General System Theory: Foundations, developments, applications*. New York: Braziller.
Wagner, Z. M. (1999). Using student journals for course evaluation. *Assessment & Evaluation in Higher Education*, *24*(3), 261–272.
Watson, A. J., & Lovelock, J. E. (1983). Biological homeostasis of the global environment: The parable of Daisyworld. *Tellus*, *35B*, 284–289.
Wilber, K. (1998). *The essential Ken Wilber*. Boston, MA: Shambala.
Wilber, K. (2006). *Integral spirituality*. Boston, MA: Integral.
Wilson, E. O. (1998). *Consilience: The unity of knowledge*. New York: Knopf.
Wolf, J., Stanton, M., & Gellott, L. (2010). Critical thinking in Physical Geography: Linking concepts of content and applicability. *Journal of Geography*, *109*(2), 43–53.

Going mobile: perspectives on aligning learning and teaching in geography

Claire H. Jarvis, Jennifer Dickie and Gavin Brown

Department of Geography, University of Leicester, UK

There has been little consideration to date regarding how we might best adjust our assessment protocols so that the overall learning experience remains appropriately aligned to both content and teaching approach when adopting location-specific mobile learning. This paper explores the success of a novel strategy to design an assessment regime that captures a field experience in Dublin as a whole for both staff and students alike. Evaluation identified that the use of a combined mediascape-essay approach as a major component of the assessment successfully captured the main elements of the learning and teaching experience and facilitated deeper learning and creativity.

Introduction

In Geography, there has been a rapid increase in the use of mobile technology-assisted research (Ricketts Hein, Evans, & Jones, 2008) and learning (Lynch et al., 2008). The term "mobile technology" is widely used, and covers a wide range of hardware that can easily be used "on the move" such as digital cameras, laptops or tablets, digital sensors, simple and more complex global positioning systems (GPSs), mobile phones (or cell phones), personal digital assistants (PDAs) and mp3 players. Importantly in a geographical context, generic devices may be coupled with the location-awareness provided by GPS, either directly or indirectly. Further, we are increasingly seeing the coalescence of functionality within one highly portable device; the "smartphone", for example, refers to a unit that is currently likely to contain a camera, GPS, accelerometer and digital voice recording function, in addition to the standard functionality of a mobile telephone and PDA.

The literature reporting the use of such mobile technologies in learning, teaching and assessment in Geography and the allied subjects largely focuses on situations where mobile technologies are used by students for data collection and learning in the field (e.g. Clegg et al., 2006; Latham & McCormack, 2007; Menking & Stewart, 2007; Rogers, McCann, & Cooper, 2010; Saunders, 2011). A second and much smaller group of papers refer to mobile teaching interventions that may be location-aware in the sense that materials are delivered to a student according to their GPS coordinates (Armstrong & Bennett, 2005; Loveless, Denning, Fisher, & Higgins, 2008; Williams, Jones, Wood, & Fleuriot, 2006) or not (e.g. Butler, 2008; Jarvis & Dickie, 2010). The focus of this paper lies with the second of these categories, with a specific focus on location-aware (or context-based) mobile teaching interventions.

The journal literature regarding mobile teaching interventions explicitly within Geography focuses, in the main, on possibilities and potential. Drawing on his research work along the River Thames, Butler (2008, p. 1750) for example suggested a range of interesting ideas for students to develop their own "memoryscapes". In using this term, he refers to "outdoor trails that use recorded sound and spoken memory played on a personal stereo or mobile media to experience places in new ways". His exemplars show personal reflections, sounds and oral histories that may be downloaded onto an mp3 player to be played at certain pre-specified location points; alternatively, they may be triggered from a static web mapping interface. The Soundwalk project's Bronx Graffiti trail (Soundwalk, 2004) follows a similar approach. In his paper, Butler (2008, p. 1750) suggests that "Creating such work is an active, multi-sensory and profoundly challenging experience that can offer students the chance to master multi-media skills as well as apply theoretical understandings of the histories and geographies of place." This is an attractive goal to explore in the context of assessing undergraduate teaching that France and Wakefield (2011) touch on from the static, as opposed to mobile, perspective.

As Armstrong and Bennett (2005) highlight, using a tool that allows you to present information to an audience on the move on a device at appropriate locations *automatically* could also offer powerful teaching opportunities. Locations need not be visited in order, or even be apparent to the student until the intervention occurs. The example they use is that of a video on the subject of stream meanders; this is automatically shown to a student as they near an oxbow lake. In different contexts or areas of the geographic discipline, we could substitute this video for an audio clip asking students site-specific questions, audio clips from different local residents discussing their views on urban change in a particular area, a historical or seasonal soundscape, and still photography from similar or connected locations or times past. Critically, multimedia are *automatically triggered* on a GPS-enabled smartphone device as the holder passes through an area relating to the material being played. In this paper, we refer to this location-aware collection of information as a mediascape. Butler's (2008) approach, while offering a considerably more challenging and sensuous experience over paper tours, is more a modern implementation of a traditional, somewhat linear guide in comparison.

Taking both Butler's (2008) and Armstrong and Bennett's (2005) ideas together, and drawing on the work of Williams et al. (2006) and Loveless et al. (2008) with younger students, there is a case for both teaching *and* assessing students using a mediascape approach. Interestingly, within the geographical literature, there is an evident distinction between situations where mobile interventions are key teaching tools as proposed here and those where the form of assessment is heavily influenced by a mobile technology (e.g. France & Wakefield, 2011; Kemp, Mellor, Kotter, & Oosthoek, 2011; Mavroudi & Jöns, 2011). In a locative mobile technologies' setting, rarely are these two aspects of the learning ecosystem considered together, either in the geographical literature or beyond. Between these two poles, however, there is a widely practised arena in which students are not *taught* through mobile devices directly, but rather use them in their *learning* and to inform their assignments.

This paper explores the assessment of a human geography fieldcourse module to Dublin in which staff and students alike used mediascapes. In this case, staff-authored, location-aware mobile teaching methods were used to a wide extent to support student learning while students created group mediascapes as a key component of their assessment. This is a significant departure from previous pedagogic work in Geography regarding locative mobile technologies; in general, although there have been developments in both mobile teaching and assessment strategies, they are less commonly coupled together. The focus of this paper is to evaluate whether this explicit coupling of teaching and learning approach and assessment style gives rise to deep learning, as Biggs' (2003b) theory of constructive alignment suggests may be the case.

Fieldcourse context and approach to mobile learning

General overview

The fieldcourse aims were to explore the social construction of "Modern Ireland", both through an examination of the material rebuilding of Dublin's built environment (e.g. issues of gentrification and urban regeneration) and by exploring cultural representations of Irishness in the city. In particular, we wished students to grasp the fundamental shift in the city from Ireland as a diasporic nation to one that has, in recent years, been receiving significant numbers of immigrants and is increasingly establishing itself as a multicultural city. The emphasis of the fieldcourse is upon understanding the legacies of Ireland's post-colonial development and the transition from a rural to an urban economy, particularly through a distinct geopolitical engagement with the European Union. Within this broader context, more generic learning outcomes for the field module were to:

- encourage students' critical intellectual development through guided learning in a research environment;
- encourage personal reflection and the acquisition of subject-specific skills to enable graduates to achieve success in their career paths;
- develop advanced IT skills and familiarity with new generation mobile technologies, promoting employability;
- practice innovative research-led methodologies in the preparation for independent study;
- develop new ways of visualizing advanced theoretical concepts from urban geography.

Figure 1 summarizes our overall teaching and learning approach to this module. It is important to note that both subject and technological learning are carefully scaffolded throughout our degree. Prior to the start of the year, students had received basic training in field observation and note taking. They were also familiar with GPS phototagging, digital cameras, the use of simple Garmin GPS units, the principles of GPS and Geographical Information Science. Further, the aspects of urban and cultural geography and digital representation are considered elsewhere in the second year syllabus and developed further in the third year. In this paper, our particular focus is on the assessment of the mediascape component within our teaching and learning strategy.

Mediascape as a teaching and learning tool

In our introduction, we defined mediascape as a multimedia display on a GPS-enabled smartphone device that is *automatically triggered* as the holder passes through an area relating to the material being played.

In the case of the work described here, we drew on visual and sensory social science research methods (e.g. Pink, 2009; Rose, 2008) to gather multimedia data for Dublin as part of our initial field reconnaissance trip. During our visit we evaluated areas that we wanted students to explore in depth, and considered themes we wished them to reflect upon, questions that we might pose to them in particular places and the level of discoveries and connections that they should find out for themselves. These data and our staff audio "overlay" were then collated on a PC using the software framework "mscape" (Stenton et al., 2007).

The mscape software allowed us to designate regions or points of interest on a digital map and assign multimedia data to play at these particular locations in a very easy and intuitive manner. This software has previously been used by young children to help them

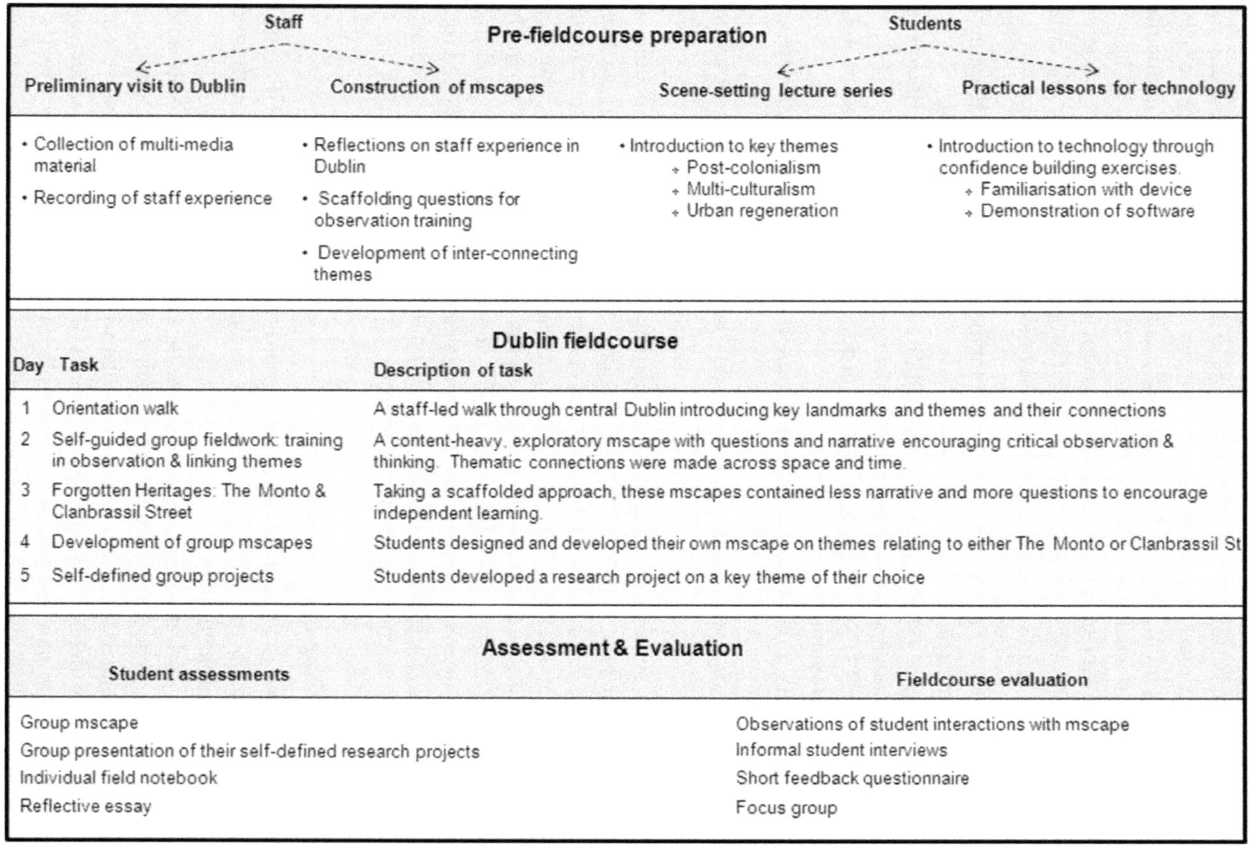

Figure 1. Fieldcourse structure, incorporating pre-field preparation and training, key elements of in-field study and assessment regime.

express a sense of place (e.g. Williams et al., 2006; Loveless et al., 2008); a very important consideration when setting the student assignment, since we did not wish to obscure subject learning by over-emphasizing technology. Subsequently, our mediascapes were then exported to a version that ran on our departmental collection of GPS-enabled smartphones. At times when no media were triggered, the mediascapes were set to show a background map of the area in which the user is situated, and so identified the students' location for them. These maps did not overtly show where multimedia data might be found. The staff group re-visited Dublin to hone the mobile mediascapes and, in particular, to ensure that the multimedia triggered at the geographical locations intended in the appropriate order.

In the fieldcourse context, in addition to a short staff-led introductory tour, mediascapes were firstly used to guide students during their initial experiences of the city and to provide training in observation both by example and situated questioning (Figure 1). There was no expectation that students should use their own GPS-enabled smartphones, although many do use such technologies routinely. The mediascapes enabled students to experience Dublin's past and present, taking them on a journey through time and space that illustrated the interweaving stories between people and places. The principal difference between the mediascape approach and a traditional guide was that students did not undertake a linear "tour" either with staff or with a fact sheet; rather, they roamed as groups within an area of the city that we bounded using a musical border, discovering answers to questions set at the start and also developing further questions of their own for later project work. They could not see the material in the mediascape in advance and did not know when it would be triggered; thus, initiative was required of the students to explore and seek out interesting geographical issues. The style of teaching was more "supported exploration" than "traditional guide", with mainstream delivery provided by machine and a much lighter staff tread than the norm. Staff did hover in the background to ensure no insurmountable issues were arising with the mediascapes and to discuss subject-related questions.

Secondly, the students themselves developed group mediascapes, which were assessed a large proportion of their research project reports. The other component of the project assessment was a more traditional reflective field diary. The research projects, designed to hone our students' ability to ask and solve focused research questions in an appropriate way, were typical of their type. Some looked to expose specific critical processes behind the material rebuilding of Dublin while others explored and challenged the cultural representations of Irishness in the city. The students made full use of field notebooks, cameras and digital recording devices during this in-field project process. The existing staff mediascapes also provided students with GPS-enabled background mapping for use in project work, and a sense of what we look for in the assignment submissions. The latter point is important, as significant student anxiety can potentially arise when introducing innovative methods of assessment for the first time unless good exemplars are provided.

What was unusual about the assessment was the representation of the project findings when compiled back home; the majority of student outputs are subsequently processed to a linear textual narrative with interspersed visual evidence. This was not the case here. When developing project outputs in mediascape, students who collected photographic, textual and sound-based materials as part of their projects were largely able to keep these formats in their submission. The affect in the findings could potentially find its way through into the project report. Textual material, however, if detailed versus (say) a newspaper headline, was better converted to an audio overlay; also, just as staff provided commentary to justify the inclusion of images or to explain the context for video footage, so students were advised to do similarly. As noted by Butler (2008), doing this well is a greater challenge than producing linear narrative. In particular, designing the order in which media would be triggered in the submissions such that the assessor could be walked through a network of nonlinear space–time links is not straightforward. Implicitly, in the process of building such mediascapes, students record their own learning and reflections on the relatedness of materials; pedagogically, the approach not only provides students with experience using leading edge active research methods but also provides an additional and more sensory means of recording an account of student learning possible in a traditional written notebook.

For the marker, the final mscapes can be viewed as an interactive map on a PC, on which both the students' highlighted regions of interest and materials can be seen in addition to a smartphone screen emulator. The assessor can move their mouse across the map, much as you might across Google maps, and this triggers sequences of connected materials of various types when regions of interest are entered. In style, this final product is more akin to a multimedia essay set in space than a website or blog. Our emphasis in marking the projects was on the intellectual aspects of the projects, not on the quality of film or audio produced as long as it was satisfactory enough in quality to be understood.

Evaluative approach

This paper discusses whether aligning assessment involving mobile technologies with teaching using mobile technologies offers our students the types of synergistic learning that Biggs' (2003b) theory suggests is possible. We explore constructive alignment in the context of new 5-day fieldcourse to Dublin with 23 second-year human geography students. This in turn requires us to think about what markers we are looking for in the students' responses to identify whether "deeper" learning and student engagement has occurred.

Constructive alignment

Biggs' (2003b) theory of constructive alignment supports the idea that prior knowledge, the process of learning as defined by learner-focused activities, and the outcome of such learning should be carefully considered together (or "aligned"). This theory has become embedded within mainstream pedagogic practice. Importantly, following the ideas of Schuell (1986) and Tyler (1949) before him, the key to Biggs' theory is that what the *student* does as part of their role in creating knowledge is key. The role of the teacher is to consider what students should do to achieve a set of learning outcomes, and then to create a supportive learning environment in which this process can occur.

The concept of constructive alignment has its critics. It is sometimes said, for example, that simply driving learning towards set "outcomes" limits the encouragement of and reward for original thinking. Nevertheless, for the "average" student, it is well documented that assessment drives student engagement and learning behaviour; thus, the quality of assessment will influence the quality of learning (Pedrosa de Jesus & Moreira, 2009). Further, outcomes that are appropriate in their specificity and that are not too narrow or dangerously broad as to encourage breadth over depth should not overly constrain creativity, but rather help structure it in a focused way. Biggs himself is clear that carefully constructed alignment is designed to deepen learning; he does not intend by alignment merely "teaching to a test" and in this sense the constructivist element of the theory in which students make sense of their world remains important. The importance of matching-up teaching and learning activities with the assessment set is further highlighted by Gibbs (1999, p. 41) who states that "Assessment is the most powerful lever teachers have to influence the way students respond to courses and behave as learners". It is clear that well-aligned assessment practices are considered key to successful, student-centred learning. In this respect, urgent attention is needed if teaching predominantly using GPS-enabled mobile technologies is to be considered a robust teaching approach.

In order to incorporate mobile technology-assisted teaching and assessment successfully into an aligned curriculum, it is important to look more closely at our motives and aspirations for incorporating the technology in the first place. We do not advocate developing assessments that focus on the students' ability to use the technology simply because we integrate it into the student experience. Instead, we believe that if there is a significant adjustment in learning and teaching style from the traditional norm, practitioners should reflect whether adjustments of assessment are needed, and if so, of what type these should be. In Biggs' terms, we need to assess the types of learning and processes that are constructed by the student in the course of the module; this is *in addition to* the subject learning outcomes more usually focused upon within both literature and practice.

In some cases, mobile technologies offer a more efficient replacement to traditional teaching and learning methods and an opportunity for students to practice basic generic skills. However, they do not otherwise challenge the overall student learning experience or outcomes. For example, field data collection methods, which may be digital, play an important role in project research work and in support of independent study. Once students have gained sufficient awareness of how to use and, critically, assess methods and devices for their "fitness for purpose" in a particular context, this knowledge can be applied in many settings. However, to alter the pattern of assessment in ways that over-emphasize the role of mobile technologies in this case would be to misalign the programme. Further, it has been argued that reflective field diaries can enhance deeper learning (Dummer, Cook, Parker, Barrett, & Hull, 2008); we have been entirely comfortable using this approach elsewhere in our curriculum as the one well suited to weighing up the relative merits of different data collection technologies. In contrast, when the objective is to enhance student activities and experiences significantly using innovative technology, the focus of the assessment should be modified to fit those criteria. This can be problematic since assessing all four learning domains that the use of mobile devices may impact upon (cognitive, affective, conative, and psychomotor; Reeves, 2006) is not straightforward and focusing our assessment strategies on what is (relatively) easy to measure is practical, replicable and equitable.

Ethnographic approach to evaluation

In this paper, we adopt an ethnographic stance in seeking to understand better the connections between the "processes" and "products" of learning; grades *per se* would not tell us whether deep or shallow learning had occurred. The case study approach is intended to be instrumental (Stake, 2005), provoking reflection on how we assess learning where field teaching is delivered (in part) on the move using digital devices; an area in which there has been little previous work either in geography or elsewhere. Materials used to investigate how the complex interplay between staff and students, and machine versus human interaction, supports synergies in learning between teaching and assessment came from a variety of sources. These included the reflective diaries of students, students' comments within a post-fieldcourse focus group and staff reflections on the style and content of the group mediascapes. In analysing these materials, our focus is on two areas. Firstly, we investigate the degree to which the assessment regime was able, through technological affordances, to convey cognitive, affective, conative and psychomotor aspects of the students' experiences. Secondly, Biggs (2003a) also argued that constructive alignment should "deepen" learning and increase student engagement, so identifying markers for deep learning was the key to our analysis.

The "deep-surface" metaphor used to distinguish between good and bad learning, respectively, is widely embedded within pedagogic rhetoric, but what should we look for as we analyse and code our evaluative materials? In essence, we need to ask ourselves whether our students are digging under the facts to make meanings from them, or merely skating over the surface using low level thinking (Biggs, 2003b). Using Biggs' (1999) SOLO taxonomy (Table 1), deepest manifestations of understanding will be seen in students looking for generalization, hypothesis generation and reflection. The ability to see relational aspects in our students' work, such as analysing, relating and applying knowledge, would support a broad case that we have made a conceptual change in our teaching approach that, with the benefit of applying constructive alignment, has made a difference to the understanding of our students. Learning that stops with an ability to make simple identification or draw-up lists, or that misses the point entirely, will suggest that the activities or processes that we have encouraged have been unsuccessful. Further, we also need to evaluate evidence with a view to seeing whether we have encouraged students "to engage the task meaningfully and appropriately" (Biggs, 2003b, p. 16) or not. As Saunders (2011, p. 186) usefully observes, there is in fact a distinction to be had between engagement *in* and engagement *with* learning. Are we "providing space for the student to become engaged *with* learning" more generally in addition to heightening the excitement of participating *in* the specific learning tasks? Issues such as independence and confidence in learning are arguably more associated with the former, while a more specific energetic consideration of urban geographies and mobile technologies might be more associated with the latter.

Findings

Did aligning assessment with teaching style offer greater student engagement?

Saunders (2011, p. 186) defines a curriculum for engagement as one which "... seeks to heighten the energy and interaction within a course, by designing challenging learning

Table 1. Biggs' SOLO taxonomy.

	Taxonomical categories	Degree of intellectual integration	Typical associated verbs
Incompetence ↓	Prestructural		Fail, incompetent, misses the point
	Unistructural	Considers one relevant aspect only	Identify, name, follow simple procedure
	Multistructural	Considers several relevant aspects, but independently	Combine, describe, enumerate, perform serial skills, list
	Relational	Considers several relevant aspects, which are integrated into a coherent structure	Analyse, apply, argue, compare/contrast, criticize, explain cause, relate, justify
	Extended abstract	Integrated understanding from one domain generalized to another domain	Create, formulate, generate, hypothesize, reflect, theorize
Competence			

Source: Biggs (1999).

opportunities that engage the student in learning, while providing space for the student to become engaged with learning".

Although we were confident that integrating location-aware mobile technology into the human geography curriculum was a useful and worthwhile goal on a number of levels, the pedagogic benefits would only be realized if there was sufficient student "engagement". A danger of using GPS-enabled devices within this cohort was that they might provoke high student anxiety, despite an initial training phase with the equipment prior to the fieldcourse, which in turn could trigger surface learning as a response. In fact, the views summarized here by three students (but gaining supportive murmurs within the focus group session) were the following:

> Although there were a few negative factors surrounding the use of the Mscapes, I believe that overall they enhanced our learning.

> There are a number of practical problems, such as relying on battery power and the low volume of audio clips. These are only minor issues, which can be overcome to a degree with preparation, and do not out-weigh the benefits of Mscape as a creative learning tool.

> There were a few practical problems with the Mscapes, most of which could not be avoided, but they were a slight drawback to using the equipment.

It does appear that the level of exposure to a range of digital representations and pervasive technologies, taken overall, was not at the expense of subject learning. Reflecting on our experiences, we believe that the intensive testing of the mscapes *in situ* was key to minimizing these issues; GPS reception and hence the suitability of the locations where we triggered material, audio quality and other factors were all adjusted during our second staff field visit. Working with digital devices is not to be undertaken lightly, and small issues arose even then, as the students note. We recommend operating as a multidisciplinary staff team while running this type of digitally enhanced teaching for the first time.

A more detailed examination of the levels of understanding achieved tells us more about engagement in the more specific learning tasks. In this case, these were to develop observation skills and proactively explore the social construction of "Modern Ireland" as part of the students' development of critical research skills. There is some evidence that the *in situ* nature of the mediascapes prompted students to start theorizing in the field. The student here implies that he/she began to move towards explanatory thinking and theorization, terms at the relational and potentially more abstract (deeper) end of Biggs' SOLO taxonomy (Table 1), although unfortunately does not provide explicit examples:

> I think this encouragement to think about the observations whilst in the field meant that more insightful analysis was possible as the longer you spend looking for answers, the more likely you are to come up with theories and ideas.

This statement speaks in favour of in-field teaching, rather than the merits of mobile technologies over traditional methods, but the key point is that learning *can* take place at the required level using the locative mobile approach.

For other students, the commentary is perhaps more relational and analytical in completeness and depth; we see students pushing beyond what they are shown to make links to other phenomena and space–time contexts about them, but not tackling theory. The mediascapes assisted with both multistructural learning (what?) and relational learning (why?) (Table 1), as we see here:

> The Mscapes which we explored in Dublin were excellent at getting us as groups to think more about what we were seeing and why things were placed where they were. They gave us that extra push to question more about the area and overall get more out of the experience.

We also see the students themselves using the term "engaged" and "enthusiastic", referring to enhanced interaction between themselves:

> Mscape was a great aid to the field work which was undertaken in Dublin, it added an extra dimension to the information we were able to gather and find out, they were simple to use and easy to get going.
>
> I think that on the whole it is a very effective piece of equipment. It engaged us with our surroundings and encouraged us to look at buildings, monuments, areas which we may not have noticed or investigated otherwise.
>
> Was a great asset, and people did not become too reliant on it, going out and collecting information beyond the context of the Mscape and the information provided... It enhanced my experience greatly.
>
> Using Mscape has made me a better, more observant and more open minded geographer.
>
> Knowing that this aid was available meant people were enthusiastic about the field work, challenging themselves and each other.

In these interactions, we see students justifying, criticizing and arguing intellectual points between themselves as a community of learners (Table 1, Relational) in addition to building multistructural understandings. This allowed staff interventions, in the main, to be deeper and more specific than they might otherwise have been. We believe this occurred since students had been drawn into prior discussion and reflection, which gave them both more understanding and also confidence to speak out. It is clear from comments to follow that this confidence is not something that students necessarily start with when placed in an unknown field situation.

As evidenced in the reflective diaries and the transcriptions of the focus group, however, the most prominent themes in terms of student engagement were the freedom and independence the mobile equipment facilitated, that is, predominantly the approach facilitated engagement *with* learning and encouraged personal development. Typical of such feedback was the comment that:

> Using the Mscape it became an interactive learning experience where we could walk, listen and take notes on what it pointed out while also then developing our own abilities to observe things independently.

By giving the students the responsibility and independence to investigate the key themes of Dublin while navigating around the city themselves, albeit with the mediascape narrative as support, the students were forced to become aware of their surroundings and create their own learning.

The initial mediascape also allowed the students to grasp the spatiality of the city and, in turn, better connect with the themes of the trip in comparison with the initial short staff-led tour. This aspect was highlighted by a student who stated the following:

> I felt like the places I'd been to I really knew, and I felt much more like I'd explored them and I knew about them" compared to the first day where she had "walked round a bit of Dublin but I didn't really know where I'd been.

The use of GPS and the integrated maps associated with the mediascapes were also seen as an added benefit to the students, for example:

> I found it easier to use the Mscape map than a paper map as it was easier to move about on the Mscape map and with a small person cursor it was easy to identify where you were.

Overall, feedback suggested that the students used these facilities to orientate themselves and navigate around the city. While subtly improving their spatial literacy skills, the mediascape and GPS gave the students more confidence to explore the city independently.

During the focus group in particular, the students discussed the merits of being given a "tutor-free" guided tour. Interestingly, they reported that group interaction and discussion flowed more freely without a member of staff present.

> I think you're more comfortable with putting (out) ideas when it's a group of students... you don't want to put out something ridiculous when the lecturer's there.

> We'd have done far less exploration wouldn't we with a staff member.

> When you're not leading yourself and you're just following someone else, you don't pay attention to where you're going.

> I think when you're looking at the map yourself and you're planning where you're going, you know what you've done a lot more and know where you've gone.

The students felt more confident vocalizing their thoughts to a student audience, whereas in other situations they would have held back in fear of making a mistake or embarrassing themselves in front of a member of staff. The only negative point they had to make on this issue was the fact that it was not possible to ask the mediascape a question. However, we did provide a question and answer session at the end of each day in addition to making informal visits to the students while they were in the project areas. In the context of Laurillard's (2002) model of discourse between staff and students, some aspects of spontaneity and immediacy as regards staff–student discussions between their relative conceptualizations were lost when using the mediascape as a "broker", but only a very small minority of students found this problematic. In sum, students "felt free to focus on the task" (Biggs, 2003a, p. 13), having been provided (quite literally, in this case) with the space to learn.

Aligning teaching and assessment

To this point, we could argue that the points we have made regarding engagement in or with learning could have occurred regardless of our assessment method and rely on a careful thought through consideration of the students learning activities while on the fieldcourse. Reflecting on our evaluation material, however, there is significant additional evidence that there are a variety of advantages of incorporating information and communication technologies (ICT) within the assessment process as well as part of the initial field teaching.

Overall, the mediascape assignment was met with a positive response from the students. The use of multimedia enabled a fuller expression of their ideas than text alone, while the mediascape format allowed the consideration of nonlinear narratives between the data from multiple sources. This has analogies with recent developments in hyperlinked digital storytelling (e.g. France & Wakefield, 2011), and should not be compared to the use of video storytelling as a means of expression for students to reflect on their project journey (e.g. Mavroudi & Jöns, 2011; Wakefield & France, 2010). The reflective essay and mediascape were considered integral to each other by both staff and students alike, and together they allowed considerably more content to be presented. In addition, as the quotes below indicate, the materials incorporated clearly show both cognitive and affective aspects of learning while picking up themes key in sensory and visual research methods:

> I definitely think the mscape was better than doing an essay...it was more emotive, you could portray things that you could never be able to write.

> To support the amount of information we had in the Mscape in an essay, it would be epically long.
>
> Being able to add in music and like other people talking, the tour guide, that added an extra dimension, it's a different way of putting things across that is much better than just reading an essay.

The freedom to communicate in novel ways and to incorporate personal experiences gave students a sense of liberation; the framework of constructive alignment, requiring us to assess learning against set outcomes, does not preclude creativity. In particular, the students enjoyed being able to communicate in less formal ways. The context surrounding the quotation below provides for an interpretation in which students appreciate freedom from the constraints of grammatical correctness and are able to include, for example, an "epic track" within a formal assessment. For those with disabilities such as dyslexia, this freedom has the potential to be particularly powerful, although in this cohort there was no scope to explore such matters in depth. In general, this novel approach resulted in a higher level of student engagement, giving the students the freedom to play with ideas in new ways and encouraging them to think outside the box.

> It's less structured as well, like you feel more liberated doing it... Like in an essay you have to worry about how someone's going to read it and things like that, whereas here you can just literally put in what you want to put in and not worry.

This issue of structure can be seen from two perspectives. In their review of curriculum design principles, Meyers and Nulty (2009) summarize material that suggests students adopting a deep learning approach form highly structured knowledge. The mediascape approach clearly does not foster this. However, it does have other advantages that reflect deeper learning consistent with a more aligned module, for example the process of thinking that assists the exploration of more elaborate nonlinear connectivity in knowledge structures. To echo Hartnell-Young and Vetere (2008, p. 291), "While we and the teachers saw value in linear narratives, which we are used to, it is a sobering thought to consider what we, as researchers, missed about learning in the rich content we saw, and what teachers might be missing every day". The format also allowed the free-thinking venturing of ideas and expression of theories, again an expression of deeper learning and a well-aligned module. We had not foreseen, for example, that students would pose questions back to us as part of their digital mediascape materials. For example:

> We also employed the use of observationally based questions which we employed to encourage independent thought and interpretation of the issues present in each area. We hoped that this multimedia narrative material would help participants to tie in themes of the area to empirical evidence as the Mscapes we used did for us.
>
> We felt it would be valuable to the audience to end this clip with a question relating to the main themes, the purpose of this is to engage the user and make them question their taken-for-granted assumptions about the area.

Looking at the following comments, it is clear that through the process of constructing their own mediascapes, the students developed their own critical (higher order) perspective on the approach as a communication tool for visualizing advanced theoretical concepts from urban geography:

> When producing an Mscape of our own our group found that our experiences with it in Dublin helped us to understand the ways in which we would like it to work and the ways in which it could be made more user friendly than it was at times during our use of it. All in all gaining experience with it first hand in the field gave us a clearer view of what we wanted to achieve with our Mscape.

> We followed the strengths of the Mscapes that we were given as a guideline for our own whilst at the same time using our own experiences to improve on areas we thought were weaker than others... Mscape as a learning tool has the potential to be extremely useful and with more use of it in the field I am sure it will become even more useful in times to come.

A number of these weaknesses related to issues of interface design that require further attention from us as staff. An inability to play a piece of multimedia multiple times without having to go physically in and out of an area is one of these; a replay button is needed. Additionally, sounding a warning just before a track or visualisation is triggered would allow students extra time to retrieve the device from their pockets.

These comments are also interesting in relation to constructive alignment and prevailing difficulties with student feedback in the sector. As Handley and Williams (2011) recently show, giving students exemplar assignments with annotated feedback prior to the submission of their final assignment was highly valued as an engagement mechanism by students. Here, by using our own staff-created mediascapes to demonstrate observation practice and also showing how we ourselves went about making sense of the city, there are echoes of such a practice. We asked questions of the students; with reflection, it is not surprising and indeed encouraging that they were developing hypotheses of their own within their mediascapes.

Collating content for the mediascape, which involved making and expressing connections between movement in the field and mental processes, appealed to a wide range of students. Aspects of technical construction, involving psychomotor skills at a different level, had a more limited appeal:

> I quite enjoyed doing our Mscape at home, but I think that's because I am quite sort of like technically minded and I sort of enjoyed seeing how it worked. Whereas I know other people in our group weren't that fussed at all about how it worked or what. They enjoyed putting the stuff into it, but the actual construction of it they wouldn't care at all really, and they still wouldn't be able to know what went into it.

> Like me, I wouldn't be able to construct one I don't think without the help of a lot of other people in the group, but then I'd still prefer to do an Mscape as opposed to write an essay. So even though I might not be able to make a great one, I'd still prefer yeah, I'd still prefer to do it.

Despite this, within a group context, students still preferred the mediascape format to an assignment assessed solely by the traditional essay. This approach offered those with different conative styles of opportunities and spaces within which they could experiment and construct their own learning. The feedback suggests that approach might not work if set as an individual assignment, but as a group submission it allowed individuals to play to their strengths and interests at different times during the project process. As one student summarized:

> Beats doing an essay any day to be honest.

In closing, it is interesting to reflect on Loveless et al.'s (2008, p. 351) experiences with mediascape and school children, in which they note that "There is an important distinction to be drawn between the authoring and "audiencing" of mediascapes". In one aspect of our work, alignment fell short; constraints regarding the cost of field time meant that students could not "walk" their own mediascapes, and those of their peers, although the materials they created back home were spatially anchored and could have been used this way. For Loveless et al. (2008), the authoring and sharing of mediascapes was the more important of the two parts. Here, however, students were reflecting intellectually and emotively about an area that is not a key place in their life. Further, while we acknowledge that developing more professional film and audio production skills in our students would be useful for their employability, we did not emphasize these aspects in our training; this would have been a key requirement if we had been assessing the effectiveness of the mediascapes for on-the-move communication to others, rather than for the shaping of intellectual ideas. In our context, it is the combined effect of audience and authorship that we believe was important in achieving our fieldcourse learning objectives.

Conclusions

In this case study, new generation digital technologies were introduced to human geography undergraduates to facilitate teaching on research themes related to the social construction of "Modern Ireland". In particular, we use context-based mobile methods in the form of what we term "mediascape" to train the students in more traditional field observation skills, an approach not widely reported upon to date in the literature concerning Geography in Higher Education.

Mindful of Biggs' (2003a, p. 26) theory of constructive alignment, which explicitly highlights that all components of the teaching and teaching "ecosystem" need to be aligned, in turn we asked students to construct their assessed group project work in the form of a mediascape. That is, we deliberately mirrored the mode of the student learning experience, and the many aspects of learning and meaning constructed by the students' activities, within the assessment process. The degree to which we achieved successful alignment within our course was evaluated using an ethnographic approach along two themes. Firstly, we asked to what extent verbs relating to deeper learning processes were expressed in students' feedback, drawing on Biggs' (1999) SOLO taxonomy. Secondly, aware that unaligned curricula lead to unmet expectations (Biggs, 2003b, p. 26) and that affect is a key aspect in learning (McConnell & van Der Hoeven Kraft, 2011), we assessed markers of student enthusiasm and engagement within the students' responses.

Our results suggest that student learning achieved reflective and on occasion sometimes abstract, integrated levels, and that engagement and enthusiasm for the context-based mobile learning was high. The approach also encouraged students to visualize and synthesize intellectually challenging theoretical concepts about the relationality of urban space (Amin & Thrift, 2002) in new ways that were technically accessible.

In general, institutional practices ensure that we link the critical components of the curriculum we teach and the subjects we assess. However, the teaching methods and assessment procedures we use are often less well aligned, and that this is particularly the case as we step out and explore novel teaching, learning and assessment concepts in Geography using mobile technologies. For example, there are examples of podcasts being used for both teaching and assessment in the literature, but not usually together; digital storytelling has been introduced as a novel assessment method but does this approach match the teaching methods used with students? Where the use of mobile technologies is a minor component of a course this may not matter, but as we move into an era of opportunity to explore more radical forms of context-based teaching supported by mobile technologies, we need to reflect on what this significant shift in teaching method means for the way we assess our students.

References

Amin, A., & Thrift, N. (2002). *Cities: Reimagining the urban*. Cambridge: Polity Press.

Armstrong, M. P., & Bennett, D. A. (2005). A manifesto on mobile computing in geographic education. *The Professional Geographer*, *57*(4), 506–515.

Biggs, J. B. (1999). *What the student does: Teaching for quality learning at university*. Buckingham, UK: Open University Press.

Biggs, J. B. (2003a). *Teaching for quality learning at university*. Buckingham, UK: Society for Research into Higher Education and Open University Press.

Biggs, J. B. (2003b). Constructing learning by aligning teaching: Constructive alignment. In J. B. Biggs (Ed.), *Teaching for quality learning at university: What the student does* (pp. 11–33). Maidenhead, UK: Open University Press.

Butler, T. (2008). Teaching and learning guide for: memoryscape: how audio walks can deepen our sense of place by integrating art, oral history and cultural geography. *Geography Compass*, *2*(5), 1750–1754.

Clegg, P., Bruciatelli, L., Domingos, F., Jones, R. R., de Donatis, M., & Wilson, R. W. (2006). Digital geological mapping with tablet PC and PDA: A comparison. *Computers & Geosciences*, *29*(4), 1682–1698.

Dummer, T., Cook, I., Parker, S., Barrett, G., & Hull, A. (2008). Promoting and assessing "deep learning" in geography fieldwork: An evaluation of reflective field diaries. *Journal of Geography in Higher Education*, *32*(3), 459–479.

France, D., & Wakefield, F. (2011). How to produce a digital story. *Journal of Geography in Higher Education*, *35*(4), 617–623.

Gibbs, G. (1999). Using assessment strategically to change the way students learn. In S. A. Brown & A. Glasner (Eds.), *Assessment matters in higher education: Choosing and using diverse approaches* (pp. 41–52). Buckingham, UK: Society for Research into Higher Education and Open University Press.

Handley, K., & Williams, L. (2011). From copying to learning: Using exemplars to engage students with assessment criteria and feedback. *Assessment & Evaluation in Higher Education*, *36*(1), 95–108.

Hartnell-Young, E., & Vetere, F. (2008). A means of personalising learning: Incorporating old and new literacies in the curriculum with mobile phones. *The Curriculum Journal*, *19*(4), 283–292.

Jarvis, C. H., & Dickie, J. (2010). Podcasts in support of experiential field learning. *Journal of Geography in Higher Education*, *34*, 173–186.

Kemp, J., Mellor, A., Kotter, R., & Oosthoek, J. W. (2011). Student-produced podcasts as an assessment tool: An example from geomorphology. *Journal of Geography in Higher Education*, *35*(4), 1–14.

Latham, A., & McCormack, D. P. (2007). Digital photography and web-based assignments in an urban field course: Snapshots from Berlin. *Journal of Geography in Higher Education*, *31*(2), 241–256.

Laurillard, D. (2002). *Rethinking university teaching*. Oxford: Routledge Falmer.

Loveless, A., Denning, T., Fisher, T., & Higgins, C. (2008). Create-a-scape: Mediascapes and curriculum integration. *Education and Information Technologies*, *13*(4), 345–355.

Lynch, K., Bednarz, B., Boxall, J., Chalmers, L., France, D., & Kesby, J. (2008). E-learning for geography's teaching and learning spaces. *Journal of Geography in Higher Education*, *32*(1), 135–149.

Mavroudi, E., & Jöns, H. (2011). Video documentaries in the assessment of human geography field courses. *Journal of Geography in Higher Education*, *35*(4), 579–598.

McConnell, D. A., & van Der Hoeven Kraft, K. J. (2011). Affective domain and student learning in the geosciences. *Journal of Geoscience Education*, *59*(3), 106–110.

Menking, K., & Stewart, M. E. (2007). Using mobile mapping to determine rates of meander migration in an undergraduate geomorphology course. *Journal of Geoscience Education*, *55*(2), 147–151.

Meyers, N. M., & Nulty, D. D. (2009). How to use (five) curriculum design principles to align authentic learning environments, assessments, students' approaches to thinking and learning outcomes. *Assessment & Evaluation in Higher Education*, *34*(5), 565–577.

Pedrosa de Jesus, H., & Moreira, A. C. (2009). The role of students' questions in aligning teaching, learning and assessment: A case study from undergraduate sciences. *Assessment & Evaluation in Higher Education*, *34*(2), 193–208.

Pink, S. (2009). *Doing sensory ethnography*. London: Sage.

Reeves, T. C. (2006). How do you know they are learning? The importance of alignment in higher education. *International Journal of Learning Technology*, *2*(4), 294–309.

Ricketts Hein, J., Evans, J., & Jones, P. (2008). Mobile methodologies: Theory, technology and practice. *Geography Compass*, *2*(5), 1266–1285.
Rogers, D., McCann, T., & Cooper, A. (2010). Integrating paper-based habitat mapping with mobile electronic field recording procedures. *Ecological Informatics*, *5*(6), 479–483.
Rose, G. (2008). Using photographs as illustrations in human geography. *Journal of Geography in Higher Education*, *32*(1), 151–160.
Saunders, A. (2011). Exhibiting the field for learning: Telling New York's stories. *Journal of Geography in Higher Education*, *35*(2), 185–197.
Schuell, T. J. (1986). Cognitive conceptions of learning. *Review of Educational Research*, *56*(4), 411–436.
Soundwalk (2004). Bronx graffiti trail. Available at http://www.soundwalk.com/#/TOURS/bronxgraffiti/ (accessed November 2011).
Stake, R. (2005). Qualitative case studies. In N. K. Denzin & Y. S. Lincoln (Eds.), *The sage handbook of qualitative research* (pp. 443–466). Thousand Oaks, CA: Sage.
Stenton, S. P., Hull, R., Goddi, P. M., Reid, J. E., Clayton, B. J., Melamed, T. J., & Wee, S. (2007). Mediascapes: Context-aware multimedia experiences. *Multimedia, IEEE*, *14*(3), 98–105.
Tyler, R. W. (1949). *Basic principles of curriculum and instruction*. Chicago, IL: University of Chicago Press.
Wakefield, K., & France, D. (2010). Briging digital stories into assessment. *Planet*, *23*, 63–67.
Williams, M., Jones, O., Wood, L., & Fleuriot, C. (2006). Investigating new wireless technologies and their potential impact on children's spatiality: A role for GIS. *Transactions in GIS*, *10*(1), 87–102.

Considerations of How to Study Learning Processes when Students use GIS as an Instrument for Developing Spatial Thinking Skills

LENE MØLLER MADSEN & CAMILLA RUMP
Department of Science Education, University of Copenhagen, Denmark

ABSTRACT *Within the last 30 years, geographical information systems (GIS) have been used increasingly in the training of geographers. On the basis of the philosophy of technology and instrumental genesis, we sketch how the use of instruments interacts with learning processes and outline how this can be studied. We empirically analyse students' learning processes and the influences of teaching practice in an introductory course in GIS. We show that students have different strategies for creating their personal instrument for spatial thinking and how teaching interacts with the students' learning processes. Finally, we discuss how GIS may gradually alter future professional development of geographers.*

Introduction

Within the last 30 years, geographical information systems (GIS) have been introduced and integrated into the curricula of geographers in many countries. Different types and levels of GIS courses have been included in a wide number of geography university curricula and at various levels of lower and upper secondary schools. Today, GIS can be said to be an important part of the training of geographers and an integrated part of thinking about geography in the wider public sphere (Catling & Taylor, 2007).

There are many challenges involved in including GIS in the geography curricula at different educational levels; for example the technical impediments that stem from the complexity of software, the lack of experienced teachers and unclear curriculum demands (Lloyd, 2001; Rød *et al.*, 2010). Different examples have been given on how to integrate GIS in relation to specific geography subjects such as heritage landscapes (Summerby-Murray, 2001), different teaching methods such as problem-based learning (Drennon, 2005), self-learning modules (Zerger *et al.*, 2002) and the use of a multimedia classroom (Deadman *et al.*, 2000).

GIS have been linked to the concept of spatial thinking in relation to students' development of spatial abilities. This has been done by claiming that GIS can develop students abilities to solve spatial problems, reason spatially and improve their map-reading skills (e.g. Albert & Golledge, 1996; Hall-Wallace & McAuliffe, 2002), and it has recently been shown empirically that GIS learning improves learners' spatial thinking ability (Lee & Bednarz, 2009). At the curriculum level, support systems such as GIS have also been linked to the goal of developing competent spatial thinkers, most noticeably in the publication *Learning to Think Spatially* by the US National Academy of Sciences (National Research Council, 2006). Here, it is stated that

> GIS are only one of a suite of support systems, high and low tech in nature, that must be brought to bear in fostering an understanding for and appreciation of spatial thinking. All students deserve and need the opportunity to be challenged, to be supported and to become critical spatial thinkers (National Research Council, 2006, p. 241).

This has fostered optimism among geographers as spatial thinking is now seen as an integrator and facilitator for problem solving and a missing link across the curriculum in a wider educational sphere and not only by geographers for whom spatial thinking has long been a central issue (Catling & Taylor, 2007). Furthermore, it gives spatial support systems a prominent status as stated by Sarah Bednarz: "Spatial thinking, supported by geospatial technologies, can—and should—form the foundation of innovative geography curriculum" (Bednarz, 2007, p. 13). In *Learning to Think Spatially*, discussion of the support system has also been given focus. In reflecting on using GIS to think spatially, Lyn Malone concludes: "The power of this learning experience is the result of an inquiry process that is supported by a tool (in this case GIS)" (National Research Council, 2006, p. 241).

Following this, GIS have been and are often still seen as a technical expansion of the toolbox of geographers. However, along the way this toolbox has been more and more tightly linked to the educational request of creating competent spatial thinkers within the educational system. There is no clear consensus about a definition of spatial thinking within research communities, but this has not hindered educational discussions of how to foster spatial literacy. This discrepancy is not our concern in this paper. Our interest is in learning processes involving GIS as an instrument for learning and leading students to become spatial thinkers. We do not claim, however, that GIS should be the only way for geographers to become competent spatial thinkers. There indeed exist and should exist a wide variety of ways and instruments for geographers to foster their spatial abilities, spatial reasoning, spatial cognition, spatial intelligence and spatial concepts making. Here, we just mention one other central instrument: maps that have been extensively discussed in MacEachren (2004). We are, nevertheless, in this paper, only interested in the relation between GIS and spatial thinking.

Using an instrument in learning processes affects more than the technical expansion it involves—it affects not only *how* we learn but also potentially *what* we learn (Verbeek, 2000). Important questions in relation to GIS and spatial thinking are the following: How do students learn to see through the instrument of GIS and understand the spatiality it represents? and Does teaching practice influence this learning process? These are issues addressed in this paper. Furthermore, we hope to initiate a debate on how GIS may alter the ways we as geographers think spatially because using an instrument is not neutral, but influences the way students perceive and understand geographical issues. Thereby it may in the long run potentially modify the professional outlook of the discipline. The paper contributes to the literature by addressing the issue of learning through an instrument—an issue not often dealt with in the literature on GIS and educational issues. Furthermore, the paper links theoretical discussions of how to study learning processes when using an instrument from other disciplines in the field of geography and thereby build a bridge between educational traditions within different areas of research.

The paper is organized as follows: first, we outline our notion of spatial thinking and GIS. Second, on the basis of the philosophy of technology, we unfold how GIS can be interpreted as an instrument for learning and outline the instrumental genesis approach to studying students' learning processes when using instruments. The main part of the paper then is a case study of how students are introduced to and learn GIS at an introductory course at the University of Copenhagen. Through the analysis, we reflect upon the students' learning strategies for creating a personal instrument, how the style of teaching modifies the creation of the personal instrument, and how it relates to the students' different learning strategies. On the basis of the case study, we further reflect upon how the use of GIS in learning to think spatially might influence students' spatial skills and hence the professional outlook of geography. Finally, we discuss the research implications of our study.

Spatial Thinking and GIS

Through time, geographers have used a variety of artefacts such as maps, globes and planimeters as professional tools for spatial thinking. Each of these artefacts has in one way or the other shaped our ways of being and becoming geographers. They have shaped not only how we learn but also what we learn. New types of artefacts are constantly introduced into the profession and some have more profound and persistent effects than others. One example of this is the introduction of GIS, which has its origin in the 1960s with the Canadian Geographic Information system developed for mapping purposes (Goodchild, 1995). Several definitions of GIS exist which have evolved in different fields and disciplines, from the toolbox-oriented definition given in Burrough (1986) to definitions concerned with the interpretation of GIS such as Kemp and Goodchild (1991) and Kemp *et al.* (1992) that focus on a distinction between learning about GIS and learning with GIS.

Although many definitions exist, most in one way or other identify GIS as a tool for collecting, managing, analysing and presenting spatial data. This is also true for the definition provided recently by the National Research Council:

> Geographical Information Systems (GIS) is an integrated system of hardware, software, and procedures designed to support the collection, management, manipulation, analysis, modelling, and display of spatially referenced data about Earth's surface in order to solve complex planning and management problems (National Research Council, 2006, p. 18).

The use of GIS has increased in research and in education at geography departments worldwide in the last 30 years, and in recent years this process has accelerated. GIS had become more and more integrated into the field of geography, and this has given rise to discussions of whether GIS should be admitted as part of the intellectual core of geography (Sui, 1995; Drennon, 2005). However, these discussions have largely been bypassed by the real world integration of GIS in geography curricula and by pressure from job markets (Kemp *et al.*, 1992).

There is no clear consensus about spatial thinking, and many related concepts are in use (Lee & Bednarz, 2009). In practice within the field of geography, spatial thinking in one form or other underlies a significant amount of geographic learning such as the use of maps, graphs, images, diagrams, models and visualizations (Bednarz, 2007). According to *Learning to Think Spatially*, there are three types of spatial thinking: thinking in space, thinking about space and thinking with space (National Research Council, 2006). Thinking in space is used to perform actions and requires thinking in a real-world context, such as taking the bus to university, walking the dog or even packing a suitcase. Thinking about space is focused on a scientific understanding of the nature, structure and function of phenomena; for example the structure of the atom or the structure of the Earth. Thinking about space involves thinking about the ways in which the 'world' works. Thinking with space is the most abstract form of spatial thinking and deals with the conversion of some data relationship between objects into locations and, therefore, arrangements of the objects in a space (e.g. population pyramid). These three types of spatial thinking are highly interwoven. Thinking in space for instance promotes thinking about space, whereas the use of space as a cognitive strategy adds power and aids in thinking in and about space (Bednarz, 2007).

GIS have in various forms been linked to spatial thinking as discussed in the introduction. If we follow the definition of spatial thinking as mentioned above, the activity of learning GIS involves and develops all three types of spatial thinking. However, the focus is most often the abstract form of spatial thinking, which is to teach students to think with space. An example of how spatial thinking unfolds in relation to GIS is given in Figure 1. To undertake, read and understand the analysis in Figure 1, you need to be a spatial thinker. Being familiar with the rules of spatial thinking, you will be able to read the map and understand the analysis, but without these skills you may just see a picture with a circle and names on it. Furthermore, to analyse, communicate and judge the value of this, you must also use your spatial skills.

GIS seen as an Instrument for becoming a Spatial Thinker

An artefact such as GIS is not neutral. By being an instrument, it shapes our perception of the world. How and with what effect this shaping takes place is not explicit or shared common knowledge, but an integrated part of a geographer's experience of being a geographer. Säljö (2003) describes how both our thinking and our actions are closely related to the specific artefact and are deeply affected by the built-in resources of this artefact. He gives the example of the pocket diary:

> We think in symbiosis with the pocket diary. We exchange meaning, content and sense between the user and the artifact in relation to reading and writing. We can metaphorically speaking park meaningful statements like information in that type of artefacts and call them to life again, when we need them (Säljö, 2003, p. 80, own translation).

He further points out that the technology behind common modern artefacts is often invisible and in fact not understandable by the user. Seen in this perspective, learning is about becoming able to use the cognitive resources as information, procedures and routines, which are implicit in the artefact (Säljö, 2003).

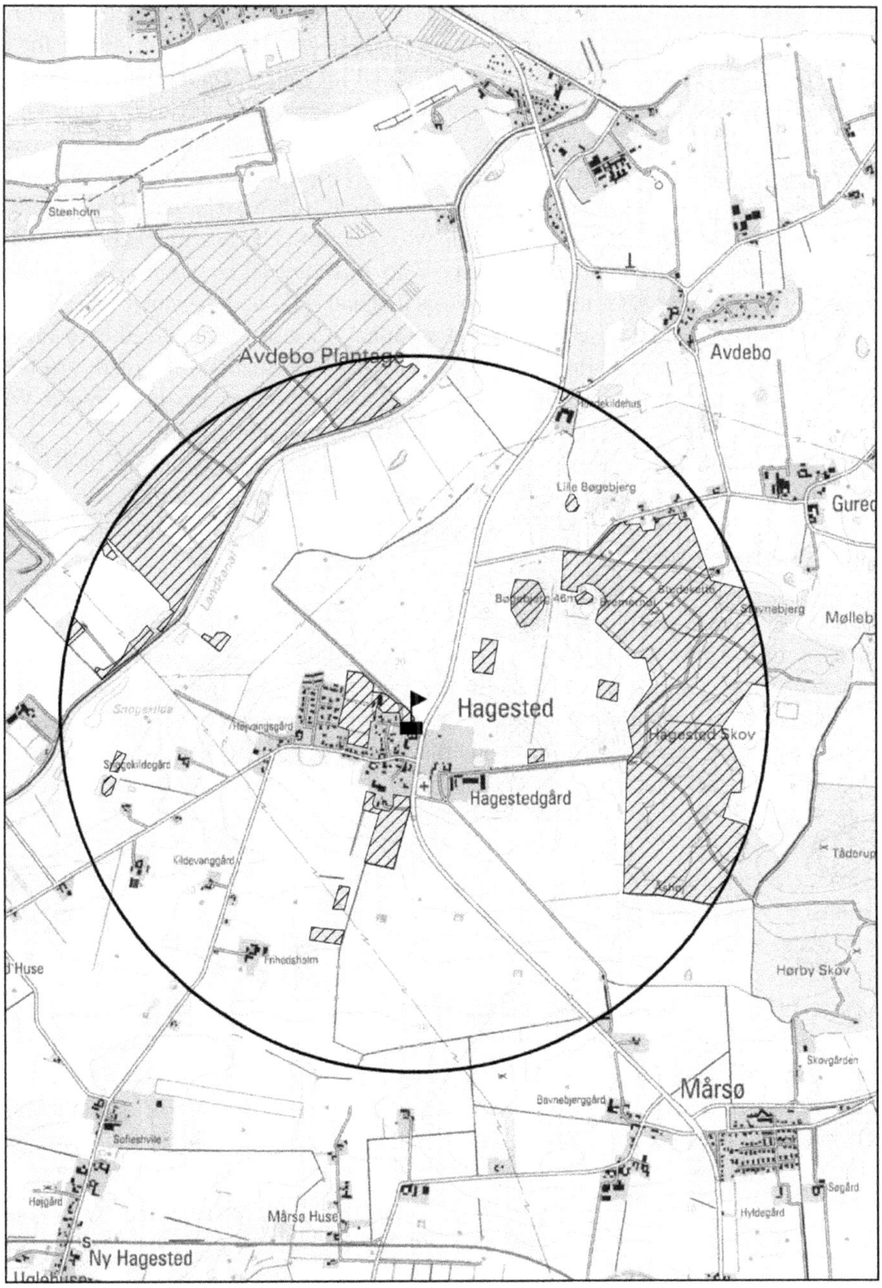

Figure 1. Spatial thinking within a GIS. Within Figure 1 is a digitalized 1:25.000 map of a rural area in Denmark. A point of interest is highlighted with a flag (a school) and the result of an analysis of the area of woodlands within a 1500 m radius from the school is shown. This analysis is made with the buffer-function within the GIS. This kind of analysis could be a result of the following exercise given to the students to train them in spatial thinking: Find the location and distance to the woodlands that is within reach of a bicycle-ride from the school of Hagested.

The relation between knowledge and scientific instruments has been an issue in the philosophy of technology and to some extent also the philosophy of science during the past 30 years in what Ihde (1991, p. 45) terms "the 'school' of instrumental realists". He argues that "Instruments form the conditions for and are the mediators of much, if not all, current scientific knowledge. They are the concrete and material operators within scientific praxis." Instruments are thus an important part of scientific knowledge and hence, the education of future scientists. It should be noted that the philosophical stance of the instrumental realists is a particular form of *realism* in the philosophy of science. The instrumental realists hold that scientific instruments can provide access to *phenomena in the world* which are otherwise hidden from us. They are entity realists in that they hold that even if 'observations' from instruments cannot be observed in the traditional sense with the naked eye, and thus potentially can be misleading artefacts produced by the instruments or be mere theoretical entities, the 'observed' objects are regularly *manipulated* to produce new phenomena. When they are so, they argue, we must accept their existence (Ihde, 1991, part V).

A 'seeing as' metaphor is strongly present in Ihde's discussions of the role of instruments in science. Indeed, the notion of seeing is not merely a metaphor to Ihde; indeed, he argues that scientific observation is perception through the so-called embodiment of instruments:

> Instruments extend the human senses, and in the context of early modern science, made its first distinctive embodiment as a thoroughly technologically embodied science possible. Galileo's use of the difficult and primitive telescope provided him with an 'artificial revelation' of phenomena never before seen nor suspected. Along with its use of a unique form of mathematization, early modern science was to henceforth differ from all its predecessors by being instrumentally embodied in technologies (Ihde, 1991, p. 74).

Ihde further argues that with some types of technology our relation is not to be seen as extensions of sensory—bodily capacities but, rather, as extensions of linguistic and interpretive capacities. He calls these *hermeneutic relations*:

> In embodiment relations the technology—instrument—is taken into perceptual and bodily experience [...]. The technology becomes a 'part' of my now extended bodily experience of the world. But in hermeneutic relations, the technology is much more 'text-like' and retains a sense of quasi-otherness not present in embodiment relations. The hermeneutic 'transparence' is thus like that of textual transparency (Ihde, 1991, p. 75).

To illustrate this, Ihde gives the example of computer printed data or a map as instruments with which we have hermeneutic relations. Hence, the GIS instrument will work for a geographer as a text works for literate persons. It is not the text which is the focus, but the content that is 'seen through' the text. In other words, you need to be able to look through the instrument (GIS) to be a competent spatial thinker.

Seen in this perspective, learning is about obtaining the necessary skills to become a 'trained scientific hermeneut' (Ihde, 1991, p. 131). It entails using the instrument to both 'write text'—to produce data, to 'read text'—in the form of output from the instrument, and in all other respects to learn to handle the instrument manually and intellectually in the scientific process. To understand how students learn to think spatially by using GIS, we, therefore, need to address the issue of the artefact as an instrument and its influence on learning.

How to look into Learning to think Spatially with GIS

To understand the process of how GIS become hermeneutic transparent for the student, we must first understand what happens in the process. Ihde (1991) uses Martin Heidegger's example of the use of a hammer:

> So long as one is using the hammer within its work context, the hammer-as-object 'withdraws' and the context of relations of the work project may remain both transparent and familiar but tacit. But what happens if the hammer is missing, or is broken, or ceases to function? Its phenomenological transparency is then changed, it becomes opaque, and the hammer may become an 'object'. That is to say, as a result of withdrawal, the hammer, even if missing, becomes not the means of achieving the work but an obstacle to its attainment (Ihde, 1991, p. 54).

As GIS as an instrument is complex and far from as straightforward in use as a hammer, the students' first encounter with GIS is comparable to the broken hammer. The GIS instrument is opaque; it is the instrument itself rather than the spatial content that is in focus. Through a hermeneutical process, or learning process, the instrument is then to become transparent. This has been described by Heelan (1972):

> Let us suppose that marks are discovered on clay tablets known to belong to an ancient but as yet undeciphered language. These marks are at first studied and compared as objects: the subject-object cut places them on the object side of the cut. Later, when fully deciphered, they are recognized as signs belonging to a natural or artificial language. When students of this language are sufficiently familiar with it, they can begin to read the signs for their meaning .... The signs, then, have come to be displaced to the subject side of the cut (Heelan, 1972, pp. 104–105).

Through education and through the continuous life-long learning of the researcher, the scientific object is then observed and studied 'directly' through the scientific instruments. The instrument becomes part of the subject because the object studied is not the scientific instrument, but the scientific object in focus. Heelan (1972) terms the process of the instrument shifting from the object side to the subject side of the subject–object cut a *hermeneutical shift*. In relation to GIS, we as professional GIS users re-experience this hermeneutical shift every time a new version is on the market; for instance the version ArcView 9.1, which had extensive changes. In such situations, we experience that the instrument is no longer transparent, and our attention is re-focused on the keys and functions that are now different. We become annoyed with having to spend time on finding out how to do the analysis we want to do—we are blinded by the instrument just like the students in their first encounter with GIS.

The theory of instrumental genesis, which has been developed within the field of educational research in mathematics (Guin & Trouche, 1999; Trouche, 2004, 2005; Guin *et al.*, 2005; Monaghan, 2005), can be used to study the hermeneutical shift in learning GIS in greater detail. The idea of instrumental genesis can be described as a process where an artefact such as GIS shifts from being an object/tool to being an instrument for the student, and the student then experience the hermeneutic shift of Heelan (1972). The process is complex and is: 'Linked to characteristics of the artefact (its potentialities and constraints) and to the subject's activity, her/his knowledge and former work methods' (Trouche, 2005, p. 144).

Within the instrument, the scholarly knowledge and the knowledge of how to use the artefact merges together. The GIS are the artefact, a material tool, while the instrument is a personal construction which develops continually, even for very experienced users (Winsløw, 2005). In relation to GIS, the idea of instrumental genesis can be described as a process, in which GIS as a physical artefact gradually becomes transparent and a personal instrument for the student.

According to the theory of instrumental genesis, the relationship between an artefact and a subject develops through two opposite processes of instrumentalization and instrumentation (Figure 2). Instrumentalization is directed towards the artefact, i.e. the subject forms usage schemes related to the artefact (artefact on the object side, opaque), whereas the instrumentation is directed from the artefact towards the subject, i.e. the student obtains experience from usage of it, which results in modification or creation of new personal schemes beyond the artefact (artefact on the subject side, transparent)—for instance, modified or new knowledge about geography. Instrumental genesis is the process in which the subject, through instrumentalization and instrumentation, builds an instrument to perform certain types of tasks (Trouche, 2005).

From the field of mathematical educational research, we know that teaching and teaching materials can, to some extent, support the development of instruments among students. A wide range of research studies have shown that the introduction of Computer Algebra Systems (CAS) in mathematics teaching changes not only the way students learn but also what they learn. For example, using CAS to differentiate $f(x) = x \sin x$ involves a totally different set of mathematical actions from what is required to differentiate this function using the standard paper-and-pencil algorithm (Monaghan, 2005). If we apply this knowledge to GIS and students' learning of spatial thinking, what is important is not

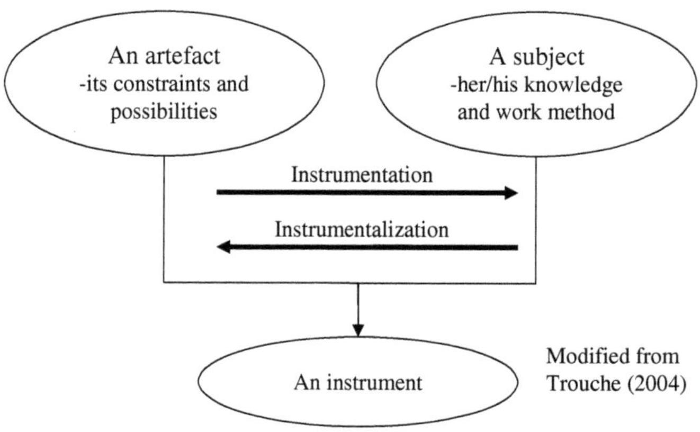

Figure 2. Creating an instrument from artefact to instrument.

that the required actions are different from spatial thinking without GIS, but to understand what the use of GIS means both for the students' personal learning process and the tasks of teachers.

An Empirical Study of Students' Development as Spatial Thinkers

In this section, we empirically show which strategies students use to learn GIS, and further, how a certain teaching practice influences the students' process of becoming spatial thinkers. This is done by analysing the group of first-year geography students at the Department of Geography and Geology, University of Copenhagen, attending the course: GIS and Cartography.

Copenhagen University is a research intensive University with approximately 35 000 students. The geography department is located at the Faculty of Science and is one out of three places you can study geography at university level in Denmark; GIS were introduced into the curriculum in the late 1980s. A course in 'Introduction to GIS' first served as a non-compulsory, technically oriented course at masters' level. Through the years, it has gradually been moved to bachelors' level. Finally, in 2002 it was placed as a mandatory introductory course for all first-year geography students at the University. The course (now named 'GIS and Cartography') gives the students an introduction and first-hand experience with GIS and is supplemented by six elective GIS courses placed later in the geography curriculum. This change also reflects the increasingly important role GIS play in the international and national communities of geographers.

The students enter the study of geography with a broad range of educational backgrounds and there are requirements of mathematical skills. All of the students have been trained in spatial thinking through their high school; however, only very few have specific experience with GIS when they enter university. A recent study has shown that the students are remarkably identical when it comes to economic status, ethnicity and family educational background, as well as region of origin (Lorenzen, 2010). Other aspects, such as the students' different IT skills, may have influenced the students' strategies. This has not been systematically tested in the present study, but could be so in the future to explore further why some students prefer certain strategies when approaching GIS.

Material and Methods

'GIS and cartography' is attended by just under 100 first-year undergraduate geography students each year. In lectures, central elements of GIS and cartography are introduced. The focus is on how different abstractions of the world can be represented within GIS through the use of different data models. This is combined with hands-on laboratory classes in which various themes of human and physical geography are treated. At the end of the course, students' theoretical knowledge of GIS and cartography is assessed in a written exam.

The course runs for 9 weeks and consists of 2 weekly lectures (each of 45 min) plus 2 weekly, 2-h computer laboratory classes in ArcGIS, version 9.1 (students are divided into four groups of separate laboratory classes). The laboratory classes, which are given by the lecturers of the course, start by introducing the day's topic. Afterwards the students work in pairs with the GIS using a 'manual' containing practical instructions and exercises. The lecturer is present throughout the classes and helps solve problems. The students also use

GIS in their parallel introductory course in human and physical geography as well as in several later courses in both human and physical geography. In these parallel and later classes the teaching and learning of GIS tends to be more and more project oriented.

In the academic year 2004–2005, 79 students were enrolled in the course in GIS and cartography. To study the students' learning experience of GIS, we collected and analysed the following material concerning this particular group of students and their experiences of learning GIS:

- Classroom observations (in 25 laboratory classes and 3 lectures).
- Qualitative interviews with teachers in 'GIS and cartography' (three interviews).
- Qualitative single and group interviews with students (six interviews with a total of nine students, during the course and eight interviews with a total of 15 students, after examination).
- Questionnaire (49 students answered the scheme equal to a response rate of 62 per cent: 27 women and 22 men. Not all students answered all questions, so exact number of answers are given in each figure shown). The questions given in the questionnaire were based on classroom observations and interviews with students as well as interviews with the teachers. It consists of both open and closed questions concerning the students' degree of participation, their experiences within the different teaching formats (laboratories, lectures and assignments), their degree of agreement with a number of statements concerning their learning practice, whose formulation was based on information from the qualitative interviews and open questions inviting them to describe their learning process. The questionnaire was electronic and handled through the university's administrative system. Mails informing about the questionnaire were sent on the day of the exam and flyers were handed out when students left the exam room. Reminders were given and the questionnaire was open for input for a 3-week period.

The methodology employed in this research is within the tradition of in-depth ethnographic work within cultural geography with a focus on qualitative and discursive research methods (e.g. Crang, 1998; Hoggart *et al.*, 2002; Pryke *et al.*, 2003). It has clear relations with phenomenographic studies as it focuses on the experiences of the students (e.g. Marton & Säljö, 1976a, 1976b). But in contrast to phenomenographic studies, our study does not have an outcome space consisting of categories of description which are logically and/or hierarchically interrelated.

The outcome space of the analysis in this paper is a typology of individuals' conceptions of their learning process in creating a personal instrument. These conceptions can vary over time and also entertain distinct conceptualizations at the same time in distinct learning situations. The developed typology is, therefore, situated within the concrete teaching and learning context, although students may come to class with individual preferences for specific approaches to learning. These preferences may or may not be realized in practice, depending on the teaching context. This is to say that we, as do Biggs and Tang (2007), take a middle position in the discussion of learning styles, in which students are either assumed to always prefer the same approach or learning approaches in which students' approaches are taken to vary fully depending on the context. The way students shape their personal instrument through their learning strategies we see as flexible and constantly under construction and re-construction by the individual student in relation to the teaching practice she/he meets, both in the different courses and at the university as a whole.

To take into account this dynamic nature of learning, the typology was developed on the basis of the students' answers to an open question in the questionnaire rather than pre-classified types asked for in a questionnaire. The question was 'How do you go about learning GIS? Please describe what you actually do, rather than what you think you ought to do.' In structuring and analysing the data, all written answers to this question were read and re-read in a process of going back and forward between the written answers and the analytical question: what learning strategies do the students use to learn GIS. This process resulted in the developed typology. In the process of constructing the typology, information from the interviews with students and classroom observations was used to shape the identified types of learning strategies. This method of analysis resembles thematic analysis described by Braun and Clarke (2006).

The Students' Learning Strategies for creating their Personal Instrument

We found that the students used five different strategies in their learning process to develop their personal instrument during the 'Introduction to GIS and cartography' course. The described strategies are the students' preferred strategies used in this specific learning situation and are shown in Figure 3.

Three learning strategies dominate the picture, namely types A, B and C, whereas types D and E were rare. Type A describes students who prioritize the task, i.e. they go through all the literature, exercises and lectures; type B refers to students who prioritize understanding in their approach to GIS and type C describes students who prefer to 'play around' with the GIS to learn. Type D students indicate that they learn GIS when they use it in the adjacent course in human and physical geography. Finally, type E students question whether they have learned GIS at all.

Examples of the three main learning strategies:

> Type A student: "I attended the lectures, wrote notes and printed out the hand-outs. Before the final exam I read all the assignments including the hand-outs. I made notes so I had an overview of the different concepts and expressions. I wrote down where in the assignments the different concepts were defined. I marked the text and made tables of content to the texts for which we did not receive tables of content.

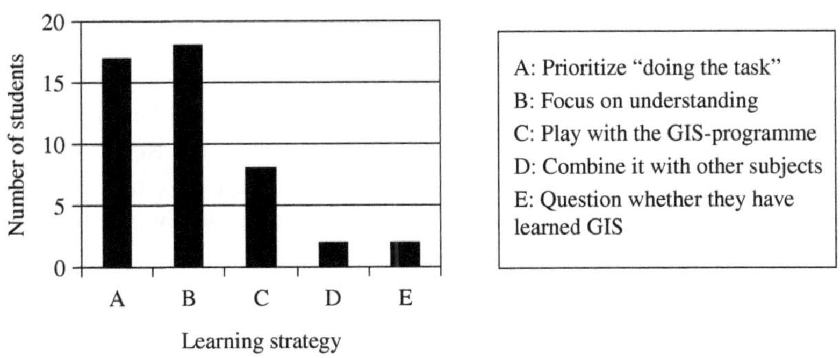

Figure 3. Students' learning strategies for creating a personal instrument ($N = 47$, three students have given two strategies, both strategies are included).

I put the texts and notes in a ring-binder organized in accordance with the lectures. I attended all the classroom exercises and the ones that we didn't finish in class we made sure we finished later."

Type B student: "I went to both the lectures and the labs. In the labs I focused on understanding what we where doing and not just going along (and finishing the exercise). I read all the course assignments, but only once.... I sat down in the exam preparation period and discussed different topics with the other students."

Type C student: "I sat many hours with ArcMap/ArcCatalog (the computer program ArcGIS) both at home and in the labs. This has given me a pretty good overview to understand what the lectures were about."

These learning strategies show that the students have different ways of developing their personal instrument and, as will be discussed later, the form of teaching in use influences this development in different ways. Type A students can be said to have a focus on intrumentalization rather than instrumentation, type B students have focus vice versa that is a focus on instrumentation rather than instrumentalization, whereas the focus of the C students may vary.

The Influence of Teaching on Students' Route to becoming Spatial Thinkers

Within the theory of instrumental genesis, it is a central idea that teaching and teaching material can, to some extent, support the students' development of personal instruments (e.g. Guin & Trouche, 1999). The external support or direction of students' instrumental genesis is called orchestration. Different types of orchestration have been described and developed in the context of CAS (Trouche, 2004).

In our study, it was evident that the type of orchestration—in this case the layout of the course and the teaching method used—had implications for the learning process. As mentioned, the experimental part of the course is organized around 2 weekly, 2-h computer laboratory classes, which are the focus in the following. The laboratory classes use a 'manual' containing practical instructions and exercises. Especially the orchestration presented by the manual had profound effects on the students' instrumental genesis. The manual can be described as a cookbook with recipe-like text that must be followed by the students to do the assignments. The following is a typical quote from the exercise manual:

> Choose Spatial Analyst → Reclassify, put Input-raster = dist_water, press Classify, press the Method-arrow and choose Equal Interval, press the Classes-arrow and enter 8 and press OK. Press the first New Values record in the Reclassify-window and replace the value 1 by 8, the value 2 by 7 and so on. Leave the field No Data untouched. Press OK and see the result which is placed in a new grid file with the name Reclass of dist_water.

This kind of manual must be followed closely to obtain a successful outcome, and does not encourage the students to explore the software as also found by Meitner *et al.* (2005). Thus, it does not support the students in constructing their own perception of the spatial issues dealt with in the specific exercise. Rather it directs the student's attention towards

GIS as an object, and hence it may not be very helpful for the students in the instrumentation process.

The orchestration created different forms of work among the students. Some students just followed the instructions and did not link their activities with their theoretical knowledge of spatial issues at first. One of the students expresses it this way in the qualitative interview:

> It is just so schematically and pedagogically put together that you are able to put yourself on autopilot—that's at least how we have experienced it—and we sit and are almost—not falling asleep, but you get very tired in your head because—and now it is very harshly said—but you just put yourself on autopilot and just do [presses with the fingers on the table] what is said in the papers [exercise manual].

Other students reported that they actively opposed this type of practice and tried to link their interaction in the exercise to their theoretical understanding, as expressed by another student:

> I often stop during the GIS-exercise and think through what I just did on the computer screen.

On the basis of the students' answers in the questionnaire, we found that these different types of responses to the teaching orchestration were related in a complex pattern to the learning strategies of the students as elaborated as follows. It appears that students with different learning strategies tend to react differently to the following two quotes given in the questionnaire: "It is important for me that I am forced to think when I do the GIS-exercises" and "I often stop during the GIS-exercise and think through what I just did on the computer screen."

A majority of students consider it important that they are forced to think when they work with the GIS exercises. This tendency is strongest among the type A students (who prioritize 'doing the task') where 65 per cent of the students agree with the quote. For type B students (who prioritize understanding), 42 per cent agree that it is important, whereas 25 per cent of Type C students (who prefer playing with the GIS) agree. These numbers indicate that A students who do not prioritize understanding as high as B students find it more important to be forced to think when they work. This indicates that A students despite prioritizing getting the work done have a high awareness of the benefits of reflecting through the exercise. Their request of being forced to think during the GIS exercises can be linked to the fact that they do not by themselves practise this as revealed by their answers to the second quote. Here, only 12 per cent of type A students agree that they often stop during the GIS exercise and think through what they just did on the computer. Instead, type A students stick with the cookbook 'recipe' to get the job done. Otherwise, type B students (those who have an understanding oriented approach) do not to the same degree find it important to be forced to think during the GIS exercises. This can be explained by the fact that more of these students practise by themselves: 37 per cent of type B students agree with the second quote that they often stop and think through what they just did in the exercise. This tells us that type B students tend to oppose the teaching orchestration with the manual: they actively stop to reflect during the exercise. Perhaps, surprising at first, type C students (who play with the GIS as a strategy of learning) seem to adapt to the orchestration by running through the manual like a cookbook recipe (13 per cent of type C students agree on the second quote and 25 per cent find it important to be forced to think during the GIS exercise). But when asked how they have learned GIS, they tell us about situations in which they have 'played' with the GIS after class. From classroom observations and interviews, we get the impression that some of the type C students use the manual quite systematically just as type A students, but at the same time they perceive this as a kind of pre-understanding. Later on, they go home and play with the computer program to gain a deeper understanding. It should be noted that all students have the opportunity to freely install the GIS software at their own computer and due to the Danish free education system and state-provided financial support, all students should be able to have computer facilities at home.

In relation to developing the students' spatial thinking, the orchestration with the manual further seems to focus students' attention onto the computer interface in classroom, thereby hindering discussions of geographical issues related to spatial thinking. By observing four laboratory classes, we found that 81 per cent (total number of questions: 105) of student questions directed at the instructor dealt with technical issues of the software or imprecise guidance in the manual (the instrumentalization process). For example:

> Why can I not save my data?
> (tries to save on the G-drive instead of the I-drive)
>
> How do I install the A/L button?
> (the area and length facility, A/L is not standard and must be installed separately).

Only 19 per cent of the questions dealt with conceptual questions regarding spatial perception (thus supporting the instrumentation process). These are questions such as

> I can't make a scale and north arrow, why?
> (It is not possible because the students are working in the system 34, a Danish technical version. A discussion begins on different representation systems).

On the basis of classroom observations and registration of dialogue between the teacher and the students, we found that the students' work with the manual redirecting the students' attention towards the computer interface, and hence the artefact as an object. The discussions between the students and the instructor are mainly concerned with the instrumentalization process, whereas the construction of GIS as a personal instrument through the instrumentation process is vaguely supported. Furthermore, the interaction with the manual draws attention away from the geographical issues that the students are supposed to see 'through' the GIS. In this way, the style of teaching affects the learning strategy of the students as they are urged to use the cookbook recipe strategy and as attention is constantly shifted to the GIS instrument itself, instead of the content that can be seen through it. The use of the manual thus seems to affect the students' instrumental genesis in a partly unconstructive way. In their learning strategy, type A students follow the orchestrated teaching and are thereby left with a low chance of creating a strong personal instrument. Type B and C students, on the other hand, have different strategies for dealing with the teaching orchestration to develop their personal instrument: type B students stop during their work with the manual and reflect on what they are doing in class while type C students play with the GIS at home and create their personal instrument outside the teaching environment.

These last two types of learning strategies result in personal instruments for the student, despite the teaching orchestration.

In the data, the following gender differences were found (Figure 4). Slightly more men than women prioritize 'doing the task' (type A). More distinctly, a larger number of women than men seek understanding (type B), whereas more men than women play with the ArcGIS program (type C) to learn GIS. If we understand type C as a 'mastering' strategy or competence strategy in which the students have a somewhat independent reaction to the teaching environment, we can see that the two types of students who oppose the teaching orchestration (type B and C) fall into two categories with the female students being more compliant within the given teaching environment and the male students being more autonomous in their learning strategy. This is similar to results of gender differences found within physics education (Hasse, 2002), in which male students are found to be more prone towards 'playing' with the laboratory settings beyond the manual prescriptions, whereas the female students stick more consciously to the manual.

Reflections on the use of GIS and the Professional Outlook of Geographers

In this section, we argue that the now widespread use of GIS in the education of future geographers has created a need for reflection within the discipline. We recognize that the map (paper or digital) plays a central role in spatial thinking and has and still have a major

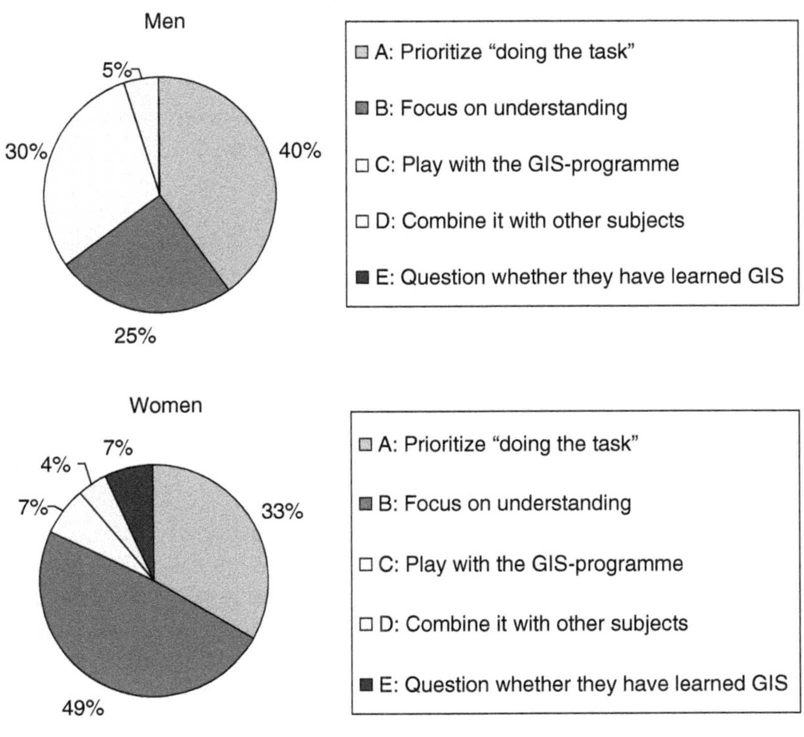

Figure 4. Students' learning strategies—men and women ($N = 47$, three students have given two strategies, both strategies are included).

influence on how we think, apply, act and define ourselves as geographers. Our claim is that using GIS to learn to think spatially is a specific way of seeing the world that further modifies and transforms the professional outlook of spatial thinking for both students and teachers/researchers and, therefore, needs attention.

First, we need to define what we mean by professional outlook. Kuhn (1970) suggests that knowledge of the rules of a discipline and of important exemplars provides a person with an outlook and an ability to act within that discipline. On the basis of this, Christiansen (2004, p. 19) defines professional outlook within a discipline as "a personal way of seeing the phenomena and situations of relevance to the profession." Hence, professional outlooks are outlooks from a particular place or perhaps from within a certain discourse of what it means to be an educated person within a particular profession. Christiansen further describes student's learning: "Understanding the fundamental rules of a scientific discipline may provide [. . .] an optic or outlook through which hypothetical situations may be envisioned by means of the rules of the discipline and the student's imagination" (Christiansen, 2004, p. 24). In this perspective, he further finds that the purpose of higher education is a matter of providing the students with professional outlooks in order for them to learn a process of discernment, and teaching becomes a question of helping students to learn to see 'through the eyes of the discipline'.

The professional outlook of a discipline is continuously contested, recreated and developed, and in these processes new instruments such as GIS play important roles. As we have shown in the analysis, students have different strategies for creating a personal instrument of spatial thinking when they are learning GIS. We have also argued in the paper that the use of instruments is not neutral and it influences the way things are perceived and understood. However, to show empirically how GIS alter the way we as geographers think spatially, and thereby transform and modify the professional outlook of geographers, is not an easy task. However, by letting the students elaborate on how they use GIS, some insights can be made. In the following quote, a student reflects on how she learned a particular task within the GIS:

> You have a road and must calculate the area that will be affected by noise in a zone 200 meters from the road. Then I think in concrete GIS: select residence and . . . then these words and that way to do it, and the structure you have trained through the laboratory classes . . . . The way you have learned to think is that you choose some things and then they must be selected, overlaid or . . . make buffers. Buffer gives sense because . . . if you just think buffer is the straight distance from a point, then it makes no sense . . . you see that picture of a circle around a point for instance and then it makes sense to a start and then you can use the theory to explain why it looks like it does.

This answer indicates that the interaction with the GIS influences the student's knowledge of spatial configuration. She thinks in terms of the functions within the GIS when she creates spatial meaning. We interpret this as an indication of a change of the professional outlook created when educating geographers using GIS. In this example, the GIS interact with geography knowledge in the process of instrumental genesis and influence spatial thinking. Yet, we need more longitudinal observational studies of the students' interaction with GIS to understand what it means for students to use GIS in a wider sense for their thinking and understanding of geography and further, to inform our knowledge of how spatial thinking and thereby the professional outlook of geographers is modified and transformed.

Conclusion

The professional outlook of a discipline—perceived as the ability to act within a discipline—is continuously created, recreated and in these processes contested and changed. In this development, instruments for doing the profession such as GIS in the field of geography play an important role because they are not neutral agents. GIS are a physical artefact that gradually becomes transparent and a personal instrument for the learner through the process of instrumental genesis. With the personal instrument, the scholarly knowledge (geography) and the knowledge of how to use the artefact (GIS) merges together. The use of GIS in geography will inevitably change the way we as geographers think spatially (in this paper illustrated by the student quote about solving a particular task). As GIS have become more and more widespread in the education of geographers, we, therefore, need to address not only the skills of students as spatial thinkers, but also the learning processes leading students to become spatial thinkers.

In a small empirical study of how students learn GIS at the University of Copenhagen, we have shown that the use of GIS is not neutral and it influences the way students perceive and understand geographical issues. Also we have shown that students have different strategies for creating their personal instrument for spatial thinking when they are learning GIS. Importantly, it was further found that the chosen teaching orchestration with a cookbook-like manual had different effects on the learning strategies of the students in their process of creating personal instruments. Type A students follow the orchestrated teaching and are left with a low chance of creating a strong personal instrument because they follow the cookbook recipe quite meticulously, causing their attention to be constantly refocused to the GIS instrument itself, instead of the result of the geographical analysis performed. Type B and C students, on the other hand, have their own strategies for dealing with the teaching orchestration to develop their personal instrument: type B students stop and reflect during their work with the manual, and type C students play with GIS at home and create their personal instrument outside the teaching environment. These last two types of learning strategies can be interpreted as a case of the students creating their personal instruments and becoming spatial thinkers despite the teaching orchestration.

The implications of these findings are threefold. First, in relation to education it is important to support the students in their process of making GIS transparent and thereby a personal instrument for spatial thinking. To understand these learning processes, we find that the theory of instrumental genesis developed in the field of mathematical educational research can be a valuable research framework. The theory has proven valuable for understanding students' use of calculators to solve mathematical problems and given rise to discussions of teaching practice within the mathematical community. We find that the theory could contribute equally to a research focus on the learning processes among students when learning GIS and give valuable findings for our geographical teaching practice.

Second, from our small empirical study, it seems that the rather 'traditional' cookbook manual used in this course is not the best way to support the learning processes of students in their first meeting with GIS. Often it is argued in favour of the use of a cookbook manual that students must have a basic knowledge of GIS before they can manage to do more project-based work. However, our results show that some students actually learn without the manual using other strategies to comprehend and develop their spatial thinking skills. The students who follow the manual meticulously state that they are almost on autopilot in the learning process and were surprised how little they could remember when we tested them (Madsen & Holm, 2006). These results are further supported by another study in which we interviewed the students later on in their study in another GIS course. Many students regardless of their former learning strategy stated that their GIS skills were relatively low entering the later course (Madsen *et al.*, 2009). We found that the cookbook style manual seems to focus the students' attention to the GIS tool itself, leaving reflections of the geography analysis untouched. Yet, one could argue that this is an introductory course in which students first have to learn basic technical skills and, therefore, need to focus on the GIS instrument; later on they can handle learning without a cookbook-like manual. This kind of argument can be based on the distinction in the literature between learning about GIS (deals with geographical data handling or geographical information management) and learning with GIS (seeks geographic knowledge and geographic intelligence) (Kemp & Goodchild, 1991; Rød *et al.*, 2010). This distinction seems to imply that students need to learn about GIS before they can learn with GIS (e.g. Andersland, 2011). Such arguments legitimate using a cookbook manual in introductory courses in GIS. However, the theory of instrumental genesis tells us that this is not how learning occurs, instead mastering a geographical analysis within the GIS is learnt when scholarly knowledge and understanding of how to use the artefact merges together and the latter becomes transparent for the learner. This is a truly complex issue and the results from our small empirical study only give us a hint about the relations. Studies of learning processes with other teaching orchestrations would be valuable. This might include manuals trying to generate geographical thinking by asking students to solve problems by applying what they have just learnt as they come to the end of each 'recipe', or using a combination of questions and a suitable GIS reference book for introductory levels.

Third, on the basis of literature from philosophy of science and by giving an example of how a student uses the terminology of GIS to support him/her in his/her spatial thinking process, we argue how GIS gradually alter the professional outlook of being a geographer and the way geographers think spatially. This constant creation and re-creation of the way we interpret the spatial world linked to the instruments we use needs attention within the discipline. We suggest that the theory of instrumental genesis can give us research ideas of how to address this.

Acknowledgements

First, we thank all the first-year students at the Department of Geography and Geology, University of Copenhagen, in the academic year 2005–2006 for their time and effort in trying to let us understand their learning processes and perceptions of the world of GIS. We thank the team of teachers at the GIS and cartography course for their enthusiasm and willingness to show and discuss their teaching practice with us. This work has been possible through a university teaching development project financed by the Faculty of Science, University of Copenhagen. The entire teaching development project is described in detail in Madsen and Holm (2006). Furthermore, we wish to thank a number of referees for their valuable comments that improved the paper significantly.

References

Albert, W. S. & Golledge, R. G. (1996) The use of spatial cognitive abilities in geographic information systems: The map overlay operation, *Transactions in GIS*, 3(1), pp. 7–21.

Andersland, S. (2011) GIS I geografifaget på ungdomstrinnet, fagdidaktiske perspektiver på å lære om og med GIS, *Doktoravhandlinger ved NTNU*, p. 12.

Bednarz, S. (2007) Mapping the way forward in an uncertain world: Spatial thinking and geography, in: S. Catling & E. Taylor (Eds) *Changing Geographies: Innovative Curricula* (Proceedings of the London Conference) pp. 13–26 (London: Herodot and International Geographical Union, Commission for Geographical Education).

Biggs, J. & Tang, C. (2007) *Teaching for Quality Learning at University*, 3rd edn (New York: Open University Press).

Braun, V. & Clarke, V. (2006) Using thematic analysis in psychology, *Qualitative Research in Psychology*, 3, pp. 77–101.

Burrough, P. A. (1986) *Principles of Geographical Information Systems for Land Resources Assessment* (Oxford: Clarendon Press).

Catling, S. & Taylor, E. (Eds) (2007) *Changing Geographies: Innovative Curricula* (London: Herodot and International Geographical Union, Commission for Geographical Education).

Christiansen, F. V. (2004) Professional outlooks in Science and Engineering, Unpublished PhD-thesis, The Technical University of Denmark.

Crang, M. (1998) *Cultural Geography* (London: Routledge).

Deadman, P., Hall, G. B., Elliot, E. & Dudycha, D. (2000) Interactive GIS instruction using a multimedia classroom, *Journal of Geography in Higher Education*, 24(3), pp. 365–380.

Drennon, C. (2005) Teaching geographic information systems in a problem-based learning environment, *Journal of Geography in Higher Education*, 29(3), pp. 385–402.

Goodchild, M. F. (1995) Geographic information systems and geographic research, in: J. Pickles (Ed.) *Ground Truth: The Social Implications of Geographic Information Systems* (New York: Guilford Press).

Guin, D. & Trouche, L. (1999) The complex process of converting tools into mathematical instruments: The case of calculators, *International Journal of Computers for Mathematical Learning*, 3(3), pp. 195–227.

Guin, D., Ruthven, K. & Trouche, L. (2005) *The Didactical Challenge of Symbolic Calculators: Turning a Computational Device into a Mathematical Instrument* (Dordrecht: Springer Science).

Hall-Wallace, M. K. & McAuliffe, C. M. (2002) Design, implementation, and evaluation of GIS-based learning materials in an introductory geosciences, *Journal of Geoscience Education*, 50(1), pp. 5–14.

Hasse, C. (2002) *Kultur i bevægelse—fra deltagerobservatoin til kulturanalyse—i det fysiske rum* [Culture on the move—from participant observation to cultural analysis—in the physical space] (Frederiksberg: Samfundslitteratur).

Heelan, P. A. (1972) Hermeneutics of experimental science in the context of the life-world, *Philosophia Matematica*, 9(2), pp. 101–144.

Hoggart, K., Lees, L. & Davies, A. (2002) *Researching Human Geography* (New York: Oxford University Press).

Ihde, D. (1991) *Instrumental Realism: The Interface between Philosophy of Science and Philosophy of Technology* (Bloomington, IN: Indiana University Press).

Kemp, K. K. & Goodchild, M. F. (1991) Developing a curriculum in geographic information systems: The national center for geographic information and analysis core project, *Journal of Geography in Higher Education*, 15(2), pp. 123–134.

Kemp, K., Goodchild, M. & Dodson, R. (1992) Teaching GIS in Geography, *The Professional Geographer*, 44(2), pp. 181–191.

Kuhn, T. S. (1970) *The Structure of Scientific Revolutions*, 2nd edn (Chicago, IL: University of Chicago Press).

Lee, J. & Bednarz, R. (2009) Effect of GIS learning on spatial thinking, *Journal of Geography in Higher Education*, 33(2), pp. 183–198.

Lloyd, W. J. (2001) Integrating GIS into the undergraduate learning environment, *Journal of Geography*, 100(5), pp. 158–163.

Lorenzen, P. (2010) Hvem er de nye studenter? Baggrund, interesse & uddannelsesstrategi—en studiestartsundersøgelse på Biologi og Geografi & Geoinformatik ved Det Naturvidenskabelige Fakultet, Københavns Universitet, Unpublished masters thesis, Copenhagen University, Specialerapport, IND's Studenterserie nr. 15 January 2010. Available at http://www.ind.ku.dk/publikationer/studenterserien/studenterserie15/ (accessed January 2010).

MacEachren, A. M. (2004) *How Maps Work, Representation, Visualization, and Design* (New York: Guilford Press).

Madsen, L. M. & Holm, C. (2006) *GIS og Kartografi—et kursus om mere end kort* [GIS and cartography—a course on more than maps] (Copenhagen: Department for Science Education, University of Copenhagen).

Madsen, L. M., Christiansen, F. V. & Rump, C. (2009) Development of personal desiderata while learning GIS, in: A. Bisel & M. U. Garib (Eds) *Proceedings of Frontiers in Science Education Research*, Farmagusta, Cyprus, 22–24 March, pp. 441–450 (Famagusta: Eastern Mediterranean University Press).

Marton, F. & Säljö, R. (1976a) On qualitative differences in learning I—outcomes and process, *British Journal of Educational Psychology*, 46, pp. 4–11.

Marton, F. & Säljö, R. (1976b) On qualitative differences in learning II—outcome as a function of the learner's conception of the task, *British Journal of Educational Psychology*, 46, pp. 115–127.

Meitner, M. J., Gonzales, J., Gandy, R. & Maedel, J. (2005) Critical thinking, knowledge retention and strife: Reflections on active-learning techniques. Paper presented at the ESRI Education Users Conference, San Diego, CA, July.

Monaghan, J. (2005) Computer algebra, instrumentation and the anthropological approach. Paper presented at the 4th CAME Symposium. Available at http://www.lonklab.ac.uk/came/events/came4/ (accessed December 2006).

National Research Council (2006) *Learning to Think Spatially: GIS as a Support System in the K-12 Curriculum* (Washington, DC: National Academies Press).

Pryke, M., Rose, G. & Whatmore, S. (Eds) (2003) *Using Social Theory, Thinking Through Research* (London: Sage Publications).

Rød, J. K., Larsen, W. & Nilsen, E. (2010) Learning geography with GIS: integrating GIS into upper secondary school geography curricular, *Norwegian Journal of Geography*, 64(1), pp. 21–35.

Sui, D. (1995) A pedagogic framework to link GIS to the intellectual core of Geography, *The Journal of Geography*, 94(6), pp. 578–591.

Summerby-Murray, R. (2001) Analysing heritage landscapes with historical GIS: Contributions from problem-based inquiry and constructivist pedagogy, *Journal of Geography in Higher Education*, 25(1), pp. 37–52.

Säljö, R. (2003) *Læring i praksis—et sociokulturelt perspektiv* [Learning in practice—a sociocultural perspective] (Copenhagen: Hans Reitzels).

Trouche, L. (2004) Managing the complexity of human/machine interactions in computerized learning environments: Guiding students' command process through instrumental orchestrations, *International Journal of Computers for Mathematical Learning*, 9(3), pp. 281–307.

Trouche, L. (2005) An instrumental approach to mathematics learning in symbolic calculator environments, in: D. Guin, K. Ruthven & L. Trouche (Eds) *The Didactical Challenges of Symbolic Calculators: Turning a Computational Device into a Mathematical Instrument*, pp. 137–162 (Dordrecht: Springer Science).

Verbeek, P. P. (2000) *What Things Do, Philosophical Reflections on Technology, Agency and Design* (Amsterdam: The Pennsylvania State University Press).

Winsløw, C. (2005) *Didaktiske Elementer—en indføring i naturfagenes og matematikkens didaktik* [Didactic elements—an introduction to the didactics of science and mathematics] (Copenhagen: Center for Naturfagenes Didaktik, Københavns Universitet).

Zerger, A., Bishop, I. D., Escobar, F. & Hunter, G. J. (2002) A self-learning multimedia approach for enriching GIS education, *Journal of Geography in Higher Education*, 26(1), pp. 67–80.

Students individual engagement in GIS

Lene Møller Madsen, Frederik Christiansen and Camilla Rump

Department of Science Education, University of Copenhagen, Østre Voldgade, Copenhagen K DK, Denmark

This paper develops two sets of concepts to theorize why students engage differently in Geographical Information Systems (GIS). These theoretical concepts are used as an analytical lens to explore empirical data on the experiences and engagement of students enrolled in an undergraduate GIS course in planning and management. The analysis shows that both the theoretical perspectives and the custom and didactical contract are important to understand students' engagement in GIS. However, it is the personal desiderata that are the key to understanding the students' different engagement. Further, a temporal dimension and contextual awareness are important in understanding students' engagement in a broader perspective.

Introduction

> I don't get the overview... I don't feel like: GIS can do this and that... I am not using it in the things I'm doing ... You need to really push yourself to use it ... it is easy not to use it (Emilie).

> I think it's extremely satisfying ... that is something that really bothers me about the university studies that you are not allowed to use your creativity. So when you suddenly get the opportunity [with GIS]...to me that's just fantastic (Maria).

The above quotes are from two students both enrolled in a university course on geographical information systems (GIS) in planning and management. They are examples of how students often engage very differently with GIS. We shall address and unfold these and other students' experiences of the specific course in GIS, as well as establish an analytical framework for understanding more generally students' different engagement in GIS. Engagement is a contested concept within educational research. It ranges from a dichotomous view that students are either engaged or not to a multidimensional view of engagement that combines diverse aspects of students' experiences. In between are notions of engagement as a measured quantity, a categorical view where engagement comprises a series of unordered categories, and a naive view where the concept of engagement is not given further scrutiny than as a well-accepted and well-understood concept (Solomonides, Reid, & Petocz, 2012). In this paper we apply a multidimensional view of engagement that: "describe various relationships between the student, study and the institution, including the campus" (Solomonides et al., 2012, p. 16) acknowledging that how engagement is created or sustained is extremely complex, and many different factors influence the development of engagement also in relation to the students perspectives of a future work.

In the research of GIS in higher education, we have within the last 30 years and with good reasons been occupied with finding and exploring new and innovative ways of promoting and teaching GIS. Learning with GIS can provide students with spatial thinking skills, giving them a tool for geographical analysis and enabling them to be critical citizens. Quite a number of papers on these issues have been published with different foci on, for example, teaching methods, learning outcomes, and interacting with communities (e.g., Barcus & Muehlenhaus, 2010; Demirci, 2011; Elwood, 2009; Lee & Bednarz, 2009; Madsen & Nielsen, 2013; Rød, Larsen, & Nilsen, 2010; Unwin, Foote, Tate, & DiBiase, 2011).

This paper is not about how to teach a course in GIS, exploring new innovative fields of teaching or concerned with the learning outcome. Instead it wonders about why it can be that students seem to experience courses in GIS so very differently as the two above quotes show. It does so with the purpose of getting a deeper understanding of students' engagement in GIS during their geography higher education. Why do some students grasp the opportunities provided by GIS, while others decide to avoid future encounters with GIS? It is based on a genuine puzzlement of why students entering a specific undergraduate GIS course at the Department of Geoscience and Natural Resource Management at the University of Copenhagen engage so differently. The GIS course in planning and management is, in terms of pedagogy, well designed. It is a hands-on, problem-based course with the central problem revolving around an authentic planning problem. It allows students to work with an open problem allowing for multiple solutions, in contrast to the often pre-structured step-by-step laboratory exercises used in many GIS courses (Madsen & Nielsen, 2013; West, 2012). Therefore, the course appears exemplary of characteristic uses of GIS in the Danish labor market (Madsen, Christiansen, & Rump, 2009), and in pedagogical organization of student activities aiming at the expected student outcomes. Despite the careful design, students engage very differently in the course.

To address this perplexity we turn to Kuhn and his description of an essential tension between convergent and divergent thinking in the practice of science (Kuhn, 1959). In his few papers on science education, we find arguments for transforming his ideas on the phenomenal world of scientists to the phenomenal world of students. However, in order to address our puzzlement more specifically to the organization of the learning process, we also draw on Balacheff (1999) and his distinction between custom and didactical contract. We develop two sets of theoretical concepts (theoretical perspective and shared and personal desiderata; custom and didactical contract) which are unfolded in the next sections. These concepts are then used in the empirical part of the paper as an analytical lens to unfold the experiences and engagement of students enrolled in the GIS course in planning and management. In this way we aim to address our research question: how can we theorize why students engage so differently in a specific GIS course?

Theoretical concepts to understand students' engagement

Theoretical perspective and shared and personal desiderata

In this section we introduce the first set of theoretical concepts for our analysis based on ideas from Thomas Kuhn's philosophy of science. Using Kuhn's theory may at first seem an odd choice, as Kuhn's main concern was with scientific development and our concern is with student development. We find, however, that Kuhn's discussion of values (Kuhn, 1977) and cognitive model for theory choice (Kuhn, 1959, 1974) have a relatively clear analog in students' choices to pursue or not to pursue GIS. We will return to this analogy below. Further, we find that Kuhn's theory provides a fruitful starting point for studying students' engagement because it provides a unique combination of sociological, cognitive, and phenomenological insights into a relatively coherent theoretical position.

The first theoretical concept that we want to describe is what we have termed *theoretical perspective*. By "theoretical perspective" we refer to paradigmatic ways of conceiving problems and phenomena within the field. In a number of his writings, Kuhn developed a philosophical and cognitive theory that could explain the apparent paradox of how educations that were by and large convergent in nature (Kuhn, 1959) could still lead to tradition-shattering periods of divergence (Kuhn, 1970, 1974, 1977). Kuhn's idea was that the problem-solving activities that take up most of science education provide the students with an arsenal of exemplars, on the basis of which he or she may proceed to new (but similar) problems. However, this "reasoning by exemplars" is *not* a rule-governed activity, but based on fundamental perceptions of similarity or dissimilarity with previously encountered problems. The absence of rules mean that even though students learn to solve problems and classify phenomena (i.e., make judgments that Y is an instance of X) in fairly much the same way, they do not do so because of a mutual agreement on the necessary and sufficient conditions for X membership. This means that in "normal" situations, the scientist works within the framework of the paradigm, extending his/her knowledge from exemplars to new phenomena and domains. Mostly, scientists will arrive at the same judgments, even though the processes leading to the judgments need not be the same for different individuals. This process of "reasoning from exemplars" we will term "theoretical perspective".

However, using the theoretical perspective is not enough to arrive at judgments in all cases. In situations where scientists are to choose between two rivaling theories, they cannot make use of their normal scientific methods and techniques. The paradigmatic "way of seeing" phenomena that the scientist has been socialized into does simply not suffice to determine the answer to which theory to choose. In such cases the scientist has to rely on more general *values* (Kuhn, 1977). Kuhn identifies five such general values: *accuracy* in accounting for phenomena, *consistency* (both internally), and with respect to other generally accepted theories, *broadness of scope*, *simplicity*, and *fruitfulness*. Kuhn regards these values as relatively stable over time, and a part of the cultural heritage of the natural sciences as such. Thus, these values are shared by all and have later been termed *desiderata* (Hempel, 1988; the term was later adopted by Kuhn).

The desiderata are quite imprecise individually. For instance, two individuals who both subscribe to the same value of accuracy may still come to disagree about whether one theory is more accurate than another. In addition, there is no fixed way in which to determine how the different desiderata should be weighed with respect to each other in specific evaluations of theory (Kuhn, 1977). The scientist's personal experiences, preferences, and values will determine how the scientist will end up weighing the desiderata. Thus, there are actually two sets of values at play here. One set is what we will term *shared desiderata*, and designates values which are shared by members of the scientific community (e.g., the list of values above). The other set of values are determined by the scientist's *personal* background and affinities, and we shall term these *personal desiderata* (see Hoyningen-Huene, 1993, pp. 150–151).

Thus, overall we distinguish between two different ways of arriving at judgments. The first has to do with what is determined by the paradigm and results in judgments that are typically unanimous within the scientific community. This is what we have termed the theoretical perspective. The other is based on desiderata that can be either *shared* within the community or *personal*. Desiderata based judgments will generally differ from individual to individual in the community.

Let us now use Kuhn's description of scientists practice to explore our aim of understanding students' engagement in GIS. We will of course necessarily have to reinterpret what we mean by the terms theoretical perspective and shared and personal desiderata now that we are considering students rather than scientists. In the student setting, the *theoretical perspective* involves the types of judgments that are facilitated or made possible on the basis of the knowledge acquired through "reasoning by exemplars" through their education. That is, having followed a course in GIS, a student will be able to make specific types of analyses based on the types of analyses encountered previously, and (ideally at least) extend this knowledge to new situations. By *shared desiderata* we refer to values and norms which are shared in the specific student body. For instance, some student environments are very competitive, while others more oriented toward student collaboration. In some environments student cherish qualitative research, in others quantitative. In some environments, education is perceived by most students as a service provided to them, in others as something that the students essentially contribute to, etc. Finally, the personal desiderata have to do with the values pertaining to the individual rather than the student body. Examples could be choosing to work with a particular region because you lived there as a child or because you want to move there, choosing elective courses because you want to strengthen your knowledge within subjects in which you have a personal interest, etc. In the following we shall designate the theoretical perspective T, shared desiderata S, and personal desiderata P.

We turn now to how these three different perspectives may have different roles to play at different times based on the context. Table 1 gives a schematic overview of the influence of the concepts (theoretical perspective, shared and personal desiderata) during a specific course for an individual student. A student in the situation of choosing an elective course (for example, in GIS) relies mainly upon shared S and personal desiderata P, whereas prior theoretical knowledge is bracketed or suppressed $[T]$, because it is of little help in the concrete situation. When working in a course, the theoretical perspective becomes important and is added to or developed. This is indicated by the apostrophe T (T'). At this time the personal desiderata become bracketed due to the convergent nature of science education. There is little room for personal judgments, they have to be suppressed. However, even though P is suppressed in the course of work, the result after the course is nonetheless a development both in personal desiderata P' and in the theoretical knowledge $[T']$. The meaning of S and S' refers to custom, and custom plus didactical contract, respectively. Further explanation is given in the following sections.

In many situations during their studies, students make choices about how they are going to specialize, one of them being when they are choosing elective courses. As with the scientist facing a choice of theory, the student relies in such situations mainly on shared and personal desiderata and only to a limited degree on the theoretical perspective in these situations of choice. The theoretical perspective simply does not provide answers to the question "which course to choose", and cannot provide valuable justifications.

Table 1. Schematic view of the development of theoretical perspectives, shared and personal desiderata for a prototypical student in a GIS course.

	Theoretical perspective	Shared desiderata	Personal desiderata
Electing a course	$[T]$	S	P
Working in a course	T'	S'	$[P]$
After a course	$[T']$	S	P'

Therefore, in situations of electing a course, we may say that the theoretical perspective is suppressed ([T] in Table 1). As the student begins working in a course, however, the theoretical perspective becomes crucial, whereas the personal desiderata are bracketed (T and [P] in Table 1) because the student will have to accept and subscribe to the teaching and learning activities in the course, and more or less leave behind his or her personal preferences and interests. The level of suppression of P depends on the students' ability to relate their personal preferences and interest to the teaching and learning activities, e.g., free choices of topics. However, due to the convergent nature of science suppression will occur. After the course, both the theoretical perspective and the personal desiderata will have been changed ([T'] and P' in Table 1). Because of the limited scope of the theoretical perspective, we have again suppressed it as before the course. The distinction between S and S' needs further elaboration in terms of another set of concepts which can extend the understanding of shared desiderata; namely custom and didactical contract.

Custom and didactical contract

The second set of theoretical concepts that we address is that of custom and didactical contract. Above we distinguished between shared desiderata and personal desiderata. Based on Kuhn's description of the shared desiderata as relatively stable over time, we here denote shared desiderata as the students' shared and general values, norms, beliefs, practices, and behavior in relation to their education. These are in play and used when a student chooses a course (S in Table 1). In accordance with Kuhn's description of shared desiderata, Balacheff (1999) uses the term custom as: "a set of obligatory practices... established as such by their use and which, in the majority of cases, is established implicitly" (Balacheff, 1999, p. 25). He further argues that customs are not confined to a particular class but extend to particular levels of schooling (for example, customs related to 7th grade geography teaching). Thus, Balacheff's concept of custom and our concept of shared desiderata are interchangeable. In addition, Balacheff (1999) makes a distinction between custom and didactical contract, which we can use to understand what happens when a student starts working in a course. In such situations the student subscribes to both the custom/shared desiderata and to a set of *explicit* but local and temporary rules governing the teaching activities in the specific course. These rules relate mainly to the teaching and learning activities and assessment of the course (e.g., deliveries, work formats, and examination) and are termed the didactical contract. This contract does not overrule the general and tacit shared desiderata S, but adds to them (S' in Table 1). The didactical contract is explicit, has a local character, and is negotiated for a particular task that requires the rules for the social functioning of the class to be defined locally and in a new way. Thus, the students come to the class with certain shared desiderata, but must then subscribe to a specific didactical contract involving acceptance of specific types of teaching and learning activities defined by the teacher (though sometimes in negotiation with the students). However, given the temporary nature of the didactical contract, the student will rely on the custom only after the course.

Consequently, in our analogy to Kuhn's description of scientists and scientific development, we may define three evaluative moments that come into play in different ways at different times as an essential tension in students development; namely the theoretical perspective T, the shared desiderata S including the distinction between custom and didactical contract, and the personal desiderata P. In the next sections, we shall outline how we can use these concepts to understand students' engagement in relation to a specific GIS course.

The empirical setting of the study

The learning environment

At the Department of Geoscience and Natural Resource Management at the University of Copenhagen, teaching in GIS has been part of the curriculum for human and physical geographers since the late 1980s. The development of different GIS courses as well as the present teaching practice has been addressed in Madsen and Rump (2012) and Madsen and Winsløw (2009).

The GIS course in planning and management is taken by around 30 students each year and is an optional course within the M.Sc. program in Geography and Geoinformatics. In Denmark there is no tradition of employing students with only a bachelor degree. Hence very few students leave the university with a bachelor degree; most graduate with a Master's degree. The GIS course is part of the elective courses within the first 3 years of the 5-year M.Sc. program which means that the geography students attending the course may not yet have chosen whether they are to become either human or physical geographers. However, the course can also be taken by students from other studies (for example, biology or natural resource management) as either a bachelor or a master course and further by people already having a master as a paid course (open university course). It is a 7-week course taking up half of the students' study activities in the period. The course introduces central GIS tools for working with planning and management issues such as network analysis, digital records, and web-based communication systems (ArcGIS, version 9.1 was used). The course curriculum is organized around the students' group based problem-solving activities. The students in the course work on three consecutive assignments in which central functions of GIS are explored in relation to semi-authentic problems. The last assignment is somewhat larger than the other two, and constitutes the basis for the oral examination. The topic of this assignment is a question of where to place a new institution (for example, kindergarten, school, and shopping mall), given the distribution of existing institutions and other relevant variables such as municipal objectives from district plans and population density in the municipality. The analysis should be based on data-sets provided to the students, and different types of GIS analyses should be made. The students' proposal should be outlined in a WebGIS for location of the institution, so decision-makers and/or citizens could get an impression of the different aspects of the problem by adding or subtracting different layers, for instance buffer zones based on noise data, or relevant age distributions.

Participants and interviews

In a specific year between 2002 and 2012 (in order to secure anonymity of the participants, the exact year is not given), we followed eight students participating in the GIS course in planning and management. Observations in the classroom were made twice in the beginning of the course and contacts with the students were established. The eight students were selected with the purpose of securing a diverse group of students in order to optimize the opportunity to learn the most from the case as discussed in Stake (2005). The group of interviewed students consisted of five men and three women. Six were geography students of which three were either taking quite a number of courses or a whole master in other disciplines (political sciences, economics, and development studies). One student was from biology and one was from geography at one of the other universities in the Copenhagen area.

We performed individual qualitative interviews with each student on two occasions. The first round was made before the students' commenced working on the final assignment, and the second interview shortly after the oral examination in the course. The interviews were conducted with an open interview guide. In the first interview, students were asked to describe why they had chosen the course, their conception of the problem that they were going to work with, and were asked to consider its relevance for their future use of GIS. In the second interview, the students were asked to explain how they had solved the problem in the final assignment and what their general conceptions of the course, the problem, and the examination were. Further, they were asked to elaborate on their motivation for working with GIS, and how they saw GIS as part of their future professional life. These questions were used because we wanted them to inform our two sets of theoretical concepts (shared and personal desiderata and custom and didactical contract) to understand students' engagement as described in the following analytical approach. Each interview took around 1 h. All eight students passed the course with grades ranging from A to C on the European Credit Transfer and Accumulation System scale, that is, all students performed well in the course.

All 16 interviews were made in Danish, taped, and subsequently transcribed with acceptance of the interviewees. We have chosen verbatim translations rather than linguistically correct ones. All interviews were performed by the first two authors together; one of us being an insider to the subject matter of GIS and geography as a discipline and one of us being an outsider (Adriansen & Madsen, 2009).

The analytical approach

The objective of the empirical part of this paper is to investigate the interplay between students' expression of the influence of their theoretical perspective T, shared desiderata S (including the distinction between custom and didactical contract), and personal desiderata P while taking the GIS course, and how the interactions between these three elements affected the students' engagement in using GIS for geographical problem solving.

Having interviews early on in the course made it possible for us to understand the students' personal desiderata P, their preliminary conceptions of T in relation to GIS, their understanding of custom S, the didactical contract S', and their engagement in relation to using GIS. Further, having interviews after the examination with the same students enabled us to analyze their understanding of the relationship between their experience with the teaching and learning activities S' and resulting content learning T', how their personal desiderata had been changed as a result of the course (from $[P]$ to P'), and how their engagement in relation to GIS had possibly changed during the course. Based on our multidimensional view on engagement, we have for each interview estimated the subject's engagement in relation to using GIS in their future work, irrespective of whether it is related to their study or a future job situation. Here we have been inspired by self-efficacy beliefs and outcome expectancy beliefs in relation to using GIS, i.e., the student's beliefs that he or she will be able to accomplish specific tasks involving GIS, and the conception of being in an environment that is conducive to using these capabilities (Bandura, 1982). In estimating the level of engagement, we have aimed at not having a precise measure of their engagement, but in making a distinction between low, medium, or high engagement based on a holistic (in the sense of taking as many factors into considerations as possible) assessment of each interview. The engagement scores in each interview transcription have been assigned independently by the authors and discussed until a consensus has been reached.

Table 2. Eight students' way through the undergraduate GIS course in planning and management described in terms of engagement and relationships between S, P, and T (see text for description of letters).

Student	Engagement during course	Engagement after course	Relationships between S, P, and T
Stories of T' influencing P'			
Søren	High	High	$T' + [P]$
Thomas	High	High	$T' + [P]$
Emilie	Medium	Low	$T - [P]$
Anette	High	Medium	$T - [P]$
Carl	High	Medium	$T - [P]$
Story of external factor E and T' influencing P'			
Charlotte	Low	High	$T + E + [P]$
Stories of S' influencing P'			
Maria	Medium	High	$S' + [P]$
Ida	Medium	High	$S' + [P]$

Findings

Table 2 shows the results of our analysis of the interviews with students. It shows the eight students' way through the course described in terms of engagement and relationships between S, P, and T. The symbols $+$ and $-$ signify a positive or a negative reinforcement between two elements. Thus, $S' + [P]$ means positive reinforcement between didactical contract and suppressed personal preferences. The E for Charlotte denotes external factors of relevance.

The relevance of geographical content: five stories of $\mathbf{T}'$ *influencing* $\mathbf{P}'$

Although they have quite different backgrounds, Søren and Thomas share many views on GIS and the role they see GIS playing in their future. Both are just finishing their studies, and both see a GIS-related job in a municipality (administrative unit in Denmark) as an attractive job opportunity. Both find the course and the assignments in the course highly relevant for them and in general. Søren has a genuine interest in the geographical planning issues involved and has chosen to look at the placement of a new school. He is excited about the possibilities for modeling the future that the GIS program provides. He also thinks that the problem resembles closely the types of problems you would get if working in a municipality, and finds this motivating. Thomas has chosen to look at the placement of a shopping mall instead. His justification for this is based on his reading of actual local development plans he found at the municipalities' website on his own initiative. He also finds the authenticity of the problem important: "I think it is a funny … lets-pretend-it-is- … It works much better than just following some tutorial. Yes I like that".

Thus, for both Søren and Thomas, there was a very good match between their personal desiderata P and the theoretical perspective adopted in the course T'. Both liked the way the teaching content was organized into a semi-authentic planning problem that students had to work on S', because it matched well with their job aspirations P.

In contrast to Thomas and Søren who are just finishing their studies, Emilie is in her third year of study. Emilie also had specific goals with the course based on her personal desiderata P. She wanted to be better at using GIS, so she could use it in her bachelor project about coast lines in the Wadden Sea. However, while she does find the authentic problem a relevant planning problem, Emilie feels that working with this problem does not help her to use GIS in her Bachelor project or elsewhere: "I had thought that it would be broader . . . it is all very focused on this specific problem about Tårnby municipality, and it's all right, but it is just a small niche of what GIS can do". In the second interview, Emilie seems more reluctant toward using GIS because of its technicality and says that she would need urging and guidance in order to use GIS in the future. Thus for Emilie the reinforcement between personal desiderata P and the theoretical perspective used in the course T' was negative.

Anette wants to finish her bachelor in geography. She has strong political interests, and she has just applied for a job in one of the unions. Her interest in GIS is that it would be a good tool to know whether she wants to work with planning in the future. However, during the GIS course she has realized that she wants to work with the political level and not the practical level at which the course operates. She finds it unsatisfying that there is so little focus on the planning and analysis behind the problem, and so much focus on the visual aspects of GIS. She finds the authentic problem about Tårnby municipality too simplistic. After the examination she tells us that she did not get the job in the union and when she looks back at the GIS course she thinks that it is nice that she now knows something about the potential uses of GIS, but she does not expect to make use of them herself in her further studies or after graduation, as she expects to operate at the political level rather than the technical. So also for Anette, the match between P and T' turns out to be negative.

The negative relationship between P and T' is also the case for Carl, another bachelor student. He enters the GIS course with no specific ideas about this particular GIS course, but with a strong interest in GIS in general. He basically wants to take every GIS-related course available, and he has some ideas about creating GIS-based software applications in the future. Some years ago he studied computer science but got fed up by the abstract work and changed to geography in order to work with some more for him real problems. He sees GIS as a possibility to combine his interest in computers with authentic problem solving. However, he finds the planning problem in the GIS course quite boring. However, because of his interest in GIS he uses the course to explore the functionalities of the software. His final assignment displays this – on the one hand he has used functionalities in GIS not used by any of the other participants, but on the other hand there is a relatively glaring mistake in his assignment. At the examination he got a lower grade, and in the final interview he is having doubts about whether GIS can be used to bridge his interest between computer and real problems.

For Carl, Anette, and Emilie, the theoretical perspective T' did not appeal to them, and they did not feel that the course helped them pursue their interests in GIS. The course did not help them develop their engagement in relation to GIS, and it has changed their personal desiderata in distancing them from using GIS. Emilie could see the idea in working with the semi-authentic problem, but felt it to be somewhat irrelevant. Anette felt that it was too simplistic and did not reflect the complexity of political decision-making. Carl had real difficulties in seeing the idea in the assignment at all.

How external factors have relevance: a story of how external factors **E** *and* $\mathbf{T}'$ *positively reinforce* $\mathbf{P}'$

Charlotte has taken the GIS course because she has heard it was an easy course to pass. She is a geography student and sees GIS as a tool: "it is good to have" but she does not find it intrinsically engaging. She does not link GIS to geographical problems. She talks enthusiastically about her job where she works with a program named *GeoGIS*. She has been typing in data into a database, but does not see that it has any relationship with GIS. She finds the authentic problem in the course okay (authentic because it is a planning problem that a municipality is likely to deal with), but does not see it as a real problem for her (since for her to be "real" it must be personally relevant). In the second interview, she has just been defending her bachelor project with good results and is very happy. She has decided to use some more time on her job and she has had some conversations about her future work tasks. In this process she has realized that there is a GIS function attached to the database she has been working with, and that *GeoGIS* is in fact a GIS. She is now fascinated with how GIS can be used to solve geographical problems (in this case soil pollution). Charlotte has during the time of the course engaged in GIS due to a combination of the course and her job functions. Thus, working with the course content T' has made her see and pursue new opportunities in her job situation, and in turn changed her personal perspective on GIS. Thus, Charlotte's story is similar to Søren's and Thomas's stories, with the important difference being that the theoretical perspective T' has been strengthened by factors outside the course while the course is running.

The relevance of geographical methods: two stories of $\mathbf{S'}$ *influencing* $\mathbf{P'}$

Ida and Maria both liked the GIS course a lot and both are really engaged in GIS. Maria is actually studying biology, but has previously taken other GIS courses. In general she has found it difficult to keep her motivation for studying, but the GIS courses have been something of a revelation to her. She wants to take all the GIS courses available and she has no specific expectations for this course. She finds GIS very appealing: "the fact that you can work creatively and get a product and all the time follow the process". It is a great challenge to work with and she finds it great fun. Her assignment is quite a work of art. Taking GIS courses has meant that Maria has found the joy in attending class again. In relation to the authentic problem, she does not find the planning problem in itself very appealing but the tools and the intentions with the course she finds highly relevant, it is the way of learning in the GIS courses that appeals to her (the relationship between P and S').

Ida's story is similar. Ida is a geography student in the third year and she has strong interests in economy. She says about the way of learning: "I think it is fun to analyze and see ... we can sit with it for hours ... clicking the layers on and off and putting photos underneath... and I could easily do it just for fun". In the second interview, Ida describes why she would like a job where GIS plays a role:

> I would like that ... I think it is really fun to work with, so if I could sit with this much of the day, that would be lovely ... I also love to sit and work in Excel and make graphs and such things ... If there is something where I can use something with some numbers, I'd like that.

While the first quote is quite geographically oriented, the second quote highlights the fundamental joy Ida has in engaging in the type of work with a computer and "something where I can use something with numbers".

For Ida and Maria, then, it is not so much the planning aspect and the specific problem that they like or can relate to (as for Søren and Thomas), but rather the way that the learning is organized and the mode of learning in the course, that is S'. They find it motivating and rewarding to transform numbers into something useful and accessible, and GIS is one instrument (among others) that allows them to do that. In that sense there is a very positive reinforcement between S' and P for these two students.

Discussion

In the following we discuss how the analysis using the introduced theoretical concepts can address our puzzlement of why students engage so differently in a specific GIS course.

First, our analysis and results show that it is mainly the shared and personal desiderata that determine in which way the students in this study are going to engage with GIS. Second, this being so, the personal desiderata are actually suppressed most of the time in the actual GIS course despite the students possibility for choosing a topic in the final assignment. Hence, in order to understand why the students engage so differently with GIS, it becomes important to consider the longitudinal interplay between theoretical perspectives, shared desiderata, personal desiderata, and possible other factors both before, during, and after a student's interaction with a specific GIS learning experience. This is unfolded in the next sections.

Theoretical perspective and proximity of goals

In terms of fulfilling the learning objectives or acquiring the intended theoretical knowledge of the GIS, all of the interviewed students solved the authentic planning problem adequately and all did well on the examination (despite the fact that one got a lower grade than the others). However, it was very different how the problem was conceived among the students, that is, how the theoretical perspective of the course influenced the students' personal desiderata. Of particular interest in this respect are the "negative" stories of Carl, Emilie, and Anette. As discussed in the introduction, the course is in many ways pedagogically well designed and authentic with respect to likely job situations. However, as we have seen, none of the three students experienced the problem as authentic or relevant for their immediate goals or their engagement in GIS. Each of them had personal and relevant ideas about potential uses of GIS for geographical problem solving, but the course did not help them pursue these goals. For these students the authentic problem tended to extinguish their interest in GIS rather than igniting it. They were not capable of relating the authentic problem to their personal desiderata, and while they all did well in the course, the course was not conducive for their engagement with GIS in the future. The curricular implications of these observations could be that it would have been more relevant to do a course which helped these students pursue their ideas about GIS use in relation to their upcoming bachelor projects.

For Charlotte, Søren, and Thomas, it was also the relationship between the theoretical perspective of the course and their personal desiderata which was crucial for their engagement in GIS. However, for them it was an igniting experience. Both Søren and Thomas were almost finishing their studies and experienced the authentic planning problem as highly relevant and were fueling their interest in GIS and their job aspirations. For Charlotte it was her job outside her study that secured a positive relationship between the theoretical perspective and her personal desiderata. For all three students it was factors outside the control of the GIS course that created their engagement.

Learning methods: customs and contracts

For Maria and Ida, what they liked about the course was not the theoretical perspective of the course and thereby the authentic planning problem, but rather their engagement in learning methods adopted in the course. They were engaged in GIS due to the relationship between their personal desiderata and the custom and didactical contract in the course. Maria feels that her potentials are realized far better in the way the learning activities are organized in the GIS course than in other courses. As she put it: "my motivation for coming to the teaching was quite different than for the teaching at the Biology Department". Ida describes how she has been taking an unusual number of courses in the political science and economy programs, for a geography student. She engages in GIS by connecting it to her interest in economy, in practice by relating her work with GIS with her work with Excel, which she much enjoys. For these two students it seems that no matter what the theoretical perspective had been in the course, they would have been engaged.

Implications for future teaching in GIS

The above analysis shows that during the GIS course, the students are suppressing their personal desiderata *P* and need to subjugate to the theoretical perspective and the shared desiderata as unfolded in the learning methods of the course. Our analysis of eight students shows that those who get engaged in GIS are those who are able to relate their initial personal desiderata with either the theoretical perspective adopted in the course or the learning methods adopted in the course. For the students who are able to relate to the theoretical perspective in the course, it is factors outside the course that are significant; either a forthcoming job or an existing study-related job. For the students who are engaged through the learning methods in the course, it is the relationship with a known learning method or an attractive unknown learning method that is significant for creating the relationship. The students who are not engaged in GIS during the course however stay on and take the examination and even do well on it. To elaborate on this is beyond the scope of this paper; however, it could be a question of deferred gratification as described in Johannsen (2012) or that the students are able to bridge the gap between their expectations to the GIS course and their experiences with the course through different negotiation strategies (Holmegaard, Madsen, & Ulriksen 2013).

The practical implications of this for teaching and learning GIS are:

(1) We need to keep teaching GIS in a pedagogically well-designed way with hands-on and authentic problem solving because it engages certain students. However, we need to address the fact that the engagement is linked to the students' personal values, meaning that elements outside the learning situation define and modify the engagement. Therefore, in order to sustain engagement, the proximity of students' goals must be closely tightened with the theoretical perspective. In our empirical study it would imply that we should either:
   (a) Consider turning the GIS course into a graduate course instead of it being an undergraduate course. In this way proximity to the goal of getting a job would be closer (because bachelor degrees are not generally seen as sufficient in the Danish labor market – in countries where bachelor degrees are a way to the labor market, this would not apply).
   (b) keep it as an undergraduate course and open up for students to work with their own ideas in relation to the content in the course.

(2) We need to allow for and fuel the creativity in learning methods used in GIS courses because it creates connections between theoretical perspectives and some students' personal values igniting an engagement in GIS.

(3) The temporal aspect of the suppressing of personal desiderata and theoretical perspectives before, during, and after the course presented in our analysis has implications for teaching as well. Most of the feedback to students about their individual learning trajectories takes place during the actual teaching. However, if

we want to influence students' long-term engagement, the time to do that might not be during courses but "in between" courses. There are many ways such guidance can be accomplished – making students keep a learning portfolio describing aims, ambitions, and engagement, and provide feedback on that, or providing student mentoring where teachers counsel students about their aspirations and how to best meet them. Such moments for meta-reflections by students might benefit immensely by keeping them "at a distance" from the didactical contracts of specific courses.

(4) But something can also be done within the course framework. Prior to the start of the course, students can be asked (e.g., as an e-learning assignment) to describe their interest in GIS, why have they chosen the course, and what are their thoughts about using GIS in the future? This will inform the teachers about the thoughts and aspirations of the students with respect to GIS, and individual needs and engagements may be met as the course progresses. Such meta-reflections could also be incorporated into the evaluation of specific courses after the examination.

Conclusion

We began this paper based on our puzzlement of why students seem to experience courses in GIS so very differently and with the purpose of getting a deeper understanding of students' engagement in GIS during their geography higher education. If we as geography faculty members agree that GIS can give students valuable competencies and see GIS as an integral part of becoming a geographer, then it seems very reasonable questions to pursue.

The analysis of 16 interviews with eight students in a GIS course shows that both the theoretical perspective (in our case capabilities obtained with respect to solving an authentic planning problem) and the learning methods in the GIS course were significant for the students' engagement in relation to GIS. However, they were not acting at the same time for individual students and they did not fulfill all students' interest in GIS. What instead was significant was the theoretical perspective and learning methods in relation to what we have called the students personal desiderata, meaning that in order for students to see problems and learning methods as authentic, they must also be seen as relevant to the student's individual and immediate and future goals and ambitions.

As for the theoretical part of this paper, we found that by developing and using the theoretical concepts of theoretical perspective, shared and personal desiderata in an empirical analysis of a specific GIS course, we became aware that studying students' engagement in relation to GIS needs a longitudinal component. At different times in relation to the students' tasks of electing a course, working in a course, and having finished a course, the perspectives have different emphasis. Thus, a temporal dimension and contextual awareness of understanding students' engagement in GIS are important.

There can be little doubt that all of the introduced concepts can be elaborated and explored further. However, our analysis of student learning in relation to the GIS course has shown that the Kuhnian analogy and the rudimentary distinction between the different types of student evaluative perspectives provide us with a useful tool for analyzing student learning trajectories and the efficacy of curriculum design.

Acknowledgments

This study is based on observations and interviews with students at the Department of Geoscience and Natural Resource Management, University of Copenhagen. We are indebted to all of them for having shared their time and thoughts with us not only in relation to the GIS course but also in relation to their personal feelings and perceptions concerning their present and future study and life trajectories. Also we want to thank the teacher for letting us observe the teaching practice, the students' engagement in the course, and discuss our queries into the course. Finally we want to thank four anonymous reviewers for their valuable comments that significantly improved the paper. All names used in the article are fictitious and all usual disclaimers apply.

References

Adriansen, H. K., & Madsen, L. M. (2009). Implications of doing insider interviews: Studying geography and geographers. *Norwegian Journal of Geography*, *93*, 145–153.

Balacheff, N. (1999). Custom and contract: Two registers of didactical interactions. *The Mathematics Educator*, *9*, 23–29.

Bandura, A. (1982). Self-efficacy mechanism in human agency. *American Psychologist*, *37*, 122–147.

Barcus, H., & Muehlenhaus, B. (2010). Bridging the academic – public divide in GIS and cartography: A framework for integrating community partnerships in the classroom. *Journal of Geography in Higher Education*, *34*, 363–378.

Demirci, A. (2011). Using geographic information systems (GIS) at schools without a computer laboratory. *Journal of Geography*, *110*, 49–59.

Elwood, S. (2009). Integrating participatory action research and GIS education: Negotiating methodologies, politics and technologies. *Journal of Geography in Higher Education*, *33*, 51–65.

Hoyningen-Huene, P. (1993). *Reconstructing scientific revolutions – Thomas S. Kuhn's philosophy of science*. Chicago, IL: University of Chicago Press.

Hempel, C. G. (1988). On the cognitive status and the rationale of scientific methodology. *Poetics Today*, *9*, 5–27.

Holmegaard, H. T., Madsen, L. M., & Ulriksen, L. (2013). A journey of negotiation and belonging: Understanding students' transitions to higher education science and engineering. *Cultural Studies in Higher Education*. doi:10.1007/s11422-013-9542-3

Johannsen, B. F. (2012). *Attrition and retention in university physics. A longitudinal qualitative study of the interaction between first year students and the study of physics* (PhD-thesis). Department of Science Education, University of Copenhagen 2012.

Kuhn, T. S. (1959). The essential tension. In CW Taylor (Ed.), *The third (1959) University of Utah research conference on the identification of scientific talent* (pp. 162–174). Salt Lake City, UK: University of Chicago Press.

Kuhn, T. S. (1970). *Structure of scientific revolution* (2nd ed.). Chicago, IL: University of Chicago Press.

Kuhn, T. S. (1974). Second thoughts on paradigms. In F. Suppe (Ed.), *The Structure and scientific theories* (pp. 459–482). Urbana: University of Illinois Press.

Kuhn, T. S. (1977). Objectivity, value judgment, and theory choice. In T. S. Kuhn (Ed.), *The essential tension* (pp. 320–339). Chicago, IL: University of Chicago Press.

Lee, J., & Bednarz, R. (2009). Effect of GIS learning on spatial thinking. *Journal of Geography in Higher Education*, *33*, 183–198.

Madsen, L. M., Christiansen, F. V., & Rump, C. Ø (2009, March 22–24). Development of personal desiderata while learning GIS. In A. Bisel & M. U. Garib (Eds.), *Frontiers in Science Education Research: Proceedings of the Frontiers in Science Education Research Conference* (pp. 441–450). Famagusta, North Cyprus: Eastern Mediterranean University Press.

Madsen, L. M., & Nielsen, T. T. (2013). Learning to do geography? University students posing questions in GIS laboratory exercises. *Norwegian Journal of Geography*, *67*, 157–161.

Madsen, L. M., & Rump, C. (2012). Considerations of how to study learning processes when students use GIS as an instrument for developing spatial thinking skills. *Journal of Geography in Higher Education*, *36*, 97–116.

Madsen, L. M., & Winsløw, C. (2009). Relations between teaching and research in physical geography and mathematics at research-intensive universities. *International Journal of Science and mathematics Education*, *7*, 741–763.

Rød, J. K., Larsen, W., & Nilsen, E. (2010). Learning geography with GIS: Integrating GIS into upper secondary school geography curricula. *Norwegian Journal of Geography*, *64*, 21–35.

Solomonides, I., Reid, A., & Petocz, P. (2012). The nature of the elephant: Metaphors for student engagement. In I. Solomonides, A. Reid, & P. Petocz (Eds.), *Engaging with learning in higher education* (pp. 475–485). Faringdon, UK: Macquarie University Research Online.

Stake, R. E. (2005). Qualitative case studies. In N. K. Denzin & Y. S. Lincoln (Eds.), *The sage handbook of qualitative research* (3rd ed.) (pp. 443–466). Thousand Oaks, CA: Sage.

Unwin, D. J., Foote, K. E., Tate, N. J., & DiBiase, D. (Eds.). (2011). *Teaching geographic information science and technology in higher education*. Chichester: John Wiley & Sons.

West, E. (2012). Where to begin? Getting started teaching GIS&T. In D. J. Unwin, K. E. Foote, N. J. Tate, & D. DiBiase (Eds.), *Teaching geographic information science and technology in higher education* (pp. 145–157). Chichester: John Wiley & Sons.

Writing in Geography: Student Attitudes and Assessment

VANESSA SLINGER-FRIEDMAN & LYNN M. PATTERSON
Department of Geography and Anthropology, Kennesaw State University, USA

ABSTRACT *This paper presents the results of two independent studies that assessed the impact of writing as a learning tool in introductory-level geography courses. Student attitudes toward writing and their performance on exams were analyzed to determine the impact of writing for comprehension and application of geographical concepts. Student perceptions were mostly positive; however, there was no significant improvement in grades. As both students and instructors view writing as valuable for learning, these results underscore the need to develop alternative methods of assessment and additional studies to quantitatively determine the impact of writing on learning.*

Introduction

As geography has become increasingly de-emphasized in the US K-12 curriculum, more students have limited geographic literacy and consequently lack the ability to think and communicate in the geographic lexicon. By the time students enroll in geography classes in post-secondary US institutions, many have the belief that geography is just memorizing maps and place names (Fournier, 2002; Mansfield, 2005). This misperception often affects their level of interest and even their comprehension of geography as a broader and more complex discipline. Another consequence is that students often fail to recognize the importance of writing as a tool to communicate with a broader audience about geographic ideas.

Students often perceive writing as an onerous task rather than as an opportunity for learning and communication. For example, teaching assistant John Isom once lamented, "They like maps ... They liked Geography. But they shied away from the writing assignments I was trying to add to the course" (Smith, 2010). However, getting students more comfortable with writing regularly within the geography discipline can increase the overall quality of their learning and help develop better communication skills (Winchell & Elder, 1992; Gedye & Chalkley, 2006). Similarly, many institutions also compartmentalize writing as a communication tool rather than a learning tool. This constraint overlooks the opportunity to recognize writing as a part of the learning process. Combined, these challenges have prompted some geography instructors to seek innovative pedagogies to stimulate interest and assist learning in geography by introducing writing as a tool for learning and communication.

Writing across the curriculum (WAC) is a process in which writing is integrated into courses within all disciplines and at all levels of the curriculum as a tool for learning and thinking (McLeod, 1987). This paper reports on two separate studies (conducted in Spring 2009 and Spring 2010) in which writing activities were used in geography classes at a large American University as part of a College of Humanities and Social Sciences WAC initiative. The objective of this paper was to identify impacts of writing-to-learn activities on student learning. Specifically, we were interested in the following questions:

- Did students perceive that writing could help them better understand geographical concepts and applications?
- Did writing help students to better understand geographical concepts and applications as manifested through improved student grades?

Background

Writing as a Learning Process

There is a pervasive perception among students, faculty, and institutions that writing belongs solely or principally in English classes (Carter, 2007). This compartmentalizing, also referred to as 'inoculation theory', supposes that a dose of writing in a first and/or second year composition class should suffice in terms of creating satisfactory writing skills (White, 1990; Daniel & Odom, 2010). This view of writing as mostly a communication tool limits the potential for using writing as a continual tool for learning.

Writing activities can be identified along a continuum from 'writing-to-learn' to 'writing-to-communicate' (Young, 2006). On one end, writing-to-learn activities are expressive and informal and focuses on process (Britton *et al.*, 1975). That is, they are integrative, interactive, and increase critical thinking (Young, 2006). These activities are often reflective and creative and have students writing about concepts to help them gain clarity for themselves. These include note card questions, journals, notes, free writes and rough drafts. For example, students may be asked to explain to themselves how course material applies to personal experience or larger issues and put their answer down on notecards for the instructor to review, but not grade. On the other end of the spectrum, writing-to-communicate exercises are typified as transactional or formal. Examples are research papers and reports in which students have to convey information to a public audience. Along the continuum are activities that blend the reflective and communicative properties of writing. These may include discussion boards and worksheets. Discussions require students to utilize concepts and apply them to particular settings and garner feedback from other students. Worksheets ask students to answer questions about concepts as applied to specific examples not given in class.

Beyond the individual learning benefits, advocates assert that writing in classes can create an interactive classroom experience. Writing stimulates students to become active and engaged participants (e.g., Stainer, 1997; Cook, 2000; McGuinness, 2009). In addition, by writing out their thoughts before, or as part of, a discussion, students gain the confidence to express themselves and are more willing to participate in class, thus resulting in much more lively and intelligent discussions (Bean, 2001; Lapadat, 2002).

Writing can help students with deep learning rather than just surface learning (e.g. Thompson *et al.*, 2005; Dummer *et al.*, 2008). As surface learning is characterized by memorization of facts, deep learning requires understanding and application of concepts and ideas (Biggs, 2003). With deep learning, writing can often help with comprehension problems when students cannot successfully articulate concepts. Thus, the process of writing enables students to explore ideas in-depth and identify questions they may have about the ideas. The 2008 National Survey of Student Engagement (NSSE) showed a positive correlation between students writing in college and 'active and collaborative learning, student–faculty interaction, and deep learning' (NSSE, 2008).

For geography, writing is an essential component of the discipline: "To make writing a component of the teaching of geography begins with the supposition that to understand geography and communicate that understanding to others requires the skills of language and provides creative opportunities for learning" (Walker, 1996, p. 158). Although geographers do communicate using other media, writing is, "the most generic skill students of geography invoke in order to share their learning with others" (Proctor *et al.*, 1995, p. 571). With this combined impact on learning and communication, geography instructors using writing concur that writing in the classroom is important to the development of students who think like geographers (Libbee & Young, 1983; Maraffa, 1985; Hooey & Bailey, 2005; McGuinness, 2009). Both employers and geography graduates rank writing, and its associated ability to allow people to think about and understand a subject more deeply, as an important skill. For example, Clark and Higgitt (1997) noted that geography graduates surveyed in their research cited: the ability to analyze, knowledge-acquisition skills, an ability to explain things, problem-solving, and writing skills among the top 13 most valued skills in terms of their career progression. Teaching geography students to express themselves clearly in written form is one of the keys to success in our discipline (Gedye & Chalkley, 2006).

Geography instructors have incorporated writing into their courses because they considered writing a good resource for helping students be more reflective and learn course content (Maraffa, 1985; Winchell & Elder, 1992; Walker, 1996; Cook, 2000; Haigh, 2001; Chappell, 2006; Dummer *et al.*, 2008; McGuinness, 2009). Early on, some of the writing techniques in geography included creative writing assignments, critical analyses of recent articles (Winchell & Elder, 1992), a pen pal letter system to exchange student-generated letters from around the world (Cunningham, 1990), the use of multimedia computer learning modules that involved students learning and collaboratively writing about human geography topics (Proctor *et al.*, 1995) and a debate format with students doing prewriting (Estaville, 1988). In the last decade, learning journals (also known as student journals, reflective journals and reflective diaries) and narratives have appeared more frequently as a mode for deep learning (Cook, 2000; Haigh, 2001; Hyers, 2001; Harrison *et al.*, 2003; Park, 2003; Dummer *et al.*, 2008; McGuinness, 2009; Summerby-Murray, 2010). Many other writing-intensive activities are available but have not yet been the subject of inquiry in the geographic literature.

Empirical analysis of writing interventions in geography has largely been concentrated on instructor and student reflections on the usefulness of writing as a learning tool. Overall, student attitudes toward journal writing and narrative inquiries as effective learning tools have been largely positive (Cook, 2000; Haigh, 2001; Hyers, 2001; Park, 2003; Savin-Baden & Van Nierkerk, 2007; McGuinness, 2009; Summerby-Murray, 2010). Student self-reported learning gains have been identified as a useful indicator of actual learning (Chesebro & McCroskey, 2000; Jones, 2003; Schmidt, 2004). However,

Harrison *et al.* (2003) point out that one of the challenges and a concern relating to the promotion of reflective learning is whether reflective learning enhances student learning through the production of improved results. Fulwiler (1984, 1988) also cautioned researchers that quantitative measures do not adequately reflect learning.

With the exception of Hooey and Bailey (2005), quantitative research on writing in geography has been limited and has not focused on the relationship between writing and test scores. Hyers' (2001) quantitative research focused on determining the extent of the relationship between journal achievement and ability, gender, age and major subject. Outside of the discipline of geography, there has been some quantitative research on writing related to improving test scores. Such studies include two psychology courses (Connor-Greene, 2000; Cisero, 2006): a study on a foreign policy analysis course (Michalak, 1989) and a study on a literature/fiction class (Seshachari, 1994). Although Seshachari (1994) saw some improvements in exam grades,[1] Michalak (1989) found that students who wrote more did not demonstrate significant improvement on their exam grades. On the other hand, Connor-Greene's (2000) and Cisero's (2006) studies, which analyzed student test grades, indicated that a journal writing assignment increased student learning through increased exam scores, yet these authors concluded with a call for more empirical research to test writing for an increase in student learning. These mixed results may be explained by the critique that traditional testing methods are often summative and may not actually measure deep learning (Ackerman, 1993; Tynjälä, 1998; Chesebro & McCroskey, 2000; Thompson *et al.*, 2005; Dummer *et al.*, 2008). We felt it was important to explore both student attitudes and student performance in exams as a means of determining learning outcomes because there is limited empirical research to test writing for its impact on student learning, there has been limited exploration of testing as a means for evaluating writing as a learning tool in geography, and because testing remains a standard method of assessment.

Methodology

In the two separate studies conducted here, we assessed the impact of writing-to-learn exercises on student attitudes and performance in our introductory-level geography classes. Hyers (2001) maintained that objective data demonstrating whether a writing intervention (student journals in his case) improves learning are difficult to obtain without a control group and a carefully designed experiment. For this reason, we used an experimental design similar to that employed by Connor-Greene (2000), Hooey and Bailey (2005), and Cisero (2006). In two introductory-level geography courses, the World Regional Geography and the Introduction to Human Geography (see Figure 1), there was an intervention (writing-intensive) class and a control class (non-writing intensive). For each study, the intervention and control classes were conducted during the same semester, unlike the aforementioned studies that compared the same courses over different semesters (Connor-Greene, 2000; Hooey & Bailey, 2005; Cisero, 2006). These introductory courses were specifically chosen to introduce writing to students early in their post-secondary education. They were also selected to explore the effects of various writing tools in different teaching media and for different learning styles (see Michalak, 1989).

The World Regional Geography course was delivered online. Online courses can allow students to be passive participants or engage students as active learners (Swan, 2001; Lapadat, 2002; Mabrito, 2004). In a well-designed online course, students are held

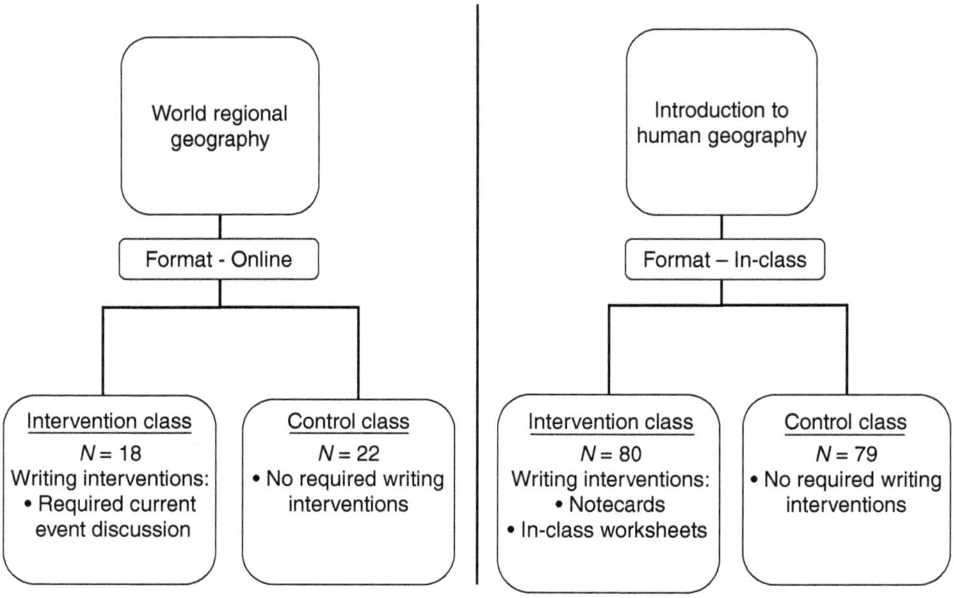

Figure 1. Structure of the two studies

accountable for participation through discussions and other interactive exercises. In this way, asynchronous discussions in online courses provide all students with a voice, are perceived as more equitable and democratic and allow students to reflect on their peer's contributions before crafting their responses (Harasim, 2000; Swan, 2001; Picciano, 2002). The Introduction to Human Geography course was a larger, lecture-style class, in which students can sometimes 'get lost in the crowd'. In-class writing, even if it is not heavily graded, can be the impetus for a discussion by eliciting a response from students and improving the quality of classroom interaction (Ventis, 2000). As such, in both of these course styles, writing was considered as a way to involve students, garner their feedback about course content, facilitate deeper learning and assess comprehension.

During the hiatus between the two studies, the two instructors conferred on how to improve the writing activities for the second study. It is important to note that these studies were not meant to compare outcomes to each other, but rather to examine how writing interventions overall might affect student attitudes and performance. Although the specific writing activities varied between classes, the major commonalities between the interventions were that both the professors used writing techniques to address the issue of how students, "learn to write and write to learn" (Fistek, 1998, p. 82). Both the courses used a combination of qualitative and quantitative data in the form of writing activities, exam scores and feedback from student surveys.

World Regional Geography Study

In the online World Regional Geography classes (Spring 2009), a total of 40 students were enrolled between the two sections. Using online lectures, readings and exercises, each week the students would study a specific region of the world. The writing-to learn activity, which was designed to reinforce geographic concepts and make geographic inquiry more relevant to the students, consisted of a weekly online written discussion of an instructor-identified current event related to the module's specific region. The instructor provided a summary of the current event with links for more information. Students were then asked to (1) specifically identify how geography played a role in the current event, (2) reference information from the course materials and (3) encourage exploration of the current events through online student-to-student discussion. Students in the intervention class (WRG1) were required to participate in these discussions as part of their grade (10 per cent of total grade). Grades were assigned based upon the number of postings (one initial comment and one follow-up per student) and each post had to include direct references from the readings, online presentations, and other related material to demonstrate that the student actually understood the material and its relationship to the current event.

Students in the control group (WRG2) were given the current event to review, but discussion was not required nor assigned any type of grade, and no guidelines were given to guide discussion. Participation was also not required nor given any type of grade. At the end of the semester, both classes were given an online final exam that included multiple-choice questions and short-essay questions based on these specific current event topics.

Introduction to Human Geography Study

In the Introduction to Human Geography course (Spring 2010), there were a total of 160 students evenly divided across the intervention and the control classes. The intervention class (IHG1) participated in writing-intensive activities to discuss the material being covered in the classroom, whereas the other control group (IHG2) was not given the writing-intensive activities. The writing-to-learn activities included notecards for students to provide feedback or answer questions on course material and in-class worksheets that reviewed course material. The notecards and the worksheets were distributed and collected during class time to encourage participation from all students and were assigned one extra credit point each for completion and relevance.

Data Sources

Multiple data sources and methods of analysis were used in both studies to assess the impact of writing as a tool for increasing comprehension and perceptions toward writing in a geography class. The first research question aimed to reveal students' perceptions of writing as a learning tool in geography classes. Student feedback for both the courses was gathered from an institutional WAC survey, and a pre- and post-intervention Writing in Geography survey was administered to the intervention and control classes of the Introduction to Human Geography course. Both the WAC survey and the Writing in Geography survey contained closed- and open-ended questions. Descriptive statistics and responses to open-ended questions are presented, and a Fisher's exact test[2] was run on the data from the Writing in Geography survey.

The second research question sought to detect quantitative improvements in student performance. Student performance data sources for the World Regional Geography course included grades for current event discussions and final exams, whereas the Introduction to Human Geography course analyzed two major exam grades. Statistical analyses of student grades for both studies were conducted using independent samples *t*-tests.

Results

Student Perceptions of Writing-to-learn for Comprehension

Overall, student surveys from both the studies revealed that writing is moderately valued as a mode to improve comprehension. The quantitative results from the WAC survey suggest that students in both the courses generally had positive perceptions of writing as a learning tool (Table 1). More than half of the students in the World Regional Geography class (WRG1) felt that the writing helped them learn course material (57.1 per cent), facilitated class discussion (64.3 per cent), helped them come with original ideas related to the material (61.5 per cent), engaged them in the material (69.2 per cent), understand the reading assignments (66.7 per cent) and also seeing the writing of others was helpful (61.5 per cent). For online classes, the facilitation of class discussion and engagement was particularly important from a pedagogical perspective as interaction in online courses can enhance student satisfaction with the class (Swan, 2001; Lapadat, 2002; Picciano, 2002; Skinner, 2007).

Likewise, with the exception of facilitating student discussion (43.5 per cent), a comparable percentage of students in the Introduction to Human Geography class (IHG1) agreed/strongly agreed that writing helped them learn the course material (57.1 per cent), come up with original ideas related to the material (58.7 per cent) and got them engaged (58.7 per cent) (Table 1). As facilitating class discussion was also one of the desired outcomes of the writing interventions in this large class, this was a disappointing, but not surprising result. So, although writing was perceived as helpful for learning content, it was not able to overcome the anonymous and sometimes imposing nature of a large class.

Qualitative comments from the WAC survey lent additional insight into the quantitative results. In the World Regional Geography class (WRG1), 7 out of 14 students who completed the WAC survey offered specific comments related to how writing improved the students ability to learn the material. Six out of these seven were positive—reflecting that writing was helpful in engagement, interest, comprehension of concepts and assistance in reading. The seventh comment was neutral. Sample comments from the World Regional Geography WAC survey were as follows:

- "Current events that related to the material was [sic] unique and fun. Thank you... The writing was fun and kept my interest".
- "The discussion posts for each region were most helpful. I found them to be beneficial in way [sic] the textbook could not".
- "I do think it's helpful to write about current events to reinforce the material".
- "... I did not particularly find either of them helpful, but I didn't find them unnecessary either".

Qualitative comments from the WAC survey conducted in the Introduction to Human Geography class (IHG1) showed that 14 out of 17 students who commented on the writing assignments as a tool for learning had positive comments. This is in contrast to 3 out of 17 students who felt that the writing experience had been negative. Samples of both sentiments can be seen in these statements:

- "I think it helps us learn and forces us to make connections".
- "I enjoyed the writing assignments in this class and I learned more about the geography of different areas [because] I had to do my own research on them".

Table 1. WAC survey, all responses

	World Regional Geography Intervention Class (response rate—72.2%)			Introduction to Human Geography Intervention Class (response rate—75%)		
	Agree/ strongly agree	Disagree/ strongly disagree	Neither agree nor disagree	Agree/ strongly agree	Disagree/ strongly disagree	Neither agree nor disagree
The writing in this class helped me learn the course material	57.1% (8)	28.6% (4)	14.3% (2)	57.1% (36)	15.9% (10)	27.0% (17)
The writing in this class facilitated student discussion	64.3% (9)	21.4% (3)	14.3% (2)	43.5% (27)	22.6% (14)	33.9% (21)
The writing in this class helped me come up with my own ideas about the material in this course	61.5% (8)	15.4% (2)	23.1% (3)	58.7% (37)	11.1% (7)	30.2% (19)
The writing in this class got me engaged with the course material	69.2% (9)	23.1% (3)	7.7% (1)	58.7% (37)	14.3% (9)	27.0% (17)
Seeing/hearing the writing of the other students in the class was helpful*	61.5% (8)	7.7% (1)	30.8% (4)	*	*	*
The writing assignments helped me understand the reading assignments*	66.7% (8)	16.7% (2)	16.7% (2)	*	*	*

* Not applicable in the Introduction to Human Geography Intervention Class (IHG1).

- "I enjoyed the writing assignments. I felt like I learned a lot by all the writing we did".
- "The writing assignments particularly the research paper was at best confusing and distracting from the course".

As part of the pre- and post-semester writing in Geography survey conducted in both the intervention and control Introduction to Human Geography classes, students were asked their opinion on whether writing in a geography course can help them better understand geographical concepts and applications. At the end of the course, positive responses from students in the intervention class (strongly agree or agree) increased dramatically from 49 per cent to 81 per cent. A Fisher's exact two-sided test showed that the difference in the responses from the pre-survey to the post-survey were statistically significant ($p = 0.0004$). Although the percentage of students in the control class who answered this same question using strongly agree or agree increased from 65 per cent to 77 per cent, a Fisher's exact test shows that this increase was not statistically significant. Two interesting points emerge here. First, there was a 16 per cent difference in initial perceptions of the value of writing as a learning tool for geography between the intervention and control class. In reviewing the characteristics of the class participants, we found that the demographics were comparable, so we do not have a rationale for this. Second, although both classes had generally positive first impressions about writing as a learning tool, these results show that the writing activities changed the intervention classes' perceptions about how writing in a geography course helped them to better understand geographical concepts and applications.

Survey data indicate that students who had been exposed to writing in these geography classes perceived WAC as a mostly positive and helpful tool for learning geographical concepts and applications. This trend is in keeping with other geographers' descriptive statistics and anecdotal conclusions about writing in geography classes (Maraffa, 1985; Winchell & Elder, 1992; Walker, 1996; Cook, 2000; Haigh, 2001; Chappell, 2006; Dummer *et al.*, 2008; McGuinness, 2009). Although these generally positive perceptions were encouraging, we also wanted to know whether the writing had an immediate impact on student learning via student performance in conventional assessment methods.

Analysis of World Regional Geography Student Performance

In the World Regional Geography writing intervention study, the hypothesis was that students enrolled in the intervention group (WRG1) would earn higher grades on the final exam than students enrolled in the control group (WRG2), particularly on those exam questions that pertained to the current event discussion. To test the hypothesis for the World Regional Geography classes, we calculated an independent samples *t*-test comparing the mean final exam score for WRG1 and for WRG2. No significant difference was found ($t\ (38) = -0.093$, $p > 0.05$). The mean of WRG1 ($m = 73.22$, $\text{sd} = 14.86$) was not significantly different from the mean of WRG2 ($m = 73.636$, $\text{sd} = 13.38$). A second independent samples *t*-test was then calculated comparing the mean aggregate scores for the exam questions related to the current event activities for WRG1 and for WRG2. No significant difference was found ($t\ (38) = -0.957$, $p > 0.05$). The mean of WRG1 ($m = 8.22$, $\text{sd} = 5.81$) was not significantly different from the mean of WRG2 ($m = 6.45$, $\text{sd} = 5.82$). Results are presented in Table 2.

Table 2. *t*-Test comparing student performance

	Intervention class mean	Control class mean	*t*	*p*
World Regional Geography				
Final exam	73.22	73.64	− 0.093	0.927
Current event questions	8.22	6.45	− 0.957	0.345
Introduction to Human Geography				
Exam I	39.48	40.34	− 1.058	0.292
Exam II	40.34	39.55	0.766	0.445
Final research papers	27.43	26.8	1.345	0.181

A third analysis looked for correlations within WRG1 to see whether there were noticeable differences between performance on the current event activity grade and on the final exam question grade. Table 3 shows the Pearson correlations as calculated examining the relationship between the current event activity and its related final exam question (Europe, Latin America and Southeast Asia). A weak correlation that was not significant was found in all three cases, respectively (r (18) = 0.429, $p > 0.05$; r (18) = 0.214, $p > 0.05$; r (18) = 0.380, $p > 0.05$).

Analysis of Introduction to Human Geography Student Performance

For the Introduction to Human Geography study, student performances on the two exams were analyzed with a similar hypothesis that students enrolled in the intervention class (IHG1) would earn higher grades in their two exams than those students enrolled in the control class (IHG2). To test the hypothesis for the Introduction to Human Geography, we calculated an independent samples *t*-test comparing the mean Exam I and Exam II scores for IHG1 and for IHG2. For the first exam, no significant difference was found (t (120) = − 1.058, $p > 0.05$). The mean of IHG1 ($m = 39.48$, sd = 4.465) was not significantly different from the mean of the IHG2 ($m = 40.34$, sd = 4.513). For the second exam, no significant difference was found (t (120) = −0.766, $p > 0.05$). The mean of IHG1 ($m = 38.81$, sd = 5.052) was not significantly different from the mean of IHG2 ($m = 39.55$, sd = 5.604). Results are presented in Table 2.

Our second research question sought to determine whether writing-to-learn activities helped to improve student exam performance in their respective geography courses. For both the World Regional Geography study and the Introduction to Human Geography study, the quantitative analyses showed no statistically significant difference in performance between the students who had the writing intervention and the control group. From these results, we cannot assert that writing activities helped students to better understand geographical concepts and applications as manifested quantitatively through improved student grades.

Discussion

As the students perceived that writing positively impacted their comprehension of the material, but the exam scores did not reflect this learning, we had to reflect upon the value of student perceptions as measures of learning in relation to traditional measures of performance (i.e. exam grades). The positive reactions of the students toward writing in

Table 3. Pearson correlations for current event grade and related final exam question

		Europe final exam question			Latin America final exam question			Southeast Asia final exam question
Europe ($N = 18$)	Pearson correlation	0.429	Latin America ($N = 18$)	Pearson correlation	0.214	Southeast Asia ($N = 18$)	Pearson correlation	0.380
	Significant (two-tailed)	0.076		Significant (two-tailed)	0.394		Significant (two-tailed)	0.119

our studies indicate that the students saw the value of writing as a cognitive learning strategy in the comprehension and application of geographic concepts. Put in context, studies about writing in geography classes have almost exclusively used student feedback as their determinant for the worth of writing as a learning tool (e.g., Cook, 2000; Haigh, 2001; Hyers, 2001; Park, 2003; Savin-Baden & Van Nierkerk, 2007; McGuinness, 2009; Summerby-Murray, 2010). Consistent with the broader literature on student perceptions of learning, there is evidence to support student perceptions as an indicator of actual learning from both process (Hawthorne, 1998) and content acquisition (Jones, 2003). Schmidt (2004, p. 463) recommends the use of student attitudes as the best alternative to traditional exam assessments and furthermore states, "as writing-to-learn activities become more commonplace, it is likely the reliability of this subscale would increase". Chesebro and McCroskey (2000, p. 301) concluded from their research that "students can provide reasonably accurate reports of the extent to which they are learning in their classrooms". At the end of her experimental study of learning experiences using writing, Tynjälä (1998, p. 227) stated that, "students' subjective learning experiences should receive much more attention in the evaluation of learning". Finally, Hawthorne (1998, p. 57) presents the value of student perceptions in a relatable way, "Just as student evaluations show something, although not everything, about teaching efficacy, student perspectives show something – if not everything – about learning efficacy".

So, if student perceptions are an important part of the picture, but not the whole picture, what are the other measures of learning in which we can determine the impact of writing on learning geographic concepts and applications? Assessment convention in higher education is often an exam. And, certainly in lower-level courses that often have high enrollments, multiple-choice and short-answer components find their way onto these exams as quick and efficient method for assessing knowledge. Given their popularity, we attempted to determine student learning using these performance measures. The results from the studies showed no statistically significant difference between the intervention and control groups in both the studies. This outcome might be explained by any one or a combination of the following: (1) writing-to-learn activities do not promote learning, (2) we did not use the best writing activity for our intended outcomes and (3) we did not ensure curricular alignment between the writing-to-learn activities and the assessment instrument. Because the results of the student perceptions surveys and supporting literature give merit to writing as a learning tool, we do not simply accept the first explanation.

The second explanation may be valid; however, writing activities were selected from researched and established WAC techniques designed to enhance learning for a variety of learning styles (Bean, 2001; Young, 2006). Sometimes it is a matter of trial and error to find the best techniques. Seshachari (1994) tried a number of different writing activities until she found one (instructor-mediated journals) that resulted in an increase in performance. This does not mean that the other writing techniques that she tried were ineffective, but perhaps only ineffective in the limited scope of improving student scores. Similarly, writing techniques may be better suited to different populations. Future research on writing methods used in geography courses, therefore, should be designed to consider other variables such as gender, ethnicity and age. This type of research among various sub-groups of students will be valuable in expanding the research done by Hyers (2001) to understand and address differences in learning styles.

The third explanation requires an evaluation of curricular alignment between the higher-order writing assignments and the assessment of learning tools used in the experimental classes, in this case, traditional multiple-choice exams and some short-answer questions. Although the questions on our more traditional mode of assessment tested the material covered in the higher-order writing activities, they did not emphasize the learning process itself and encourage the students to engage in reflective or analytical thinking. Tynjälä (1998) and Chesebro and McCroskey (2000) caution against using traditional recall tests to assess the impact of writing on student learning. A student's relationship with the content, his/her familiarity with that content and the extent to which he or she is engaged with the activity and topic can all influence the success of writing as a mode of learning (Ackerman, 1993). This concept of knowledge transmission is based on a 'constructivism' point of view that is more in line with the thought that traditional exams may not be able to determine actual changes in students' knowledge because these exams tend to test surface learning of material and memorization (Tynjälä, 1998). These traditional methods of assessment give no idea of the quality of student learning, such as the changes in students' knowledge and the development of their subject expertise. Conversely, if the intent of writing in a class is to generate higher-order learning such as critical thinking, then the mode of assessment should more closely match the intent of the writing activity. Senko *et al.* (2008) suggest that performance-approach goals, particularly multiple-choice exams, are best suited for surface learning and not the deep learning desired from writing-to-learn activities. Schmidt (2004) and Bowman (2010) stress the importance of using assessments that are appropriate for the activities and provide more complete information about what students have learned and are able to do with their knowledge; in this case, assessment should include specific opportunities for students to demonstrate deeper learning. A limitation of multiple-choice exams is that they test a student's ability to select a choice that best represents the answer rather than originating that answer on her own. Even short-answer exam questions offer limited time and space for students to fully develop their thoughts. An alternative assessment to measure the impact of the writing activity on learning would be a take-home writing assignment that tested the student's ability to analyze, explain, problem-solve and communicate their knowledge.

In our case, we organized the exam questions to closely match or mirror the content covered in the writing activities. In addition, the questions were both multiple choice and short answer. In retrospect, our exams may have focused too much on surface learning and not probed deeper learning as experienced by the students. Considering this, in future classes using writing-to-learn techniques, we will include alternative measures to evaluate student learning. Senko *et al.* (2008) recommended that exam questions should ask for students to elaborate beyond class material. For example, in the World Regional Geography class, rather than asking students to answer questions directly related to the current event class discussions, the exam question would present a different current event and ask the students to relate the appropriate geographic concepts to explain and interpret this event.

As one of the goals of writing in classes is to foster the development of expert knowledge and higher-order thinking in students, a pivotal part of the development of this expertise is "converting formal knowledge into an expert's informal knowledge and skills" (Tynjälä, 1998, p. 212). This can be done by applying formal knowledge to solve problems. In lieu of an exam, a final essay assignment for the Introduction to Human Geography course might be to take on a role and explain to an alternative audience an issue using geographic concepts in lay language (see example below).

> Imagine you are the mayor, with a degree in geography, of a small town near a large metropolitan area. The decennial census has been completed and limited data are just now being released. You realize there has likely been significant undercounting in your community. Write a speech that you will deliver at the next town-hall meeting as to the importance of the census. Discuss the purpose of the census, what undercounting is, and what the political, economic, social, fiscal and environmental impacts of undercounting are. Use Pattison's geographic traditions to frame your speech. Make recommendations to your constituents about how you will overcome the undercounting problem and why they should complete the next decennial census.

Returning to the importance of student perceptions of writing as a valuable learning tool, we were intrigued by the significant number of students who responded to the survey questions as 'neither agree nor disagree' in both surveys (Table 1). These numbers averaged 17.7 per cent for the World Regional Geography students and 31.9 per cent in the Human Geography students. These results suggest that some students (1) may still not make the connection for the potential of writing as a learning tool, (2) did not recognize writing-to-learn activities as 'writing' (as opposed to conventional essays or papers) or (3) were not affected by these particular writing activities. The survey instruments we used did not provide us with adequate data to address this point. In lieu of conjecture, we recommend that future studies of writing-to-learn as a learning tool should include interviews with students to uncover reasons behind this ambivalence.

Furthermore, although the open-ended responses in the surveys administered provided a substantial amount of qualitative feedback, we feel research would benefit from additional data from semi-structured interviews and focus groups of students who have the opportunity to reflect upon the writing assignments and give details about their experience with using writing as part of their learning process (similar to Dummer *et al.*, 2008). This is consistent with Hawthorne's (1998) recommendation for an increased use of student voice, through open-ended interviews with students, in evaluation of education programs, including those related to writing.

In summary, we recommend the continued evaluation of writing as a learning tool within Geography courses. These evaluations should include the exploration of alternative methods of evaluation other than traditional exam scores for determining the impact of writing on student learning. Finally, consistent with the existing literature, we support the use of student feedback for assessing the impact of writing activities, but recommend that more in-depth conversations be held with students to discern the extent to which students felt the writing best assisted them. We will continue to evaluate different writing activities into geography courses as a way to engage students and hopefully shift the mindset of students about the importance and relevance of writing in any discipline, especially geography.

Conclusion

Writing offers an important opportunity for increasing students' learning of the discipline of geography and better preparing geography students to enter and function in a rapidly globalizing world. At its best, writing in classes can encourage students to focus on thinking and understanding, rather than just memorizing facts (Schmidt, 2004, p. 459).

In the field of geography, this can lead to benefits including the development of higher conceptual skills that will provide working geographers with effective verbal and written communication skills. Writing activities can also be used to provide geography students with theoretical and real-world scenarios in which they can practice applying concepts and ideas learned in class through working through the writing activities. In this way, students develop the habit of thinking spatially and get the importance of understanding concepts of space, place, environment, society and representation. Although "examinations are very poor predictors of any subsequent performance, such as success at work" (Gibbs & Simpson, 2004/2005, p. 7), writing-to-learn activities in classes offer opportunities to learn a practical still that is important in the contemporary working world (see Clark & Higgitt, 1997). Therefore, writing-to-learn activities are a long-term professional commitment.

The quantitative and qualitative results from these analyses provided feedback on the impact of writing activities in the geography classroom. In both courses, we found that although student grades in the writing intervention geography classes did not improve when compared with the non-writing class, students still believed that writing helped them to better understand and engage in the course material, facilitated class discussion and helped to clarify concepts presented in readings or in-class lectures. On the basis of these findings, we conclude that writing does have a role for learning in geography courses. The exact nature of this learning merits further inquiry.

Notes

1 Seshachari did not use any statistical tests to determine the significance of the grade differences between the courses in which she used the writing intervention and courses in which she did not use the writing intervention.

2 Fisher's exact test is a statistical significance test that is used with two nominal values. The Fisher's exact test was determined to be appropriate in this situation due to the small sample size with low expected values. "Fisher's exact test is more accurate than the chi-squared test or G-test of independence when the expected numbers are small" (McDonald, 2009, pp. 70–75).

References

Ackerman, J. M. (1993) The promise of writing to learn, *Written Communication*, 10(3), pp. 334–370.

Bean, J. C. (2001) *Engaging Ideas: The Professor's Guide to Integrating Writing, Critical Thinking and Active Learning in the Classroom* (San Francisco, CA: Jossey-Bass).

Biggs, J. B. (2003) *Teaching for Quality Learning at University* (Buckingham: Society for Research into Higher Education).

Bowman, N. A. (2010) Can 1st year college students accurately report their learning and development? *American Educational Research Journal*, 47(2), pp. 466–496.

Britton, J. N., Burgess, T., Martin, N., McLeod, A. & Rosen, H. (1975) *The Development of Writing Abilities (11–18)* (London: MacMillan Educational for the Schools Council).

Carter, M. (2007) Ways of knowing, doing, and writing in the disciplines, *CCC*, 58(3), pp. 385–418.

Chappell, A. (2006) Using the 'grieving' process and learning journals to evaluate students' responses to problem-based learning in an undergraduate geography curriculum, *Journal of Geography in Higher Education*, 30(1), pp. 15–31.

Chesebro, J. L. & McCroskey, J. C. (2000) The relationship between students' reports of learning and their actual recall of lecture material: A validity test, *Communication Education*, 49(3), pp. 297–301.

Cisero, C. A. (2006) Does reflective journal writing improve course performance? *College Teaching*, 54(2), pp. 231–236.

Clark, G. & Higgitt, M. (1997) Geography and lifelong learning: A report on a survey of geography graduates, *Journal of Geography in Higher Education*, 21(2), pp. 199–213.

Connor-Greene, P. A. (2000) Making connections: Evaluating the effectiveness of journal writing in enhancing student learning, *Teaching of Psychology*, 27(1), pp. 44–46.

Cook, I. (2000) Nothing can ever be the case of 'us' and 'them' again: Exploring the politics of difference through border pedagogy and student journal writing, *Journal of Geography in Higher Education*, 24(1), pp. 13–27.

Cunningham, M. (1990) A geography boost: Writing across the curriculum, *Journal of Geography*, 89(5), p. 226.

Daniel, B. & Odom, M. L. (2010) WAC at three levels: The evolution of a program—WAC strategies to build a WAC program. Paper presented at the Writing Across the Curriculum 2010 Conference, Bloomington, IN.

Dummer, T. J. B., Cook, I. G., Parker, S. L., Barrett, G. A. & Hull, A. P. (2008) Promoting and assessing: 'Deep Learning' in Geography fieldwork: An evaluation of reflective field diaries, *Journal of Geography in Higher Education*, 32(3), pp. 459–479.

Estaville, L. (1988) Debate: A teaching strategy for geography, *Journal of Geography*, 87(1), pp. 2–4.

Fistek, M. A. (1998) Writing in the Social Science Department, in: R. Andrews (Ed.) *Journal on Writing Across the Curriculum*, 9, pp. 82–88 (Plymouth, NH: Plymouth State College).

Fournier, E. J. (2002) World regional geography and problem-based learning: Using collaborative learning groups in an introductory-level world geography course, *The Journal of General Education*, 51(4), pp. 293–305.

Fulwiler, T. (1984) How well does writing across the curriculum work? *College Composition and Communication*, 46(2), pp. 113–125.

Fulwiler, T. (1988) Evaluating writing across the curriculum programs, in: S. H. McLeod (Ed.) *Strengthening Programs for Writing Across the Curriculum. New Directions for Teaching and Learning*, no. 36, pp. 61–75 (San Francisco: Jossey-Bass).

Gedye, S. & Chalkley, B. (2006) *Employability Within Geography, Earth and Environmental Science* (Plymouth, NH: The Higher Education Academy GEES Subject Centre).

Gibbs, G. & Simpson, C. (2004/2005) Conditions under which assessment supports students' learning, *Learning and Teaching in Higher Education*, 1, pp. 3–31.

Haigh, M. J. (2001) Constructing Gaia: Using journals to foster reflective learning, *Journal of Geography in Higher Education*, 25(2), pp. 167–189.

Harasim, L. M. (2000) Shift happens: Online education as a new paradigm in learning, *The Internet and Higher Education*, 3, pp. 41–61.

Harrison, M., Short, C. & Roberts, C. (2003) Reflecting on reflective learning: The case of geography, earth and environmental sciences, *Journal of Geography in Higher Education*, 27(2), pp. 133–152.

Hawthorne, J. I. (1998) Student perceptions of the value of WAC, *Language and Learning Across the Disciplines*, 3(1), pp. 41–63. Available at http://wac.colostate.edu/llad/v3n1/hawthorne.pdf (accessed May 2011).

Hooey, C. A. & Bailey, T. J. (2005) Journal writing and the development of spatial thinking skills, *Journal of Geography*, 104(6), pp. 257–261.

Hyers, A. D. (2001) Predictable achievement patterns for student journals in introductory earth science courses, *Journal of Geography in Higher Education*, 25(1), pp. 53–66.

Jones, T. L. (2003) Self-reported Instrument for measuring student learning outcomes. Proceedings of the 2003 American Society for Engineering Education Annual Conference & Exposition, Nashville, TN.

Lapadat, J. C. (2002) Written interaction: A key component in online learning, *Journal of Computer-Mediated Communication*, 7(4). Available at http://www.ascusc.org/jcmc/vol17/issue4/lapadat.htm (accessed December 2010).

Libbee, M. & Young, D. (1983) Teaching writing in geography classes, *Journal of Geography*, 82(6), pp. 292–293.

Mabrito, M. (2004) Guidelines for establishing interactivity in online courses, *Innovate*, 1(2). Available at http://www.innovateonline.info/index.php?view=article&id=12 (accessed December 2010).

Mansfield, T. D. (2005) Geography and schools. Paper presented at the Annual General Meeting of the Canadian Association of Geographers, University of Western Ontario, London, ON.

Maraffa, T. (1985) Ungraded writing assignments in geography classes, *Journal of Geography in Higher Education*, 9(2), pp. 165–169.

McDonald, J. H. (2009) *Handbook of Biological Statistics*, 2nd edn (Baltimore, MD: Sparky House).

McGuinness, M. (2009) Putting themselves in the picture: Using reflective diaries in the teaching of feminist geography, *Journal of Geography in Higher Education*, 33(3), pp. 339–349.

McLeod, S. H. (1987) Defining writing across the curriculum, *Writing Program Administration*, 11(1–2), pp. 19–24.

Michalak, S. (1989) Writing more, learning less? *College Teaching*, 37(1), pp. 43–45.

National Survey of Student Engagement (NSSE) (2008) *Promoting Engagement for All Students: The Imperative to Look Within*. Available at http://nsse.iub.edu/NSSE_2008_Results/docs/withhold/NSSE2008_Results_revised_11-14-2008.pdf (accessed September 2010).

Park, C. (2003) Engaging students in the learning process: The learning journal, *Journal of Geography in Higher Education*, 27(2), pp. 183–199.

Picciano, A. G. (2002) Beyond student perceptions: Issues of interaction, presence, and performance in an online course, *Journal of Asynchronous Learning Networks*, 6(1), pp. 21–40.

Proctor, J. D., Sutton, P. & Michaels, G. H. (1995) Multimedia guided writing modules for introductory human geography, *Journal of Geography*, 94(6), pp. 571–577.

Savin-Baden, M. & Van Nierkerk, L. (2007) Narrative inquiry: Theory and practice, *Journal of Geography in Higher Education*, 31(3), pp. 459–472.

Schmidt, L. A. (2004) Psychometric evaluation of the writing-to-learn attitude survey, *Journal of Nursing Education*, 43(10), pp. 458–465.

Senko, C., Durik, A. M. & Harackiewicz, J. M. (2008) Historical perspectives and new directions in achievement goal theory: Understanding the effects of mastery and performance-approach goals, in: J. Y. Shah & W. L. Gardner (Eds) *Handbook of Motivation Science*, pp. 100–113 (New York: The Guilford Press).

Seshachari, N. (1994) Instructor-mediated journals: Raising critical thinking and discourse levels, *College Teaching*, 42(1), pp. 7–11.

Skinner, E. (2007) Building knowledge and community through online discussion, *Journal of Geography in Higher Education*, 31(3), pp. 381–391.

Smith, B. (2010) Surveying Worlds of Writing with Geography TAs. Available at http://mendota.english.wisc.edu/~WAC/page.jsp?id=124&c_type=article&c_id=4 (accessed August 2010).

Stainer, L. (1997) Peer assessment and group work as vehicles for student empowerment: A module evaluation, *Journal of Geography in Higher Education*, 21(1), pp. 95–98.

Summerby-Murray, R. (2010) Writing for immediacy: Narrative writing as a teaching technique in undergraduate cultural geography, *Journal of Geography in Higher Education*, 34(2), pp. 231–245.

Swan, K. (2001) Virtual interaction: Design factors affecting student satisfaction and perceived learning in asynchronous online courses, *Distance Education*, 22(2), pp. 306–331.

Thompson, G., Pilgrim, A. & Oliver, K. (2005) Self-assessment and reflective learning for first-year university geography students: A simple guide or simply misguided? *Journal of Geography in Higher Education*, 29(3), pp. 403–420.

Tynjälä, P. (1998) Writing as a tool for constructive learning: Students' learning experiences during an experiment, *Higher Education*, 36(2), pp. 209–230.

Ventis, K. (2000) Writing to discuss: Use of a clustering technique, in: L. T. Benjamin, C. L. Brewer & M. R. Hebl (Eds) *Handbook for Teaching Introductory Psychology*, Vol. 2, pp. 87–90 (New Jersey: Lawrence Erlbaum).

Walker, D. (1996) Using maps to stimulate writing in geography classrooms, *Journal of Geography*, 95(4), pp. 158–161.

White, E. (1990) The damage of innovations set adrift, *AAHE Bulletin*, 43(3), pp. 3–5.

Winchell, D. & Elder, D. (1992) Writing in the geography curriculum, *Journal of Geography*, 96(1), pp. 273–276.

Young, A. (2006) *Teaching Writing across the Curriculum*, 4th edn (Upper Saddle River, NJ: Pearson/Prentice Hall).

Capstone portfolios and geography student learning outcomes

Joann Mossa

Department of Geography, University of Florida, Gainesville, FL, USA

Due to increasing demands regarding student learning outcomes and accreditation, a capstone portfolio was added to assess critical thinking and communication skills of geography majors at a large public university in the USA. The portfolio guidelines were designed to be adaptable to a flexible curriculum where about half of the requirements within the major are structured. This paper discusses the portfolio requirements and assessment rubrics developed following multiple portfolio iterations, gives guidance to instructors, and synthesizes potential benefits for students, departments and employers.

Introduction

Academic accountability regarding student learning has been increasing and administrators at institutional, state, and national levels others are looking for evidence of student learning beyond course grades (Huisman & Currie, 2004; Middaugh, 2007). Accountability is tied with accreditation, a voluntary quality assurance process (e.g., http://www.voluntarysystem.org/) designed to help higher education institutions make curriculum and other improvements in an effort to improve student learning, yet essentially imperative in that the vast majority of academic institutions undergo this "voluntary" process. In most countries, government organizations coordinate the process; in the USA, it has involved peer review by accreditation commissions and member institutions for more than a century, with the federal government having peripheral involvement (http://www.advanc-ed.org/). Despite concerns of many critics (e.g., Huisman & Currie, 2004; McLendon, Hearn, & Deaton, 2006; Shin, 2010), regular reporting of program goals and student learning outcomes, assessment methods, results, and use of results are expected in present-day academia, one of many challenges in the changing landscape of higher education (Erickson, 2012).

Student learning outcomes fall in different categories, such as content knowledge, critical thinking skills and communication skills, and realistically, many other categories including those identified by Valid Assessment of Learning in Undergraduate Education program associated with the American Association of Colleges and Universities (http://www.aacu.org/value/rubrics/index.cfm). Content knowledge is commonly assessed through exams, with some institutions evaluating retained knowledge when students are nearing graduation. Departments can implement exams from a national organization such as the Educational Testing Service or ETS® Major Fields tests (http://www.ets.org/mft/about) or Area Concentration Achievement Test (ACAT®) (http://www.collegeoutcomes.com/lv1/faq.aspx) in the USA, although these exams are not available for geography and many other subjects. Because each institution has somewhat different course requirements, and geography degrees tend to be part structured and part flexible, one alternative is that each institution can develop their own exam with questions from several faculty regarding major areas in the structured part of the curriculum, considering diverse approaches in assessing achievement of content (Ediger, 2001).

For the other learning outcomes, such as critical thinking and communication skills, options for assessment are less clear. Consequently, many institutions are at a stalemate because of the limited examples and models for assessments. Many instructors believe that they integrate critical thinking skills into their classes, yet this is particularly challenging in large introductory classes, where lectures are the primary mode of delivery and multiple choice exams are typically used for much of the assessment. In an international survey, geography faculty identified critical thinking, framing research questions, reflectivity, and creativity as the most challenging research skills to teach early in the undergraduate curriculum (Walkington et al., 2011). Students can showcase communication skills and critical thinking skills in smaller specialized classes with projects, assignments, and thoughtful instructor feedback during the latter stages of their program, but ultimately their products and projects are dispersed and diverse, and generally not revisited or improved following the final assessment.

Capstone courses are a means to assess critical thinking and communication. Synthesis, reflection, and key geographic ideas can be an important component of capstone courses in geography (Hovorkaa, 2009), and the transition between schooling and the work world or a graduate program is an opportune time to reflect on one's skills, accomplishments, next steps and long-term goals. Prior studies in the USA have found that almost half of 707 regionally accredited colleges and universities use capstones as part of their institution's assessment program, although they are more common at smaller colleges and universities than larger ones (Berheide, 2007). Capstone courses are also in widespread use at many large institutions in the USA (Beyer, 2001; Kerrigan & Jhaj, 2007; Rhodes & Agre-Kippenhan, 2004; Rowles, Koch, Hundley, & Hamilton, 2004).

One mechanism for capturing some aspects of critical thinking and communication skills relevant to geography is a portfolio. Portfolios have been used for a variety of purposes, such as a Personal Reflective Portfolio (Kneale, 2002), a booklet that follows students throughout their programs. Students use these as a tool to identify weaknesses and to develop personal academic goals, not driven by other's agendas. This is an interesting and beneficial process where students are in geography throughout their university education, but it does not fit in well with the broader undergraduate education in the USA in which geography coursework typically constitutes between one-fourth and one-half of the higher education degree, with some students changing to the geography major as late as the third or fourth year of study. No particular instructor or tutor regularly interacts with the student to discuss his or her progress, or is assigned to review individual goals and plans, although many interact informally with an undergraduate advisor or academic coordinator or a faculty member whose academic interests align closely with theirs.

A capstone portfolio can meet the needs of assessing critical thinking skills and communication skills and fits within the synthesis purview of capstone courses. Somewhat different than a student thesis or dissertation where critical thinking, analysis, evaluation, effective communication, and independence are components (Hill, Kneale, Nicholson, Waddington, & Waverly, 2011), portfolios allow for a wide range of products to show progress and achievement (Ediger, 2001) and have high potential for creativity and innovation. Students can choose and improve their best prior work from geography and related classes, employment, internships and experiences, or create new products which integrate their learning in different classes. The portfolio requires additional work by students, but professional gains are likely from the process, particularly for those seeking employment.

For faculty reviewers involved in evaluation of learning outcomes, it requires less effort than obtaining and compiling final projects from instructors of several different classes.

In considering a geography capstone portfolio at a large public university in the USA, limited published work was available to provide guidance. Prior study of capstones in general found that students wanted more examples of capstone final products and clear grading criteria (Kerrigan & Jhaj, 2007). To address these students' needs, a mini-grant regarding Capstone Portfolios was submitted to the Association of American Geographers' (AAG) Enhancing Departments and Graduate Education (EDGE) project. One priority area of the EDGE project was the employability agenda (Gedye, Fender, & Chalkley, 2004) of the discipline. The portfolio constitutes a product that facilitates feedback regarding learning outcomes from employers. As part of a professional development course, guidance and feedback on portfolios also met another EDGE priority area, to prepare students for connections to business, government, and nonprofit employers. This paper discusses implementation, requirements and potential content of capstone portfolios. In order to produce clear grading criteria, and a stronger final product, a rubric was developed with input and discussion from faculty and students. The experiences and lessons learned are potentially beneficial for the wider geographic community, particularly departments considering the addition of a portfolio requirement.

Student learning outcomes and portfolios

Establishing student learning outcomes for all majors at our university began about 10 years ago. These were discussed in departmental faculty meetings, and were reviewed and approved by college and university committees. Subsequently, they were published in the undergraduate catalog, currently an online document which few students read in much detail. The geography major has three content learning outcomes, and one each for critical thinking and communication skills, each with subcomponents. For this paper, the focus is on the latter two outcomes. The critical thinking/analytical learning outcome for geography states that students will "analyze geographic information and apply interpretation of data toward problem solving or modeling." The communication skills learning outcome states that students will "interpret and effectively communicate information spatially, graphically and/or with statistics." These are published in the undergraduate catalog, mentioned periodically on the major's list-serve, discussed and listed on syllabi of intermediate and advanced courses, and reviewed in detail on the first day of Senior Seminar, the required departmental capstone course.

In our geography degree programs, about half of the curriculum is specified, and the other half is flexible with options in different categories. All majors take introductory physical geography and accompanying lab, a human geography course, a regional course, statistics outside the department, quantitative analysis for geographers, and two techniques courses, at least one involving Geographic Information Systems (GIS), and Senior Seminar. To assess content knowledge, an exam was conducted with input from several faculty members. A few years ago, a capstone portfolio was added to assess critical thinking and communication skills.

Portfolios have many potential functions and benefits, for the student, department, employers, and outside groups (Table 1). For students, functions of portfolios include reflection on learning and a vehicle for facilitating student employment in government, business, and other sectors. The department has a means of assessing student learning outcomes, a method for accreditation of prior learning, products to assist in departmental self-evaluation and improvement, and evidence to evaluate how student knowledge

Table 1. Benefits of geography capstone portfolios for students, departments, and employers.

Benefits for students

- Assists in reflection and self-evaluation
- Synthesizes varied, specific accomplishments
- Useful in preparing for interviews by reviewing prior work
- Convinces employers about skills, abilities, and work ethics
- Assesses one's own progress in his/her career development
- Sees and evaluates the patterns in work preferences and values
- Opportunity to improve prior work products from earlier faculty feedback

Benefits for departments

- Demonstrates and documents student learning outcomes such as critical thinking and communication
- Identifies gaps in knowledge, understanding, thinking, and communication
- Tool for potential curriculum modification, concepts, and skills reinforcement
- Helpful for recruiting new majors unaware of professional geography and its applications
- Leads to better alumni successes, improving departmental reputation
- Successful alumni have better potential to give back targeted donations to department
- Portfolios showcase student work products to employers who misunderstand geography
- The content can educate administration on geographic work and the use of technology in geography
- Useful for accreditation and outside forms of evaluation such as external reviews

Benefits for employers

- Can see competence areas of job candidates from evidence, facilitating screening
- Improves communication, by providing springboard for discussion at job interviews
- Gives insights regarding organization, creativity and work ethic of potential employees
- Sample geographic work products help educate employers who have not previously hired geographers
- Increasing awareness of geographic technology may expand capabilities of employers

progresses over time. Employers, especially those who have not typically hired geographers, gain more awareness of geography work products.

Technically, the learning outcomes can be assessed by having a faculty/administrator (Coordinator or Chair of Curriculum) compile the results of specific projects or pieces done in several courses. This is being done at some universities (Greg Pope, Montclair State University, personal communication) and is very burdensome work. At our university, this would involve cooperation from multiple instructors of GIS courses, and to improve the curriculum, it would also involve appreciable time capturing, compiling, and saving the specific work, not just project grades. The portfolio puts more responsibility on students to compile and organize, but it gives students a product that has potential to facilitate employment, allowing the department to focus on the larger issues such as curriculum needs.

Geography portfolio content and initial input

The amount of content required for the capstone portfolio was minimal, four to six items tied to student learning outcomes. Majors are reminded of this requirement through multiple modalities including the undergraduate catalog, the departmental list-serve, faculty syllabi, and in-class reminders from instructors to keep best work and instructor comments.

Appreciable thought was given to the type of content that could be included, and a draft portfolio description was reviewed at a faculty meeting and later by the first classes to complete the portfolio. Each portfolio, it was decided, should be framed with a cover page and table of contents, and captions explaining each figure. Four pieces of required content was deemed reasonable and the list of potential content was quite diverse (Table 2), and seemed to satisfy students as being inclusive of the types of work completed in varied classes. Some faculty wanted effective writers to be given more opportunity to showcase this strength and, following this input, the list of potential types of submissions demonstrating writing abilities then expanded. However, because criteria other than written communication were being evaluated, a student would need to find other examples to meet those requirements. Comparisons are facilitated with some consistent elements, thus requirements included one map, one graph, and one statistical analysis for assessing communication. Critical and analytical thinking would be assessed by one written piece and a piece that could overlap with other products involving data analysis, with the remainder of the portfolio being flexible (Table 2). Optional content could include a resume, which could be useful for an employer; a reflective piece regarding the student's professional journey and future goals, potentially useful to students composing a letter of intent or statement of purpose for graduate study; and a references section with any relevant citations or data sources.

Students had the option of submitting an original piece or improving earlier submissions, the latter potentially helpful for those students wanting to showcase their best work to employers. This practice varies somewhat from a traditional portfolio in which the original pieces remain intact. We felt it best to maintain this flexibility because some students had lost or misplaced original pieces, the software with which the work was originally done had sometimes changed or improved since the original submission, and faculty grades and comments might or might not be on a hard copy, as feedback is often given electronically.

While the portfolio was intended for program-level assessment nearing graduation, student learning is assessed at different levels, such as course, program, and institution. Some varied content was allowed from courses in other departments, especially areas related to the major. Similarly, content from projects that students created in part-time professional employment, internships, and related experiences was allowable. Students

Table 2. Examples of potential content in a geography capstone portfolio.

- Graph analyzing primary or secondary data[a]
- Map analyzing primary or secondary data, preferably in ArcGIS[a]
- Writing with critical thinking or data analysis, such as a journal article review, literature review, data analysis project, or honors thesis[a]
- Analysis of primary or secondary data using statistics or modeling[a]
- Projects that show the ability to use other software, such as AutoCAD, Matlab, SigmaPlot for varied applications in geography
- Illustration, conceptual diagram, including a Photoshop or Illustrator product
- Table, concept map, or compilation that synthesizes information or ideas from the literature or other
- Analysis of data collected in field (GPS points, surveying skills, vegetative, hydrologic, soils or other) or laboratory (sediments, soils, water, biota, etc.)
- An economic base analysis, a site location analysis, or other human geography project
- A power point poster (printed on a page for easing viewing) or presentation (in handout form) of a project associated with a class or research in geography
- Links to web pages created (with screen capture or printout of content)
- Links to videos created (e.g., scientific animation, featuring educational content)
- Blog related to the profession
- Other professional work products relevant to geography (see instructor for questions)

[a] Required of all.

could also create new products which integrated their learning in different classes. For the most part, these were products stemming from foundations obtained in the program, and the student was still responsible for tying these to learning outcomes. However, no content from outside the department was required, nor was the portfolio intended for assessment at other levels. The portfolio was also not intended to show developing geographic knowledge gained by students over time in the program, yet some aspects of this could be interpreted with content spanning several years in programs with those goals.

The rubric and portfolio evaluation

Rubrics articulate the standards by which a product or outcome will be evaluated and are a prevalent method for evaluating portfolios (Buzzetto-More & Alade, 2006). Research on rubrics has suggested that they improve consistency in scoring, better communicate expectations and performance to students, measure student progress over time, facilitate the evaluation of complex and subjective skills, and promote learning (Jonsson & Svingby, 2007). Rubrics are generally seen by students as less arbitrary than other measures, and help improve the quality of work of university students (Andrade & Du, 2005) and young students in primary or elementary school (Andrade, Du, & Wang, 2008). Also, rubrics are valid and reliable for peer evaluation, as found in a study of 1500 peer-group evaluations over three years, where students' evaluations of their peers were similar to faculty when used with an oral presentation rubric (Hafner & Hafner, 2003).

A capstone portfolio could be assessed in different ways, and several examples for evaluating honors theses and capstones were reviewed. After a few iterations with varied rubrics, a hybrid analytic–holistic rubric (Mertler, 2001) was developed, analytic in that it listed learning outcomes as individual components but holistic in that the overall performance was evaluated for each outcome (Table 3). It does not, however, have narrative descriptions of excellent and poor work for each individual attribute. The rubric was developed by dissecting the critical thinking and communication student learning outcomes into separate logical components, and giving students specific, detailed information for what constitutes quality work in each component. Based on the student learning outcomes described previously, the five categories evaluated are as follows: (1) analyzing geographic information; (2) applying data toward problem solving or modeling; (3) spatial communication/mapping; (4) graphical communication; and (5) statistical interpretation and communication. The students were tasked with reflecting, assessing, and addressing the learning outcomes each item demonstrates, although some requested feedback for verification. Their table of contents, captions, and descriptions of each entry would help make this link, describing the purpose or why it was done and the concepts and skills shown. Their description would mention where (class, other) and when the work was done, and they could leave graded content or use feedback given by the instructor to improve on the initial submission. Together, the figure and caption should also describe data sources and associated limitations, what specifically was done to manipulate or analyze the data if relevant, and what it shows or interpretation of the final product. The students were encouraged to display other skills, including organization, creativity, and reflectivity through their cover page and table of contents, portfolio formatting and appearance, transitions and the ties to our specified learning outcomes, and through a number of artistic and optional projects on the list (Table 2). The rubric developed was reviewed by the undergraduate student learning outcomes committee and the entire faculty and was given as a resource to students in class, posted on the course e-learning site, and later added to the undergraduate section of the departmental web page.

Table 3. Rubric for evaluating student learning outcomes.

Criteria and ranking	4 (Excellent)	3 (Good)	2 (Fair)	1 (Poor)
Critical and analytical thinking				
• Analyze geographic information (understands issues; evaluates geographic literature, data, methods, and assumptions; interprets patterns, relationships, differences, and trends; and synthesizes information into coherent explanations)	Meets all criteria	Meets most criteria	Meets some criteria	Rarely meets criteria
• Apply interpretation of data toward problem solving or modeling (understands problem, poses research question, gathers and organizes data, evaluates data and applies appropriate methods, and interprets and understands findings, implications and limitations)	Meets all criteria	Meets most criteria	Meets some criteria	Rarely meets criteria
Communication				
• Spatial communication/mapping (presents meaningful spatial data effectively, without distortion and chartjunk; uses appropriate map and design elements, i.e., title and sources, scale, direction, labels, inset, appropriate categories, colors, tones, symbols, line weights, fonts, legend, etc.)	Meets all criteria	Meets most criteria	Meets some criteria	Rarely meets criteria
• Graphical communication (presents meaningful graphical data effectively, without distortion and chartjunk; uses appropriate graph and design elements, i.e., title and sources, including axis labels and units, feature labels, categories, color, tones, symbols, line weights, fonts, legend, etc.)	Meets all criteria	Meets most criteria	Meets some criteria	Rarely meets criteria
• Statistical interpretation and communication (understands variables, data, hypotheses, correctly chooses statistical tests or models, and interprets and communicates statistical findings)	Meets all criteria	Meets most criteria	Meets some criteria	Rarely meets criteria

To better comprehend concepts of spatial and graphical visual communication, a presentation was given in the capstone class. The presentation consisted of a few dozen graphs and maps from varied sources, many from journal articles and professional organizations. As a group, we discussed what we liked about each map, graph, or image and ways in which they could be improved visually and scientifically. We compared these with the rubric and discussed whether these met all, most, or some criteria or rarely met criteria listed (Table 3). Students were quite critical and collectively came up with several suggestions for improving the communication of each piece, even for items produced by highly reputable scientific and professional organizations. The presentation was posted on the class e-learning website, so that students could review and reflect further on examples discussed. Students were also given citations for helpful books involving visual communication (e.g., Tufte, 2001) and links to helpful websites for graph design (http://www.sealthreinhold.com/tuftes-rules/rule_three.php), map design (http://www.gsd.harvard.edu/gis/manual/style/), and website design (http://webdesign.tutsplus.com/articles/understanding-visual-hierarchy-in-web-design–webdesign-84), the latter not required but a skill that some wanted to demonstrate. A few portfolio examples from recent graduates, with their permissions, were accessible through our course e-learning site.

Compilation of the portfolio can be an incremental or comprehensive process, and there are advantages and disadvantages to each. If the pieces of the portfolio are submitted for evaluation throughout a term, advantages include early feedback, better understanding of the quality required, and a dispersed workload for the students, which might make the project seem less daunting. Some universities have adopted this model (Burrell Montz, East Carolina University, personal communication). Alternatively, the entire portfolio could be completed a few weeks before the end of the term, allowing for instructor feedback on a draft and avoiding the pressures of final exams. Knowing that some of the seniors were concurrently enrolled in classes where their submissions might be coming from final projects and labs, the comprehensive approach was favored for our program. Submission of a complete product is also more likely to result in a consistent appearance (e.g., same font types and sizes, other aspects of formatting), which in turn might appear more professional to employers.

Because there are advantages to both the paper and electronic format, and because computers are much more widely available and used than when Kneale (2002) found that some greatly leaned toward the paper format, it was recommended that students have copies in both formats. Paper might be useful for in-person interviews but the departmental submissions were in an electronic form such as a .doc, .docx, or .pdf file so that we could better retain records, due to storage limitations, and facilitate review by the faculty committee. Electronic files were organized into folders by the students' last name and the semester that they were enrolled in the capstone course. Through viewing submissions, the student learning outcomes committee, comprised of faculty representing different subfields, can examine and reflect on the sample portfolios and provide feedback on the content, process and next steps, and make recommendations for curriculum changes and skills and approaches that need more emphasis in the classroom.

Portfolio considerations feedback and prospective changes

Through several semesters of the portfolio requirement, we learned from student ideas, frustrations, and the faculty viewpoint. Students came up with a variety of creative ideas in compiling the portfolio, one of which was to make digital photographs of lab assignments with their camera/phone. Some students expressed frustration that they had saved prior work as.mxd (ArcGIS® Map Document) files which developed broken links when examined on a different computer. They recommended that we advise future students to save and export files in.pdf or.jpg formats. Instructors of GIS classes have been informed of this concern and are now recommending that students save portfolio work in different formats. From the faculty viewpoint, it was more work to manage portfolios that consisted of multiple files, thus we advocated the use of Adobe Professional® to merge small pdfs into a larger portfolio. This software was available in computer labs in our department, but also very affordable through a student discount at the university.

Grading is one of the major considerations in portfolios. Some instructors might feel strongly that the work is already graded, so it should not be graded again. Others might feel that the effort put into the product should be acknowledged and part of a course grade. Our choices were somewhat constrained by our university's Office of Institutional Assessment which preferred that assessments be a component of course grades. In addition, we wanted students to put some effort into the process, which would be less likely to happen without any recognition of that effort. With a rubric, it is possible to give a letter or numerical grade or give an satisfactory/unsatisfactory grade as part of the capstone assignment grade, letting the other assignments decide the course grading outcome. With the latter approach, a student could receive an unsatisfactory on the portfolio but still pass the course if the portfolio comprises only a small portion of the class grade. Ultimately, we felt the latter was best practice for us. Because this was a course requirement, students who did not submit a portfolio by the semester deadline received an "I" or incomplete for the course which could later be changed to a letter grade. Depending upon the institutional requirements, each department might have differing considerations or constraints for choosing their own best practice.

Students generally understood the expectations, were able to meet them, and had done projects in their classes that produced potential content. Most students mentioned that they would show all or parts of the portfolio to potential employers. At the end of each semester, we gave opportunity for input and gradually improved the communication processes, reminders, and refined the content requirements and rubric. In past years, some students complained that this requirement was not publicized, but objections subsided with repeated announcements over subsequent semesters.

At other institutions where a capstone course comprises more credit hours, options include changing the number of required entries or having students give an oral presentation about their work. We went with a small minimum number because this was a 1-credit semester course and also to simplify the process for students. Benefits of oral presentation are that it represents an additional form of communication that could be assessed and improved; this option would give students a chance to develop communication skills, and promote peer learning and feedback, particularly if done in a constructive way. This was not implemented in our class due to this only being a 1-credit semester course not focused exclusively on assessment and having other goals and assignments toward professional development.

Having been through the process a few times, we have also learned that there is a need to reinforce the fact that all majors have learning outcomes and highlight those in geography. In addition, there is a need to repeatedly inform and remind students to maintain content from classes with project work for the capstone portfolio. It is best practice that both learning outcomes and reminders of capstone assessments be incorporated into all course syllabi, that instructors give reminders about retaining and backing up project work on multiple media, and that students receive periodic communication through email. Instructors, both faculty and graduate teaching assistants, are given reminders to prompt students. Also, to increase communication and visibility, we have placed the capstone portfolio guidelines, associated suggestions, and rubric on our departmental webpage.

Based on what we have learned from student feedback, we anticipate some changes in curriculum. Some students suggested more classes in geospatial technologies, including more sequencing with the advanced classes which focus exclusively on projects. To a large extent, this echoes the plans of the department, partly held back by needing additional faculty in this area. Others wanted to see more classes with final projects, some with GIS as a prerequisite. This would work for some students, but not others, as many students declare the major after they have third-year status, not taking GIS until their last year, and thus not likely continuing with subsequent coursework in GIS. In recent years, more of our advanced and specialized classes include an opportunity or requirement for projects involving geospatial technologies. Another area where we anticipate changes is increasing data analysis and graphing in human geography classes. Some students focused in this area mentioned having minimal background because of the specific courses they took. Those in physical geography had abundant samples involving climatology, geomorphology and hydrology so we did not anticipate a need for these students. We are developing a matrix of our coursework to track which typically have projects connected to learning outcomes, so students know which are content-only courses or courses that mesh content learning with projects involving critical thinking and communication.

Conclusions

In this changing higher education environment, geography departments are mandated to assess student learning outcomes. To assess communication and critical thinking skills and learning outcomes, we have implemented a capstone portfolio requirement. The current portfolio requires only four to six pieces, in which a student ties his/her pieces from course projects to learning outcomes specified for our major. Through review of the portfolios, and related student feedback, the department has learned some ways in which the curriculum and specific course requirements can be enhanced, modified, and improved. Despite the portfolio compilation being additional work for students, many reported that it has facilitated conversations with prospective employers. Through feedback from employers, reviews from the faculty, and concerns from the students, we hope to continue to improve the capstone portfolio process, the geography curriculum and applied work in specific courses, and the needs of employers who recruit and hire our majors.

Acknowledgements

The author is grateful to Burrell Montz of East Carolina University for discussion and resource exchange regarding capstone portfolios, and two anonymous reviewers for their helpful comments.

Funding

This project was funded by the Association of American Geographers' (AAG) Enhancing Departments and Graduate Education (EDGE) project, sponsored by the National Science Foundation [grant number DRL-0910041].

References

Andrade, H., & Du, Y. (2005). Student perspectives on rubric-referenced assessment. *Practical Assessment, Research and Evaluation*, *10*, p. 11. ISSN 1531-7714.

Andrade, H. L., Du, Y., & Wang, X. (2008). Putting rubrics to the test: The effect of a model, criteria generation, and rubric-referenced self-assessment on elementary school students' writing. *Educational Measurement: Issues and Practice*, *27*, 3–13.

Berheide, C. W. (2007). Doing less work, collecting better data: using capstone courses to assess learning. *Peer Review*, *9*, 27–30.

Beyer, C. H. (2001). Assessment in the majors, 2000. Retrieved from http://www.washington.edu/oea/pdfs/reports/OEAReport0103.pdf

Buzzetto-More, N., & Alade, A. (2006). Best practices in e-assessment. *Journal of Information Technology Education*, *5*, 251–269. Retrieved from http://jite.org/documents/Vol5/v5p251-269Buzzetto152.pdf

Ediger, M. (2001). Assessment, geography, and the student. *Journal of Instructional Psychology*, *28*, 150–154.

Erickson, R. A. (2012). Geography and the changing landscape of higher education. *Journal of Geography in Higher Education*, *36*, 9–24.

Gedye, S., Fender, E., & Chalkley, B. (2004). Students' undergraduate expectations and post-graduation experiences of the value of a degree. *Journal of Geography in Higher Education*, *28*, 381–396.

Hafner, J., & Hafner, P. (2003). Quantitative analysis of the rubric as an assessment tool: An empirical study of student peer-group rating. *International Journal of Science Education*, *25*, 1509–1528.

Hill, J., Kneale, P., Nicholson, D., Waddington, S., & Waverly, R. (2011). Re-framing the geography dissertation: A consideration of alternative, innovative and creative approaches. *Journal of Geography in Higher Education*, *35*, 331–349.

Hovorkaa, A. J. (2009). A capstone course of "geographic ideas". *Journal of Geography*, *108*, 252–258.

Huisman, J., & Currie, J. (2004). Accountability in higher education: Bridge over troubled water? *Higher Education*, *48*, 529–551.

Jonsson, A., & Svingby, G. (2007). The use of scoring rubrics: Reliability, validity and educational consequences. *Educational Research Review*, *2*, 130–144.

Kerrigan, S., & Jhaj, S. (2007). Assessing general education capstone courses: An in-depth look at a nationally recognized capstone assessment model. *Peer Review*, *9*, 13–16.

Kneale, P. (2002). Developing and embedding reflective portfolios in geography. *Journal of Geography in Higher Education*, *26*, 81–94.

McLendon, M. K., Hearn, J. C., & Deaton, R. (2006). Called to account: Analyzing the origins and spread of state performance-accountability policies for higher education. *Educational Evaluation and Policy Analysis*, *28*(1), 1–24.

Mertler, C. A. (2001). Designing scoring rubrics for your classroom. *Practical Assessment, Research & Evaluation*, *7*. Retrieved from http://PAREonline.net/getvn.asp?v=7&n=25

Middaugh, M. F. (2007). Creating a culture of evidence: Academic accountability at the institutional level. *New Directions for Higher Education*, *140*, 15–28.

Rhodes, T. L., & Agre-Kippenhan, S. (2004). A multiplicity of learning: Capstones at Portland State University. *Assessment Update: Progress, Trends, and Practices in Higher Education*, *16*, 4–5, 12.

Rowles, C. J., Koch, D. C., Hundley, S. P., & Hamilton, S. J. (2004). Toward a model for capstone experiences: Mountaintops, magnets, and mandates. *Assessment Update: Progress, Trends, and Practices in Higher Education*, *16*(1), 1–2, 13–15.

Shin, J. C. (2010). Impacts of performance-based accountability on institutional performance in the US. *Higher Education*, *60*, 47–68.

Tufte, E. R. (2001). *The visual display of quantitative information* (2nd ed.). Cheshire, CT: Graphics Press.

Walkington, H., Griffin, A. L., Keys-Mathews, L., Metoyer, S. K., Miller, W. E., Baker, R., & France, D. (2011). Embedding research-based learning early in the undergraduate geography curriculum. *Journal of Geography in Higher Education*, *35*, 315–330.

SYMPOSIUM

Re-framing the Geography Dissertation: A Consideration of Alternative, Innovative and Creative Approaches

JENNIFER HILL*, PAULINE KNEALE**, DAWN NICHOLSON[†], SHELAGH WADDINGTON[‡] & WAVERLY RAY[^]

*Department of Geography and Environmental Management, University of West of England, UK, **Higher Education Subject Centre for Geography, Earth and Environmental Sciences, University of Plymouth, England, UK, [†]Department of Environmental and Geographical Sciences, Manchester Metropolitan University, UK, [‡]Department of Geography, National University of Ireland, Ireland, [^]Department of Geography, Texas State University – San Marcos, CA, USA

ABSTRACT *This paper reviews the opportunities and challenges for re-framing the purpose, process, product and assessment of final-year geography dissertations. It argues that the academic centralities of critical thinking, analysis, evaluation, effective communication and independence must be retained, but that the traditional format limits creativity and innovation. Re-imagining capstone projects has implications for students, faculty, departments and institutions, but greater diversity could enhance its relevance to students and employers, better aligning the student experience with the academic interests and future career demands of the 21st century graduate.*

Context

The final year of many degree programmes, internationally, involves some form of capstone project. In UK and Irish Honours programmes, the traditionally recognized format is an individual research activity assessed by a written dissertation of 8000–12 000 words (Nicholson *et al.*, 2010). This dissertation should enable students to demonstrate competence in a range of subject-based and transferable skills, and Honours-level knowledge and understanding (QAA, 2007; Harrison & Whalley, 2008), although its effectiveness has been questioned (Greenbank & Penketh, 2009). It is regarded as an indicator of deep learning, independent research and valuable training for further academic study (James, 1998), but is it a relic of a previous academic era?

There is evidence that the dissertation neither encourages students to be stakeholders in university research, nor supports them in becoming, even peripherally, members of disciplinary research communities (Healey & Jenkins, 2009a). Various studies indicate that undergraduates believe themselves to be recipients of rather than producers of research (Jenkins *et al.*, 1998; Zamorski, 2002; Brew, 2006). While Harrison and Whalley (2008) highlighted a number of positive issues around dissertations, they also reported concerns about a wide range of aspects of the process, noting that: "As a consequence, the quality of the dissertation is often less than it could have been" (Harrison & Whalley, 2008, p. 407).

Arguably, the final-year dissertation in its UK and Irish form has remained somewhat static while undergraduate programme objectives increasingly emphasize graduate employability, skills mobility and civic responsibility, alongside subject-based knowledge and understanding (Hennemann & Liefner, 2010). Graduates are expected to demonstrate autonomy, communicate clearly, engender and respond to change, be socially, economically and politically aware of local, national and global issues, and be prepared for portfolio careers (Harvey *et al.*, 1997). Additional pressures come from the rise in dominance of the global knowledge economy, reduced financial support for institutions and rapid advances in information technology (Terenzini & Pascarella, 1998; Haigh, 2002; Simplicio, 2007; Lynch *et al.*, 2008). We consider that all undergraduate Honours geography students should engage with primary research (Healey & Jenkins, 2009b), but we argue that higher-level cognitive, affective and inter-personal engagement is unlikely to be successful unless this research is relevant for the student, aligning with their academic and career interests, and is managed effectively under the constraints identified (Jenkins, 2003; Jenkins *et al.*, 2003, 2007; Turner *et al.*, 2008; Healey & Jenkins, 2009b; Jenkins & Healey, 2010).

In this paper, we argue that the dissertation process has limited long-term benefits for students individually, and it neither fulfils the broader ambitions of the research-teaching agenda nor the knowledge-transfer agenda of many departments. The authors consider it timely to examine alternative forms of the dissertation, offering the possibility of engaging with different pedagogic processes, student–student and student–supervisor relationships, and final summative products. As such, options are explored, already established in geography and other disciplines, which would present fit-for-purpose, relevant capstone experiences for geography students as alternatives to the traditional student-selected solo research project. The challenge for the geography community is to consider creatively the purpose, process, product and assessment of final-year research projects to align with the aspirations of 21st century geographers.

Opportunities and Challenges Exemplified

Changing Purpose?

A traditional research project prepares the minority of very able students for further academic study. However, for the majority of students of all abilities seeking employment outside higher education (HE), it may not instil the core personal and transferable skills required by many employers. Additionally, Troyer (1993, p. 247, cited in Hovorka, 2009) asks, "What last academic experience can [faculty] provide graduating seniors that will be valuable for citizenship in the human community?" Arguably, the dissertation could function as a broader socializing agent, instilling a sense of self and civic responsibility.

In addition to disciplinary understanding, a broader range of final-year undergraduate capstone project styles might better inspire and motivate their authors, preparing the students for further study or employment.

This paper does not seek to dilute the rigour of the dissertation product or experience. Concepts that define the rationale and goals of the majority of final-year courses; critical thinking, synthesis, evaluation and self-authorship (the capacity to internally define one's beliefs, identity and social relations *sensu* Baxter Magolda, 1999, 2009), student ownership, responsibility and engagement must remain central to the experience, with ample opportunity for critical reflection. It is suggested that more relevant and creative opportunities might offer students greater opportunity to engage in meaningful collaboration and to synthesize learning from personal, academic and professional contexts. This paper presents examples of different styles of capstone projects to meet these goals.

Changing Process?

We have identified eight styles of capstone project for geography, their challenges and opportunities, ranging from the traditional to the more innovative (Table 1).

Evolving the traditional dissertation. The undergraduate geography dissertation in the UK and Ireland is normally assessed by a written report presenting the outcomes of independent research, sometimes with an oral or poster presentation at an interim stage or alongside final submission (Nicholson *et al.*, 2010). Occasionally, outstanding students are encouraged to seek formal publication of their work, possibly jointly authored with their supervisor. Publication of final-year research project work is becoming more common through online student research journals, e.g. Reinvention (Available at http://www2.warwick.ac.uk/fac/cross_fac/iatl/ejournal/) and The Plymouth Student Scientist (Available at http://www.theplymouthstudentscientist.org.uk/index.php/pss).

Submitting dissertation research as an academic journal paper, rather than in thesis format, adds a layer of academic credibility making the process more authentic (Nicholson, in press). In part, this is because undergraduates become familiar with the styles and foci of journal articles but rarely read a thesis, and shorter word limits focus thoughts and writing. The attractions of publishing include the opportunity to disseminate work to a wider audience, to complete the research cycle, to participate in a 'real-world' research experience, to develop transferable skills and to reward good and innovative research practice (Walkington, 2008; Nicholson, in press). It can also contribute to career-development opportunities, initiate understanding of publication processes (Walkington & Jenkins, 2008) and improve motivation to perform at the highest level (Charlesworth & Foster, 1996).

The Boyer Commission (1998) stimulated a plethora of undergraduate research journals in the USA, but there are far fewer examples in the UK and Ireland (see Walkington & Jenkins, 2008). UK examples include Charlesworth and Foster (1996) who created a student journal for articles from a second-year course in hydrology. Walkington (2008) piloted a national online undergraduate geography journal *Geoverse*, developed from an in-house e-journal at Oxford Brookes University called *Geoversity*. More recently, Nicholson (in press) evaluates a simulation of the professional publication process in a final-year field-based module at Manchester Metropolitan University.

Walkington and Jenkins (2008) present a series of strategies to facilitate the publication of undergraduate research, recommending that dissemination is embedded within the

Table 1. Styles of capstone project

	Product typical and additional	Skills	Issues	Advantages	Example geography references	Examples from other disciplines
0. The traditional dissertation	Thesis 8–12 000 words	Academic writing. Information literacy.	Recipients rather than producers of research? Fit-for-purpose? Non-collaborative. Assessment consistency.	Traditional product is understood and accepted. Training for budding academics.	Harrison and Whalley (2008) Nicholson *et al.* (2010)	
1. Evolving the traditional dissertation	Journal or e-journal publication. Reduced word length. Oral presentation. Draft thesis proposal.	Academic writing. Information literacy. Publishing. Refereeing.	Workload to manage journal. Consistency of product and assessment.	Students as stakeholders. Publication for CV. Understanding the publication process. Authentic research experience.	Charlesworth and Foster (1996) Walkington (2008) Walkington and Jenkins (2008)	
2. Academic apprentice—working with university research groups	Report on research results in thesis or journal paper. Oral report on findings. Co-authorship of research-group output.	Academic writing. Information literacy. Research skills awareness. Laboratory and/or field note book.	Lack of freedom to develop own project ideas. Staff time for training. Ethical issues. Impact on overall project if unsuccessful	Developmental approach. Academic community experience. Support within research team. Development of practical research skills. Understanding the academic research process.	Hovorka (2009)	Amin and Roberts (2008)—generic Hodgson *et al.* (2008)—health and social care Whitmeyer *et al.* (2008)—geology and engineering
3. Group Research and cross-disciplinary collaboration	Group report writing. Individual report section. Reflection on research process: learning journal or reflective diary. Solution to real-life problem. Demonstration or presentation of product.	Academic writing Information literacy. Team working. Reflection. Collaboration in authorship. Peer assessment. Negotiation. Ideas exchange. Awareness of group skills.	Group dynamics. Rewarding equity of effort. Staff time to provide adequate support. Administrative challenge for cross-disciplinary work.	Team support. More complex projects are possible. Shared tasks can support mixed learning styles. Resolves lone working concerns.		Aller *et al.* (2008)—engineering and technology- Brackin and Gibson (2005)—mechanical engineering Bruhn and Camp (2004)—computer science Lynch *et al.* (2007)—information technology Maleki (2009)—engineering and technology Rohleder *et al.* (2008)—psychology and social work Rowles *et al.* (2004)—generic

4. Business aware graduates	Business plan for a new business or substantial redevelopment of an established business. Business pitch presentation.	Team working. Business awareness. Negotiation skills. Entrepreneur or employer engagement. Consultancy skills.	Identification of appropriate new businesses. Ensuring equity of challenge.	Employability skills. Business and consultancy processes awareness.	Chalkley and Gibson (2009) and Gemmel (1995)	Fisher (2007)—business Tatner and Tierney (2008)—bioscience
5. Workplace, service learning and charity sector projects	Client-focused report. Product—client requirement. Business style presentation. Reflection on research process: learning journal or reflective diary.	Academic and business writing. Information literacy. Commercial and community awareness. Business practices. Project planning. Negotiation skills. Client or community engagement.	Group dynamics. Sourcing appropriate placements. Ensuring appropriate employer involvement. Ensuring output is 'professional' standard. Comparability of projects.	Employment relevance. Induction into professional or business community. Active citizenship. Experience in community and political groups.	Bednarz *et al.* (2008); Chalkley and Harwood (1998); Nicholson *et al.* (2010) and Waddington (2001)	Leidig *et al.* (2006)—information systems, Talley (1997)—electrical engineering and Wallen and Pandit (2009)—engineering
6. Vertical project—involving students from all years	Thesis 10–12 000 words. Academic journal or e-journal paper Oral presentation. Portfolio for pre-final-year work.	Academic writing. Peer assessment. Information literacy. Team working and mentoring. Negotiation. Ideas exchange. Awareness of group skills.	Logistics for timing, student availability and group allocation. Group dynamics. Rewarding equity of effort.	Creating communities of practice. Progression built in Year 1 students get an immediate insight into research processes.		Chang (2005)—science and technology McEwen (1994)—engineering Pahl (2008)—environmental management and planning
7. Creative forms—art and design, photography and video, online technologies	Art, video, film and web resource. Critical evaluation and reflection on product and the creative process. Learning journal. Reflective diary.	Film—storyboard, editing and writing. Use of technology. Exhibition of products. Option for general public viewing. Self and peer assessment. Contributions to blogs, wikis, forums, etc. Field investigation.	Ensuring comparability in assessment criteria and standards. Technology problems.	Facilitates different learning styles. Variety of assessment and skills development. Ability to use media in different ways. Links between modules.	Latham and McCormack (2007)	Dougherty and Parfitt (2009)—engineering McGrann *et al.* (2005)—engineering MacKeogh (2006)—psychology
8. Mapping products and visualization	Map and related products Critical evaluation and reflection on product and the development process.	Use of technology. Exhibition of products. Option for general public viewing. Mapping. Data handling. GIS, remote sensing and surveying.	Technology problems. Project comparability—challenge level	Development of wide variety of skills. Ability to use media in different ways. Experience in community. Employment relevance.		Swanson and Bampton (2009)—geology Whitmeyer *et al.* (2008)—geology and engineering

requirements for such projects. Undergraduate research journals could be mainstreamed into assessment, acting as stepping stones to postgraduate research and research careers for academically able and motivated students. Evolving dissertation formats from the thesis to a journal format, with possible e-journal publication, should motivate students to perform at a higher level.

Although there are potential implications for staff workload in managing an undergraduate journal, changing the product to a shorter article may reduce assessment time. There may also be issues relating to product quality and whether inexperienced undergraduates are capable of achieving the standard for 'publication'. However, the success of in-house journals that publish material from all stages of a degree programme (e.g. *Geoversity*) indicates such difficulties can be overcome by setting appropriate standards and guidelines.

Academic apprentice model—working with university research groups. Most academics work in collaborative research groups. Working in relative isolation with occasional direction from a supervisor does not provide final-year students with a positive introduction to such academic communities of practice (Lave & Wenger, 1991; Wenger, 1998; Brew, 2003), nor does it introduce them to presenting results in an academic environment (Wagner *et al.*, 2007; Whitmeyer *et al.*, 2008). Amin and Roberts (2008) recognize that there is scope for undergraduate projects to engage students more integrally with research group discussions. These groups create and share knowledge, exchange ideas, collaborate and learn from one another reciprocally through participation (Lave & Wenger, 1991; Boud *et al.*, 1999).

A cognitive apprenticeship model lays emphasis initially on sharing and passing on what is already known before constructing new knowledge. Students might simply observe and perform basic research tasks in their first year of studies as they learn how the group operates and how they can participate. Subsequently, they might undertake more challenging problem-solving tasks, collaboratively sharing knowledge to enhance performance. There could be an extension to the model found in the physical sciences in the UK, where students often work on projects that are derived from faculty and graduate-student research interests (Stefani *et al.*, 1997). Thus, undergraduates could build collaboratively on the skills they have developed earlier in their studies, they could be integrated into cutting-edge research projects and they could potentially work in more depth than the current dissertation model permits (Rowley & Slack, 2004). It is important, however, for students undertaking faculty-directed research to be involved centrally in decisions throughout the research process, so that they are realizing desirable learning outcomes, including elements of self-directed learning and decision-making.

Empowering students to become active participants in a research community would encourage relationship building between students and faculty in which undergraduates could be 'co-researchers', jointly constructing knowledge with their supervisors in a negotiated form of independent learning. This aligns with the concepts of students' voices in education (Oldfather, 1995; Fielding, 2001, 2004; Hadfield & Hawe, 2001; Arnot *et al.*, 2004) and, more particularly, with the concept of radical collegiality, where groups of staff and students are genuinely committed to formulating and solving problems within mutually supportive frameworks of democratic practice (Fielding, 1999). Such partnerships should offer new understandings and insights to faculty and students and should allow students to research more involved and complex issues. This should be more inspiring and motivating, with students gaining a sense of intellectual agency and legitimacy. But there are ethical issues in negotiated pedagogies, relating largely to equity and power relationships (Le Heron *et al.*, 2006), for example, the negative connotations inherent in employing students as 'menial' research labourers versus the opportunity for formative collaboration to bring about closer faculty and student engagement (Askins, 2008). There are also issues for supervisors in a relationship, which demand empathy for students as reciprocal co-researchers. Providing a balance between student autonomy and staff support is challenging but essential. There may also be issues relating to the staff time required to provide this support and to the potential effects on the overall project if the student is unable to complete their work. Finally, there are challenges for institutions concerning the requirement to accept students as equal partners in research, developing an 'inclusive emancipatory community' (Fielding, 2001, p. 130).

Group research and cross-disciplinary collaborations. Many US institutions encourage capstone courses that offer experiences beyond the independent research project (Healey & Jenkins, 2009b). This may involve group work, including external partners, aligning with the collaborative nature of many faculty research projects (Table 1). Hovorka (2009) reported that 43 per cent of departments provided group field courses or internships. Group and cross-discipline projects also mirror organizations where individuals work together, applying differing expertise to a problem and possessing increased capacity and resources over solo researchers to tackle more complex projects (Lynch *et al.*, 2007; Maleki, 2009). English and Computer Technology students at Indiana University–Purdue University, Indianapolis, collaborate on a capstone multimedia project. They are charged with envisioning, developing, presenting and evaluating web-based projects, cross-pollinating one another's learning with their respective disciplinary expertise as the project progresses (Rowles *et al.*, 2004). Rohleder *et al.* (2008) discuss issues in collaboration across two universities.

It is common practice in engineering disciplines to place students in teams to solve design and development problems from an early stage. Students build relationships with industry and are inspired through practical insights into solving current problems (Dutson *et al.*, 1997; Bruhn & Camp, 2004; Brackin & Gibson, 2005). At the College of Engineering, University of Maryland, USA, student teams undertake 4-year, student-initiated research projects, where they analyse and propose solutions to societal problems, generally involving a significant technology focus (Gemstone, 2010). Team members work as a coordinated group, investigating their project from the perspective of different disciplines, with guidance from a faculty mentor. In the final year, student teams present their research to experts in the field or external agencies and write a team thesis (Healey & Jenkins, 2009a).

This approach is inspiring for students because they are able to tackle complex problems and address 'live' issues beyond the scope of a solo investigator (Aller *et al.*, 2008). In the context of geography dissertations, group field research could also offer a solution to health and safety concerns around lone working. While such collaborations have obvious potential, it must be acknowledged that implementation may be complex. From an administrative viewpoint, collaboration even within a single discipline can often be challenging—cross-disciplinary and cross-institutional work is likely to pose even more difficulties without considerable time investment by supervisors and a flexible approach within modular structures at institutional level.

Business aware graduates. Business programmes typically have a business plan dissertation in which students research and present a plan either for a new business or to reinvigorate a current business. In each case, the student gains considerable insight into business processes, strategy and awareness of enterprise and entrepreneurship (Fisher, 2007; Chalkley & Gibson, 2009).

The capstone Environmental Enterprise Project for the BA in Environment and Business at the University of Leeds asks students to validate an environmental enterprise idea, and to develop a detailed action plan to take it to market, demonstrating how to develop an environmental innovation idea into a feasible enterprise. The assessment is primarily based on a report and the presentation of the plan to a hypothetical investment panel, which includes an entrepreneurial mentor and a venture capitalist.

Students of geography may engage in business and enterprise through the *sustainability entrepreneurship* concept: "Sustainability entrepreneurship is interdisciplinary by nature, drawing on knowledge from a range of research areas, including conventional entrepreneurship studies, business and environment studies, and newer research into environmental and social entrepreneurship" (Tilley & Parrish, 2009, p. 5). This work might be undertaken in the context of the Education for Sustainable Development agenda (e.g. Higgitt, 2006), directly linking business awareness with the elements of environmental geography.

Projects that mirror or simulate consultancy processes introduce students to relevant professional business practices and teach them about the nature of collaborative working. At the University of Aberdeen, Scotland, the final-year course in Applied and Engineering Geomorphology involves a group simulation requiring classroom, field and library work (Gemmel, 1995). Student teams work together to prepare a proposal and compete to 'win' a tender from the client assessors. Thus, proposals are evaluated against the assessment criteria and against each other. This simulates the real-world situation in which companies enter into tender competitions to win a contract, a process that all students could benefit from experiencing. One time-consuming issue with such projects is likely to be the need to source appropriate projects on an on-going basis as the work requires an appropriate level of challenge within a genuine (or at least authentic) scenario.

Workplace, service learning and charity sector projects. Dorsey (2001) explored issues around service learning to link undergraduates and faculty with local community members and professionals, such as planners, to tackle community projects (Table 1). Work-based individual and collaborative projects are well established in a range of HE institutions, such as the Work-Place Cooperative Project at the University of Leeds (Hogg, 1998). However, such opportunities are usually optional, and additional to the final-year dissertation, rather than as an alternative to it.

Electrical Engineering students at Texas A&M University, who design robots for their capstone project, benefit from the experience of mentoring local high-school students in robotics competitions as part of their capstone project work (Talley, 1997). This links the University to school education in a very practical way. Similarly, Kostelnick *et al.* (2009) integrate students into Geographical Information System (GIS) projects with a range of communities from the academic to the civic.

Team-based problem-solving is used in the final-year capstone course at the University of Queensland's Faculty of Natural Resources, Agriculture and Veterinary Science, to give students the experience of research-based consultancy. An interdisciplinary staff team solicits suitable 'problems' and clients among their contacts, including government agencies, non-governmental organizations, land-care groups and the private sector. The clients attend an initial briefing with their student 'consultants', watch presentations at the end of the first and second semesters and receive the final reports. They liaise with the students throughout the year, usually by phone and email. A flexible programme of lectures in the first semester provides the students with the skills they need to perform their tasks (Healey & Jenkins, 2009a).

There are a variety of ways in which this approach could be embedded into geography programmes. At Bournemouth University, for example, Shah and Treby (2006) describe a community-based project that gives geography students the chance to collaborate on live research activities and work placements. At Liverpool John Moores University, a small number of geography and earth science students submit a report concerning their work-based learning experience rather than producing a traditional dissertation. A survey of geography departments in the UK (Nicholson *et al.*, 2010) indicates that several institutions have the facility for dissertations to be built around work placements and work-based activities, but that few students opt for this route.

Whether these projects are solo or team-based depends on the circumstances. Three significant advantages of undertaking team projects in the community are that more complex issues can be tackled, more individuals can be involved and safety matters concerning lone working can be ameliorated. Group working does, however, have implications for assessment (Table 2) and, as with all group projects, group dynamics can have a significant influence on outcomes if not carefully managed.

Vertical projects—involving students from different years. At Glasgow Caledonian University, the School of the Built and Natural Environment offers a Vertical Project in which first-, second- and third-year students collaborate on a project in small groups. Students gradually progress from completing allocated tasks in Year 1, such as data collection, to taking on leadership and project management roles in Year 3 (Pahl, 2008). Thus, students are gradually integrated into the research process and into the community of practice of their subject. Similarly, Civil Engineering at Rhode Island University runs a design experience with undergraduate students from different years playing different roles (McEwen, 1994). This approach provides students with opportunities to work on current problems with practising engineers and groups of colleagues in a way that mirrors the professional design office.

In an interesting variation on this theme, Chang (2005) uses knowledge from one cohort and passes it to students in succeeding years; "each year students receive a body of work produced by the previous group of students and make improvements and additions to it; this process can be repeated until publishable materials are produced" (Chang, 2005, p. 387). The University of Leeds Innocence Project for volunteer law students combines both approaches, involving students passing to subsequent cohorts case histories of those claiming wrongful imprisonment (McCartney, 2006).

Proposals to operationalize opportunities in these styles are likely to meet pragmatic opposition based on practicalities in timetabling, but they do present interesting role models for growing and deepening understanding. Similar initiatives would have educational benefits through making students aware of other disciplinary concerns and real-world issues, as well as providing opportunities for developing mentoring skills.

Table 2. Assessment criteria for Honours dissertations (based on Hand & Clewes, 2000; Nicholson *et al.*, 2010)

Fundamentals of the dissertation
- Evidence of originality and perceptiveness
- Clarity of aims and topic identification
- Evidence of reading and awareness of literature
- Quality of research design and methodology
- Awareness of any shortcomings of design and methodology
- Quality of data
- Presentation, analysis, evaluation, synthesis and interpretation of data
- Conceptual awareness and theoretical understanding
- Sustained argument
- Logical organization
- Findings and conclusions justified and contextualized in the literature
- Where appropriate, improvements or further developments of study

Presentation
- Standard of presentation, use of English language and structure
- Use of complex academic terminology
- Correct use of referencing conventions
- Coherent integration of illustrative materials

Administrative
- Conduct including engagement with administrative processes
- Assessment of risks and ethical considerations
- Compliance with requirements

Independence
- Ability to work independently
- Exercise of personal initiative and responsibility
- Conduct and competence during practical work
- Cognitive, intellectual, practical and personal skills
- Appropriate and correct use of ICT applications
- Reflective, critically evaluating own performance and personal development

The 'X Factor'
- Demonstrable critical ability
- Creative thinking
- Flair and innovation

Creative forms—art and design, photography and video, online technologies. There is potential for capstone projects in which a creative 'product' is presented alongside some form of critical commentary or narrative, an approach that is common in the creative disciplines. The BA (Hons) Fashion, Design and Technology at Manchester Metropolitan University includes a 40-credit Major Project, where students design, make and show a 30-piece garment collection, in association with a 20-credit Personal Monograph. The monograph is amply illustrated with photographs taken at every stage in the process and provides: a formative Statement of Intent for the Major Project; a summative, critical and highly reflective commentary and justification of the design process and influences upon it; the development of design ideas including the making of garments; and an evaluation of complex technical processes. As a demonstration of advanced independent research work linking to underpinning theory and published sources, the monograph has the look and feel of a traditional Honours project. Moreover, the monograph provides a commentary on a body of creative work completed in a separate major project, thereby making explicit links between modules.

McGrann *et al.* (2005) discuss the process of connecting engineering and design students in a capstone project at Binghamton University, New York, to create an outdoor sculpture to be placed at the entrance of the Discovery Centre in the city. The authors highlight the opportunities and challenges afforded through working in multi-disciplinary teams. These include developing an effective dialogue using a language understood by all participants, appreciating and fulfilling professional responsibilities, and maintaining motivation across all participating students in order to successfully meet aggressive deadlines.

As a discipline, geography also offers a variety of opportunities for students to develop and apply creative skills to real-world issues. Examples include environmental design, urban regeneration planning, cartography and GIS, environmental and landscape art, and landscape interpretation.

Assessment of a level three urban geography fieldwork course in Geography, University of Southampton, is based on the use of digital photography and the creation of group websites by teams of students (Latham & McCormack, 2007). While in the field, in addition to notetaking, students also use diagrams, sketches, digital photographs and video material to record their impressions, which are then built into their websites.

The Geography Department of the National University of Ireland, Maynooth, has adapted a previously successful distance-learning approach to support final-year dissertation students through online supervision (MacKeogh, 2006). Students contribute to discussion fora, submit individual research proposals, and discuss and critique other proposals. Assessment is on the basis of online participation, literature review, an action plan, a reflective diary and completion of a research ethics form. Subsequently, students collect data and prepare their report and presentation, supporting each other using the online fora.

Products such as these will always need to be accompanied by critical, evaluative and reflective commentary. There may also be technical and logistical issues for some types of project, institutions and individuals. Nevertheless, these examples encourage thinking about creative ways of representing geographical knowledge.

Mapping products and visualization. Effective mapping and data visualization are fundamental to the 'business of geography', and making a creative mapping product is a wholly appropriate outcome for a geography capstone project. Mapping may seem initially to be a simple display process, but it has the potential to be as highly creative as the products referred to in Section 'Creative forms—art and design, photography and video, online technologies' (Table 1). Assessment processes based on field notebooks, draft and final mapping products are already available in GIS degree programmes. Whitmeyer *et al.* (2008) link geology and engineering students from two institutions for a field mapping project: "this experiment in cross-disciplinary undergraduate research was a big success, . . . enthusiastic interchange of expertise between undergraduate geology and engineering students . . . produced new, cutting-edge methods for visualizing geologic data and maps." Swanson and Bampton (2009) report on a programme linking undergraduates with academics to map the geology of the Maine coast. Eight weeks of mapping gives students field, GIS and digital mapping experience. The outcomes are presented as posters at the Northeast Geological Society of America Conference, giving the students an insight into academic research culture.

Changing Product?

The undergraduate research project in the UK and Ireland generally culminates in a written dissertation that is read only by the student, the supervisor and the second marker (Harrison & Whalley, 2008). In contrast, the concept of radical collegiality encourages widespread communication, with students informing each other and their faculty of their research findings. There are opportunities to rethink the format of research dissemination to integrate students into research communities. These include, for example, interim posters and/or oral presentations of research work, final-year conference presentations and research journals. Dissemination could also extend to professional audiences via presentations and displays.

A number of creative examples of alternative products can be found in the USA, both within and beyond the boundaries of the discipline. Troyer (1993), for example, offers a template for a metacognitive portfolio of students' work discussing their intellectual development as college students; Elliott *et al.* (1998) introduce an 'active-learning' project in which senior economics majors research topics linking works of distinguished economists; and Hovorka (2009) describes a classroom-based geography capstone course that explores 10 key geographic ideas. Course assignments include a group presentation in which students argue why a chosen geographic idea stands as the most significant one for the discipline and the 'real world'; a portfolio of personal reflections on the evolution of geographic ideas and how their knowledge of these ideas will serve them in the future; and a paper reflecting on a pivotal idea encountered during undergraduate studies and how this idea changed their perspective on the world.

At Alaska Pacific University, students in all disciplines present their capstone project research as a professional public presentation to faculty, students and the members of the public. Other classes are cancelled so that the student body may attend. The final assessment takes account of a 40–60-page research paper, the quality of the presentation and the handling of questions. While numbers of graduating students are small (approximately 70 per year), the idea of a public presentation as part of the final-year project could be adapted by larger departments and institutions (Healey & Jenkins, 2009a).

There are some examples of assessment diversity already. The MA Academic Practice students at Manchester Metropolitan University can choose whether the 60-credit (600 hour) project is assessed by a 8000–10 000-word dissertation, published papers or an *Equivalent Product*. The *Equivalent Product*, echoing Style 7 (Table 1), may involve an artefact such as CD or DVD, film, boxed game or other physical output, a recording of a performance, an event or a teaching resource (e.g. reusable learning object, web-based learning module or resource). Students seek approval of their product idea during the formative project proposal stage. The product is accompanied by a written commentary that must include an abstract, a rationale for the work, aims, a research methodology if appropriate, a reference list and conclusions.

At the University of Kent, final-year bioscience students can take a Science Communication project, which involves writing a 6000-word report on a controversial or poorly understood area of science (Lloyd, 2008). The students make an assessed oral presentation on their chosen issue in a local school and prepare a magazine article, museum display or website communicating the research issue to a wider public. This style of activity would be highly relevant for students seeking a teaching career.

Most of the 'product' options presented would require the application of skills that are already provided and developed elsewhere within a geography programme. These might reasonably include skills in fieldwork, GIS and digital mapping, together with generic employability skills such as group working, Information and Communication Technologies (ICT), oral presentations and posters. However, successful adoption of alternative, creative and innovative approaches to geography dissertations may require some additional skills development and preparation. For instance, students may need training in the peer-review process for submission to a student journal; practice in collaborative enquiry in multi-disciplinary groups; the development of digital photographic or video-editing skills; development of entrepreneurial skills; and training in approaches to mentoring. In addition, there may be a need for a departmental or institutional change of mindset that facilitates collaboration with academic and research communities and increases administrative flexibility within modularized programmes.

Changing Assessment?

The capstone project can represent a significant component of the degree assessment; up to one-third of final-year marks in the UK (Nicholson *et al.*, 2010). Summative assessment typically combines a number of elements including interim or progress reports, interim or final presentations and the written report (Harrison & Whalley, 2008), with the written element accounting for 70–100 per cent (Nicholson *et al.*, 2010). Some of the options for capstone products highlighted above would require more than a single output and considerable care would need to be taken to ensure appropriate output weighting and the application of consistent criteria within cohorts.

In the UK and Ireland, capstone projects are generally marked by two people. In departments with cohorts of 200 or more final-year students, this can involve 30 or more assessors, making consistency of assessment between markers challenging (Saunders & Davis, 1998; Webster *et al.*, 2000; Pepper *et al.*, 2001). Inter-marker variability is ameliorated through the use of carefully designed assessment criteria, alongside clear marking procedures, that are debated and revised periodically (Saunders & Davis, 1998; Pathirage *et al.*, 2007). The use of benchmark statements as a foundation for the development of assessment criteria ensures some degree of consistency in standards across the HE sector, but there is considerable variation in approaches to developing criteria at a local level (Nicholson *et al.*, 2010). Methods include teaching team discussions, working parties, formal review, consultation with external examiners and evolution through time as a product of shared experience and tradition. Assessment criteria for geography dissertations may be discipline specific or generic, having been developed at the institutional level. There is much commonality in assessment criteria across disciplines, including, for example, biological sciences Honours projects (Stefani *et al.*, 1997), social sciences including geography (Webster *et al.*, 2000; Pepper *et al.*, 2001), business and history (Hand & Clewes, 2000; Woolf, 2004). The data in Table 2 summarize the attributes and skills frequently occurring in learning outcomes and marking criteria for dissertations (Hand & Clewes, 2000; Nicholson *et al.*, 2010). Skills derived from team working and effective collaboration are conspicuously absent.

Woolf (2004), considering assessment criteria for final-year Honours projects in business and history programmes, noted the need for improved shared understanding by staff and students of the language used in criteria and the ways in which criteria are applied in marking. In a similar study of dissertation assessment in social sciences and law, Webster *et al.* (2000) revealed considerable ambiguity over the meaning and application of criteria among academic staff. The central role of Honours dissertations and significant contribution towards the final-year marks (Nicholson *et al.*, 2010) makes effective implementation of assessment criteria particularly important (Pepper *et al.*, 2001). For 'unconventional' dissertations, it becomes even more important to reach agreement on treatment (Webster *et al.*, 2000).

Broadening the style of projects will inevitably give rise to new challenges for fair and objective assessment, which may present difficulties with rigorous application of criteria (Table 2). Pathirage *et al.* (2007) highlight the apparent autonomy versus accountability dichotomy between work that is intended to represent individuality and creativity, against the necessity to implement assessment criteria—referred to as 'straight-jacketing' by Balla and Boyle (1994).

Arguably, these issues are already addressed in other disciplines, engineering and the creative arts in particular. Adopting procedures used in departments elsewhere in a university can obviate challenges about comparability and university standards. Assessment criteria will need to cater for the differences inherent in independent versus collaborative constructions of knowledge and in the differing levels of student engagement with the research process. Students working in research groups, with support from research fellows and professors, should be expected to work at a more advanced and deep level than entirely student collaborative groups and this will need to be reflected in assessment criteria. Where students are co-researching with supervisors, co-publishing and co-presenting, peer review of student articles, as a measure of assessment, could be formalized. Assessment criteria will also need to cater for increased variety in the nature of the capstone product to ensure consistency of standards and appropriate rewarding of student effort and achievement. Consistent assessment criteria could be developed for both undergraduate and taught postgraduate levels to clarify both criteria and the 'higher postgraduate' standards or demands of a dissertation submitted at the more advanced level. Students should be able to see the 'step up' to M level and rise to the challenge. Consistency across institutions will continue to be an issue, but in terms of the development of research skills, these could be included in the assessment process through a recognized Research Skills Development Framework (see Willison & O'Regan, 2007). The examples given from many disciplines reviewed above indicate the acceptability of alternatives to the traditional format.

Conclusions

Ideally a final-year research project should inspire and challenge a student to think in new ways, to apply their expertise to explore and solve problems, and to share that thinking and new experience with peers. Currently, geography dissertations in UK and Irish Higher Education Institutions (HEIs) largely constrain students to a particular process and product. We argue for broadening the format to challenge students to demonstrate their higher-order cognitive, affective and inter-personal learning skills in ways that align with their ambitions and interests (agreeing with Healey, 2010). For students who anticipate a research career, the traditional dissertation is appropriate, but could be an enhanced experience, where the student works with an active research group or produces a research-relevant product. Students seeking careers outside the Academy should be able to opt for projects that develop their understanding of workplace, charity or professional sectors, as well as developing the critical thinking, synthesis and evaluation skills that are central to all research projects. These approaches should enable students to become active participants in either academic or business and commercial communities of practice, countering Lave and Wenger's (1991) and Wenger's (1998) view that the sense of community is missing from the general student experience. More challenging is the concept of the capstone project emerging from collaborative research across years, disciplines and institutions, and the creation of student–student and student–faculty communities of practice.

Changing the format of the dissertation has implications across the curriculum. Collaborative research ideas, protocols and etiquettes would need thought and introduction from Year 1. But the benefits for research outputs, personal development and profiles are potentially huge. Where Year 1 students join academic research teams, faculty may more easily see students as young members of departments and disciplines. It could encourage mentoring to be facilitated across the student years and help to develop effective team working and writing skills. This model has the potential to work well in applied projects undertaken with external stakeholders to develop outputs that offer real impact, and build stable, long-term academy–industry links. The Business Plan capstone project already exists in some enterprise and business programmes, but it could be more widely adopted. For students intent on starting their own business, it provides an opportunity to focus on business processes, and to research and evaluate options in depth.

We recognize that any capstone product must engage students in academic or professional writing, synthesis, analysis and evaluation, and provide opportunities to develop and demonstrate creativity at an appropriate level. The contention here is that products that facilitate student work within communities of practice provide an experience of the reality of research in the Academy and the real world, which is predominantly team centred and collaborative. The ambition is for comprehensive and robust capstone learning experiences that more closely relate to processes of both knowledge creation and knowledge dissemination in the disciplines and professions. However, barriers that potentially inhibit the delivery of more creative, research-oriented capstone experiences include curriculum constructions, insufficient faculty resource allocation, static institutional values and culture, inadequate supporting infrastructure and ethical considerations. Thus, good practice must be both defined and managed appropriately, along with a desirable end state, such that newly evolved research experiences engage students in academic communities of practice and dissemination. Students should be empowered as active agents in the research process and not be alienated to independent research anchored by isolated meetings with supervisors.

Re-imagining capstone projects relies on supervisor development and shared practice, sometimes lacking in institutions (Rowley & Slack, 2004). Students must be offered opportunities to reflect upon and report their experiences, thereby informing curriculum development. A positive outcome could be a more integrated approach to planning and support for teaching, research and knowledge transfer, which emphasizes their complementary nature. The Honours dissertation should retain its central place in the undergraduate curriculum, but greater diversity could enhance its relevance to students and employers, better aligning the student experience with the academic interests and career aspirations of the 21st century geography graduate.

References

Aller, B. M., Lyth, D. M. & Mallak, L. A. (2008) Capstone project team formation: Mingling increases performance and motivation, *Decision Sciences Journal of Innovation Education*, 6, pp. 503–507.

Amin, A. & Roberts, J. (2008) Knowing in action: Beyond communities of practice, *Research Policy*, 37, pp. 353–369.

Arnot, M., McIntyre, D., Peddar, D. & Reay, D. (2004) *Consultation in the Classroom: Developing Dialogue about Teaching and Learning* (Cambridge: Pearson).

Askins, K. (2008) In and beyond the classroom: Research ethics and participatory pedagogies, *Area*, 40, pp. 500–509.

Balla, J. & Boyle, P. (1994) Assessment of student performance: A framework for improving practice, *Assessment and Evaluation in Higher Education*, 19, pp. 17–28.

Baxter Magolda, M. (1999) *Creating Contexts for Learning and Self-authorship: Constructive Developmental Pedagogy* (San Francisco, CA: Jossey-Bass).

Baxter Magolda, M. (2009) Educating for self-authorship: Learning partnerships to achieve complex outcomes, in: C. Kreber (Ed.) *The University and its Disciplines: Teaching and Learning within and beyond Disciplinary Boundaries*, pp. 143–156 (Oxford: Routledge).

Bednarz, S., Chalkley, B., Fletcher, S., Hay, I., Le Heron, E., Mohan, A. & Trafford, J. (2008) Community engagement for student learning in geography, *Journal of Geography in Higher Education*, 32, pp. 87–100.

Boud, D., Cohen, R. & Sampson, J. (1999) Peer learning and assessment, *Assessment in Higher Education*, 24, pp. 413–426.

Boyer Commission on Educating Undergraduates in the Research University (1998) *Reinventing Undergraduate Education: A Blueprint for America's Research Universities* (Stony Brook, NY: State University of New York at Stony Brook).

Brackin, M. P. & Gibson, J. D. (2005) Capstone design projects with industry: Emphasizing teaming and management tools, *ASEE Annual Conference and Exposition, Conference Proceedings*, pp. 1477–1484.

Brew, A. (2003) Teaching and research: New relationships and their implications for inquiry-based teaching and learning in higher education, *Higher Education Research and Development*, 22, pp. 3–18.

Brew, A. (2006) *Research and Teaching beyond the Divide* (London: Palgrave MacMillan).

Bruhn, R. & Camp, J. (2004) Capstone course creates useful business products and corporate-ready students, *SIGCSE Bulletin (Association for Computing Machinery, Special Interest Group on Computer Science Education)*, 36, pp. 87–92.

Chalkley, B. & Gibson, K. (2009) Enterprise education in geography: The case of the Plymouth dragons, *Planet*, 21, pp. 43–46.

Chalkley, B. & Harwood, J. (1998) *Transferable Skills and Work-based Learning in Geography* (Cheltenham: Geography Discipline Network).

Chang, H. (2005) Turning an undergraduate class into a professional research community, *Teaching in Higher Education*, 10, pp. 387–394.

Charlesworth, S. M. & Foster, I. D. L. (1996) 'Water and environmental systems': Achieving student-centred learning objectives with an undergraduate journal, *Journal of Geography in Higher Education*, 20, pp. 45–54.

Dorsey, B. (2001) Linking theories of service-learning and undergraduate geography education, *Journal of Geography*, 100, pp. 124–132.

Dougherty, J. & Parfitt, K. (2009) Framework for teaching engineering capstone design courses with emphasis on application of Internet-based technologies, *Journal of Architectural Engineering*, 15, pp. 4–9.

Dutson, A. J., Todd, R. H., Magleby, S. P. & Sorenson, C. D. (1997) A review of literature on teaching engineering design through project-oriented capstone courses, *Journal of Engineering Education*, 86, pp. 17–28.

Elliott, D., Meisel, J. & Richards, W. (1998) The senior project: Using the literature of distinguished economists, *Journal of Economic Education*, 29, pp. 312–320.

Fielding, M. (1999) Radical collegiality: Affirming teaching as an inclusive professional practice, *Australian Educational Researcher*, 26, pp. 1–34.

Fielding, M. (2001) Students as radical agents of change, *Journal of Educational Change*, 2, pp. 123–141.

Fielding, M. (2004) Transformative approaches to student voice: Theoretical underpinnings, recalcitrant realities, *British Educational Research Journal*, 30, pp. 295–311.

Fisher, C. (2007) *Researching and Writing a Dissertation: A Guidebook for Business Students*, 2nd edn (Harlow: Pearson Education Ltd).

Gemmel, A. (1995) 'Competitive' simulation in the teaching of applied geomorphology: An experiment, *Journal of Geography in Higher Education*, 19, pp. 29–39.

Gemstone (2010) *The Gemstone Program* at the University of Maryland. Available at http://www.gemstone.umd.edu/about/index.html (accessed 1 July 2010).

Greenbank, P. & Penketh, C. (2009) Student autonomy and reflections on researching and writing the undergraduate dissertation, *Journal of Further and Higher Education*, 33, pp. 463–472.

Hadfield, M. & Hawe, K. (2001) 'Voice', young people and action research, *Educational Action Research*, 9, pp. 85–99.

Haigh, M. J. (2002) Internationalisation of the curriculum: Designing inclusive education for a small world, *Journal of Geography in Higher Education*, 26, pp. 49–66.

Hand, L. & Clewes, D. (2000) Marking the difference: An investigation of the criteria used for assessing undergraduate dissertations in a business school, *Assessment and Evaluation in Higher Education*, 25, pp. 5–21.

Harrison, M. E. & Whalley, W. B. (2008) Undertaking a dissertation from start to finish: The process and product, *Journal of Geography in Higher Education*, 32, pp. 401–418.

Harvey, L., Moon, S. & Geall, V. (1997) *Graduates' Work: Organisational Change and Student Attributes* (Birmingham: Centre for Research into Quality, University of Birmingham).

Healey, M. (2010) Making creative-hops through developing creative honours projects: A proposal. Paper presented at the SRHE Student Experience Conference on Rethinking the Dissertation: The Role of Undergraduate Research in Higher Education. UCLan, Preston, 9 June.

Healey, M. & Jenkins, A. (2009a) *Developing Undergraduate Research and Inquiry* (York: HE Academy).

Healey, M. & Jenkins, A. (2009b) Developing students as researchers, in: S. K. Haslett & H. Rolands (Eds) *Linking Research and Teaching and in Higher Education. Proceedings of the Newport Nexus Conference.* Centre for Excellence in Learning and Teaching, University of Wales, Wales. Special Publication No. 1, pp. 7–11.

Hennemann, S. & Liefner, I. (2010) Employability of German geography graduates: The mismatch between knowledge acquired and competences required, *Journal of Geography in Higher Education*, 34, pp. 215–230.

Higgitt, D. (2006) Finding space for education for sustainable development in the enterprise economy, *Journal of Geography in Higher Education*, 30, pp. 251–262.

Hodgson, D., May, S. & Marks-Maran, D. (2008) Promoting the development of a supportive learning environment through action research from the 'middle out', *Educational Action Research*, 16, pp. 531–544.

Hogg, J. (1998) Evaluation of the work-place cooperative project in geography degrees at the University of Leeds, *Journal of Geography in Higher Education*, 22, pp. 126–135.

Hovorka, A. J. (2009) A capstone course of 'geographic ideas', *Journal of Geography*, 108, pp. 252–258.

James, P. (1998) Progressive development of deep learning skills through undergraduate and postgraduate dissertations, *Educational Studies*, 24, pp. 95–105.

Jenkins, A. (2003) Designing a curriculum that values a research based approach to student learning in geography, earth and environmental sciences (GEES), *Planet*, Special Edition 5, pp. 2–5.

Jenkins, A., Blackman, T., Lindsay, R. & Paton-Saltzberg, R. (1998) Teaching and research: Student perspectives and policy implications, *Studies in Higher Education*, 23, pp. 127–141.

Jenkins, A., Breen, R. & Lindsay, R. (2003) *Re-shaping Higher Education: Linking Teaching and Research* (London: Routledge).

Jenkins, A. & Healey, M. (2010) Undergraduate research and international initiatives to link teaching and research, *Council on Undergraduate Research Quarterly*, 30, pp. 36–42.

Jenkins, A., Healey, M. & Zetter, R. (2007) *Linking Teaching and Research in Disciplines and Departments* (York: Higher Education Academy).

Kostelnick, J. C., Rowley, R. J., McDermott, D. & Bowen, C. (2009) Developing a GIS program at a tribal college, *Journal of Geography*, 108, pp. 68–77.

Latham, A. & McCormack, D. (2007) Digital photography and web-based assignments in an urban field course: Snapshots from Berlin, *Journal of Geography in Higher Education*, 31, pp. 241–256.

Lave, J. & Wenger, E. (1991) *Situated Learning: Legitimate Peripheral Participation* (Cambridge: Cambridge University Press).

Le Heron, R., Baker, R. & McEwen, L. (2006) Co-learning: Re-linking research and teaching geography, *Journal of Geography in Higher Education*, 30(1), pp. 77–87.

Leidig, P. M., Ferguson, R. & Leidig, J. (2006) The use of community-based non-profit organizations in information systems capstone projects, *Working Group Reports on ITiCSE on Innovation and Technology in Computer Science Education,* 11th Annual Joint Conference Integrating Technology into Computer Science Education, Università di Bologna, Bologna, Italy. pp. 148–152.

Lloyd, D. (2008) Final year research projects in communicating science, in: M. Luck (Ed.) *Student Research Projects: Guidance on Practice in the Biosciences*, Teaching Bioscience Enhancing Learning Series (Leeds: The Higher Education Academy Centre for Bioscience), pp. 54–56.

Lynch, K., Bednarz, B., Boxall, J., Chalmers, L., France, D. & Kesby, J. (2008) E-learning for geography's teaching and learning spaces, *Journal of Geography in Higher Education*, 32, pp. 135–149.

Lynch, K., Heinze, A. & Scott, E. (2007) Information technology team projects in higher education: An international viewpoint, *Journal of Information Technology Education*, 6, pp. 181–198.

MacKeogh, K. (2006) Supervising undergraduate research using online and peer supervision, in: M. Huba (Ed.) *Proceedings of the 7th International Virtual University Conference*, 14–15 December 2006, Bratislava. TU Bratislava. Available at http://doras.dcu.ie/82/ (accessed 1 July 2010).

McCartney, C. (2006) The University of Leeds Innocence Project. Available at http://www.law.leeds.ac.uk/prospective-students/undergraduates/extra/innocence-project.php (accessed 1 July 2010).

McEwen, E. E. (1994) Integrated capstone design experience, *Journal of Professional Issues in Engineering Education and Practice*, 120, pp. 212–220.

McGrann, R., Mackiewicz, G., Walsh, J., Williams, K., Griffin, J., Hobbs, Y. & Crocker, M. (2005) Capstone design: Sculpture and structure, *ASEE Annual Conference and Exposition, Conference Proceedings*, Portland, Oregon. pp. 1485–1491.

Maleki, R. A. (2009) Business and industry project-based capstone courses: A reflection on the performance of student teams, *Industry and Higher Education*, 23, pp. 103–110.

Nicholson, D. T. (in press) Embedding research in a field-based module through peer review and assessment for learning, *Journal of Geography in Higher Education*. DOI: 10.1080/03098265.2011.552104.

Nicholson, D. T., Harrison, M. E. & Whalley, W. B. (2010) Assessment criteria and standards of the Geography Dissertation in the UK, *Planet*, 23, pp. 18–24.

Oldfather, P. (1995) Songs 'come back to most of them': Students' experiences as researchers, *Theory into Practice*, 34, pp. 131–137.

Pahl, O. (2008) Case study 8: The Vertical Project, Glasgow Caledonian University, in: F. M. Black & J. MacKenzie (Eds) *Quality Enhancement Themes: The First Year Experience*, pp. 43–45 (Glasgow: Quality Assurance Agency for Higher Education).

Pathirage, C., Haigh, R., Amaratunga, D. & Baldry, D. (2007) Enhancing the quality and consistency of undergraduate dissertation assessment, *Quality Assurance in Education*, 15, pp. 271–286.

Pepper, D., Webster, F. & Jenkins, A. (2001) Benchmarking in geography: Some implications for assessing the dissertations in the undergraduate curriculum, *Journal of Geography in Higher Education*, 25, pp. 23–35.

QAA (2007) *Geography, and Earth Sciences, Environmental Sciences and Environmental Studies Benchmark Statement* (Gloucester: Quality Assurance Agency).

Rohleder, P., Swartz, L., Bozalek, V., Carolissen, R. & Leibowitz, B. (2008) Community, self and identity: Participatory action research and the creation of a virtual community across two South African universities, *Teaching in Higher Education*, 13, pp. 131–143.

Rowles, C. J., Koch, D. C., Hundley, S. P. & Hamilton, S. J. (2004) Towards a model for capstone experiences: Mountaintops, magnets and mandates, *Assessment Update: Progress, Trends and Practices in Higher Education*, 16, pp. 1–2 and 13–15.

Rowley, J. & Slack, F. (2004) What is the future for undergraduate dissertations? *Education and Training*, 46, pp. 176–181.

Saunders, M. N. K. & Davis, S. M. (1998) The use of assessment criteria to ensure consistency of marking: Some implications for good practice, *Quality Assurance in Education*, 6, pp. 162–171.

Shah, A. & Treby, E. (2006) Using a community based project to link teaching and research: The bourne stream partnership, *Journal of Geography in Higher Education*, 30, pp. 33–48.

Simplicio, S. C. (2007) *Educating the 21st Century Student* (Bloomington: AuthorHouse).

Stefani, L. A., Tariq, V. N., Heylings, D. J. A. & Butcher, A. C. (1997) A comparison of tutor and student conceptions of undergraduate research project work, *Assessment and Evaluation in Higher Education*, 22, pp. 271–288.

Swanson, M. T. & Bampton, M. (2009) Integrated digital mapping in geologic field research: An adventure-based approach to teaching new geospatial technologies in an REU Site Program, in: S. J. Whitmeyer, D. W. Mogk, and E. J. Pyle (Eds.) *Field Geology Education: Historical Perspectives and Modern Approaches: Geological Society of America Special Paper* 461, pp. 117–133.

Talley, T. J. (1997) Using a state-wide high school robotics competition as the basis for a capstone design experience in electrical engineering, *Proceedings—Frontiers in Education Conference*, 2, pp. 775–777.

Tatner, M. & Tierney, A. (2008) Case study 6: Commercial projects for final year bioscience students, in: M. Luck (Ed.) *Student Research Projects: Guidance on Practice in the Biosciences*, pp. 51–53 (Leeds: HEA Subject Centre for Bioscience).

Terenzini, P. T. & Pascarella, E. T. (1998) Studying college students in the 21st century: Meeting new challenges, *The Review of Higher Education*, 21, pp. 151–165.

Tilley, F. & Parrish, B. D. (2009) Introduction to sustainability entrepreneurship research, *Greener Management International*, 55, pp. 5–11.

Troyer, R. J. (1993) Comments on the capstone course, *Teaching Sociology*, 21, pp. 246–249.

Turner, N., Wuetherick, B. & Healey, M. (2008) International perspectives on student perceptions of research: The role of academic development in implementing research-based teaching and learning in higher education, *International Journal of Academic Development*, 13, pp. 199–211.

Waddington, S. (2001) Working with the community: Improving the learning experience for large classes, *Journal of Geography in Higher Education*, 25, pp. 67–82.

Wagner, R., Anderson, J. L., Cao, W., Gao, Y., Ikeda, T., Jacobs, R., Johanesen, K., Mai, J., Memeti, V., Padilla, A., Paterson, S. R., Seyum, S., Shimono, S., Thomas, T., Thompson, J. & Zhang, T. (2007) Geologic wonders of Yosemite at two miles high: An undergraduate, learner-centered, team research program at the University of Southern California, American Geophysical Union, Fall Meeting 2007, abstract #ED21A-0082. Available at http://adsabs.harvard.edu/abs/2007AGUFMED21A0082W

Walkington, H. (2008) Geoverse: Piloting a national e-journal of undergraduate research in geography, *Planet*, 20, pp. 41–46.

Walkington, H. & Jenkins, A. (2008) Embedding undergraduate research publication in the student learning experience: Ten suggested strategies, *Brookes E-journal of Learning and Teaching*, 2(3). Available at http://bejlt.brookes.ac.uk/article/embedding_undergraduate_research_publication_in_the_student_learning_experi/

Wallen, M. R. & Pandit, A. S. (2009) Encouraging undergraduate engineering students towards civic engagement, *European Journal of Engineering Education*, 34, pp. 141–148.

Webster, F., Pepper, D. & Jenkins, A. (2000) Assessing the undergraduate dissertation, *Assessment and Evaluation in Higher Education*, 25, pp. 71–80.

Wenger, E. (1998) *Communities of Practice: Learning, Meaning and Identity* (Cambridge: Cambridge University Press).

Whitmeyer, S. J., de Paor, D. G., Nicoletti, J., Rivera, M., Santangelo, B. & Daniels, J. (2008) Cross-disciplinary undergraduate research: A case study in digital mapping, western Ireland, American Geophysical Union, Fall Meeting 2008, abstract #ED52A-04. Available at http://adsabs.harvard.edu/abs/2008AGUFMED52A.04W

Willison, J. & O'Regan, K. (2007) Commonly known, commonly not known, totally unknown: A framework for students becoming researchers, *Higher Education Research and Development*, 26, pp. 393–409.

Woolf, H. (2004) Assessment criteria: Reflections on current practices, *Assessment and Evaluation in Higher Education*, 29, pp. 479–493.

Zamorski, B. (2002) Research-led teaching and learning in higher education, *Teaching in Higher Education*, 7, pp. 411–427.

SYMPOSIUM

Embedding Research-Based Learning Early in the Undergraduate Geography Curriculum

HELEN WALKINGTON*, AMY L. GRIFFIN**, LISA KEYS-MATHEWS†, SANDRA K. METOYER‡, WENDY E. MILLER^, RICHARD BAKER§ & DEREK FRANCE¶

*Department of Anthropology and Geography, Oxford Brookes University, UK, **School of Physical, Environmental and Mathematical Sciences, University of New South Wales, Australia, †Department of Geography, University of North Alabama, USA, ‡Department of Geography, Texas A&M University, USA, ^Department of Geography, State University of New York College at Cortland, USA, §Fenner School of Environment and Society, The Australian National University, Australia, ¶Department of Geography and Development Studies, University of Chester, UK

ABSTRACT *This article considers the rationale for embedding research and enquiry skills early in the undergraduate geography curriculum and for making these skills explicit to students. A survey of 52 international geography faculty identified critical thinking, framing research questions, reflectivity and creativity as the most challenging research skills to teach early in the undergraduate curriculum. This article provides a range of practical examples illustrating research skill teaching from geography courses internationally. The case studies demonstrate that by embedding research skill development early, scaffolding provided throughout a degree programme can support geography students as they become producers of knowledge.*

Introduction

The Boyer Commission's (1998) call for greater undergraduate engagement in research has been heard by many university systems around the world. A body of literature has documented the advantages to students of engaging with research (McGuinness & Simm, 2003; Seymour *et al.*, 2004; Healey, 2005a, 2005b; Healey & Jenkins, 2006, 2009; Findley *et al.*, 2007; Garde-Hansen & Calvert, 2007; Hodge *et al.*, 2007, 2008; Jenkins *et al.*, 2007; Karukstis & Elgren, 2007; DeHaan, 2009; Elsen *et al.*, 2009). There is a growing recognition that research skills are essential for all students because knowing how to critically evaluate information and to enquire is of increasing importance (Brew, 2006), and research skills are required for graduates to function effectively in an increasingly complex world (Bar-Yam, 2004; Harris, 2007; Mitchell, 2009). As Scott notes (2002, p. 13):

> In a 'knowledge society' all students – certainly all graduates – have to be researchers. Not only are they engaged in the production of knowledge; they must also be educated to cope with the risks and uncertainties generated by the advance of science.

However, Harrison and Whalley (2008) have questioned the extent to which undergraduate students are equipped to engage in dissertation research as a result of their preceding years in undergraduate education, and Lovitts (2005) has even highlighted the lack of preparedness of postgraduate level students for thesis research. Students struggle with research when they do finally engage with it because no explicit attention was given in earlier years for developing the skills needed for success in research (Evans & Witkosky, 2004; DeHaan, 2009).

This article explores the skills which students need to become effective researchers, and provides case study examples of how skills teaching can be embedded into geography curricula *before* students are expected to engage in self-directed honours-level research. As the case studies we present highlight, research skills can take many forms and different universities have approached developing these skills in a range of ways, including building student skills in critiquing the research of others and communicating research findings to new audiences. These skills and practices are based on the broad definition of student research given by Willison and O'Regan (2007, p. 394) as:

> a continuum of knowledge production, from knowledge new to the learner to knowledge new to humankind, moving from the commonly known, to the commonly not known, to the totally unknown.

This inclusive definition of research opens up pedagogic spaces to engage students in the research culture of our universities. Universities can offer transformative opportunities through increased student collaboration and dialogue (often enhanced by learning technologies), in the process of developing research skills.

For the purpose of integrating research and teaching, we advocate an alternative view of research as a lifelong-learning endeavour and as a continuum that ranges from developing research skills to the production of research (Willison & O'Regan, 2007). Awareness of the research process gained in undergraduate courses can enormously enhance student skills in critically evaluating the research of others. As one Australian student noted after undertaking field-based research in a second-year field course

> I will never read a journal article with the same naivety I once had, learning first hand and having the opportunity to conduct such raw primary research will be a valued experience in my future academic career.

Undergraduate Research Skills

Research skills can be generic (such as communication and information handling) or discipline-specific (such as systems thinking and spatial analysis using GIS). Knight and Yorke (2004) argue that it is preferable to teach even the more generic skills by embedding them into the curriculum, rather than delivering them as a discrete module or course; in this way they can be delivered throughout a programme, over a longer period of time.

Key research skills have been identified and categorized in previous work on the research process (Hackling & Fairbrother, 1996; National Research Council, 2006; Kuhn & Pease, 2008). In general, they align with the six facets of Willison & O'Regan's (2007) Research Skill Development Framework, which include generating and framing questions, collecting information/data, evaluating information, organizing information, analysing and synthesizing information and communicating results. Although this sequence of research skills mirrors the research process, research skill acquisition does not have to follow a linear progression. For example, students do not need to generate questions in order to collect data. Likewise, students can practise framing questions without any assumption that answers will be provided or pursued. Students often find their first experience of research a daunting process, so by mapping skills development across a degree programme, as well as by identifying the skills that will be taught, practised and assessed at the module/course level; student-learning goals are clarified and help to support self-reflection.

An International Network for Learning and Teaching Workshop, April 2010

A workshop organized by the International Network for Learning and Teaching (INLT) in Geography, held in Washington DC, during April 2010 on the theme 'Changing Geography in Higher Education: Towards a Scholarship of Academic Practice', was attended by the authors and provided the opportunity to gather data from 52 International workshop participants on the undergraduate research skills, which they as faculty found most challenging to teach early in the undergraduate curriculum and the barriers they encountered.

A world café format (Brown & Isaacs, 2005) was adopted to collect the data over 1 hour during which participants circulated and visited tables focused on different questions. Our table was covered with a paper tablecloth (for note-making) and dedicated to the following questions:

- Which research skills do students need before they can successfully complete a capstone research project?
- Which of these do you find most challenging to teach early in the curriculum?
- Why are these skills the most challenging?

Members of our research team stayed at the table to record discussions on the tablecloth. Participants were, therefore, able to join in conversations and read notes left by previous participants and record their agreement, new ideas or make additional comments. In this way, a variety of perspectives are quickly developed around a set of questions. The method was used to elicit a range of ideas and to look for emerging patterns and consensus to inform our thinking. After the data collection, we grouped the skills into five sets collaboratively as a research team.

Table 1 categorizes the research skills identified by the INLT participants as those necessary for undergraduate success with research projects in their final year of undergraduate study. We also asked participants to identify the skills they found the most difficult to teach and to provide reasons for this. The skills consistently highlighted as most difficult to teach were critical thinking, asking and framing questions, reflectivity, creativity and understanding the research process. Interestingly, our discussions with

Table 1 The skills described by 52 INLT 2010 workshop participants as those necessary for undergraduate students to become effective researchers

Skill sets	Component skills
Communication skills	Graphicacy; Present and communicate an argument
Understanding the research process (3)	Understanding that knowledge is constructed, understanding the value of a null result, acknowledging the work of others (avoiding plagiarism), health and safety issues related to research, understanding ethical research practice
Critical thinking (8)	Asking and framing questions (7), Developing an argument (2), choosing appropriate evidence, exploring evidence and critically evaluating information, problem-solving
Emotional intelligence (6)	Creativity (6), reflectivity (meta-cognition) (6), curiosity (2), adaptability (2), collaboration skills, ability to take on constructive criticism, self-belief and self-efficacy, self-authorship, patience
Analytical skills	Synthesizing information (2), information literacy; exploring evidence

Note: The component skills are presented within five broad skill sets. Numbers in brackets are based on participants highlighting these as particularly difficult to teach.

participants revealed a link between the skills that are difficult to teach and those that are the most important for undergraduate students to develop.

We suggest that the first two skill sets described in Table 1 (communication skills and understanding the research process) support students' development of the latter three (critical thinking, analytical skills and emotional intelligence). This is reflected by one participant noting:

> Critical thinking is the most complex of [the] skills listed, hence most difficult to achieve.

The mastery of higher order skills (i.e. our skill sets) requires the ability to deploy a number of stand-alone research skills.

Our conceptualization of research skills is that they exist along a continuum from simple to complex. Simpler research skills are less interdependent, whereas complex research skills are at the interface between three broad and overlapping skill sets: Critical thinking, analytical skills and emotional intelligence. Table 2 identifies the research skills and the extent to which they are interdependent under these broad skill sets. For example, the ability to take on board constructive criticism (point 8 in Table 2) requires emotional intelligence and critical thinking; in order to develop an argument (point 9), both critical thinking and analytical skills are required, and so forth. Some skills, such as framing and asking good geographical research questions, problem solving, reflectivity and self-authorship, require a combination of all three skill sets (points 13–16). Table 2, therefore, reveals how higher order skills, such as critical thinking, can be broken down into constituent skills, which can be developed in the curriculum.

Table 2 Research skills needed for independent research tabulated under three overlapping and complementary skill sets: Critical thinking, emotional intelligence and analytical skills

	Critical thinking	Emotional intelligence	Analytical skills	Example provided
1. Choosing appropriate evidence	■			(c) (d)
2. Curiosity		■		(a)
3. Collaboration skills		■		(b) (f) (g)
4. Self-belief/self-efficacy		■		(f)
5. Patience		■		
6. Adaptability		■		(f)
7. Information literacy			■	(d)
8. Ability to take on constructive criticism	■	■		(b) (f) (g)
9. Develop an argument	■		■	(d)
10. Critically evaluating information and evidence	■		■	(c) (d)
11. Synthesizing information	■		■	(c) (d)
12. Creativity		■	■	
13. Asking or framing questions	■	■	■	(a) (d) (g)
14. Problem solving	■	■	■	(c) (d) (f)
15. Reflectivity (meta-cognition)	■	■	■	(e) (f) (g)
16. Self-authorship	■	■	■	(f) (g)

Note: Shading indicates the skill set(s) to which each component research skill contributes, with the most interdependent at the bottom of the table. Lowercase letters refer to examples in the Case Studies section: (a) framing research questions, (b) developing constructive criticism, (c) critical thinking, (d) understanding the research process, (e) teaching reflective learning, (f) self authorship, (g) scaffolded research skills development.

The liquid café format provided us with a list of shared reasons that faculty gave for finding it challenging to teach research skills early in the curriculum. These were conceptualized as barriers and relate to challenges posed by students, faculty and institutional structures. The barriers were not quantified, as participants only added new ideas and notes to the discussion; however, there was general agreement that large class size was the most significant challenge (see also Kindon & Elwood, 2009). In addition, participants noted faculty disinterest, institutional time constraints, content assessment methods (particularly compulsory final exams), lack of confidence on the part of both faculty and students, and a loss of student curiosity because of passive student-learning approaches (Scheyvens *et al.*, 2008). The discussions revealed that these barriers were often interrelated.

In terms of barriers relating to the specific skill set of *critical thinking*, participants noted that it is "Hard to explain explicitly to students what critical thinking is and how to do it" and that "Students often struggle with words like analysis, etc. but the challenge for us is how do you explain and illustrate what 'analysis' or the larger 'critical thinking' entails." Related to this was the skill of *exploring evidence* where faculty noted that "Students tend to 'believe' everything they read."

In terms of *Creativity* faculty suggested "You have it or you don't" and that "Creativity is stifled due to deadlines and timetable [which are] not fluid enough", acknowledging also that it is "also hard to be transparent about creativity in assessment criteria." *Reflectivity* caused similar problems for faculty "To be reflective on one's position in the world and to know oneself is an art." The intangibility of research skills such as creativity and reflectivity was also related to *adaptability*, which was seen to be "part of 'the person'" rather than something that can be taught.

Recognizing barriers is the first step in overcoming them. We were able to identify aspects of our own practice which addressed some of the barriers identified, and, therefore, offer our own examples in the case studies section that we have mapped on to specific skills in Table 2.

The experience of acquiring research skills can vary between individuals. For many geography students, fieldwork represents a time when a greater understanding of the research process begins to develop (Panelli & Welch, 2005). 'Slow learning' is often required for the development of complex understanding, and slowly learnt academic skills need rehearsal and practice throughout a programme (Knight & Yorke, 2004). This slow learning is best achieved through integrated and coherent progression based on the outcomes at the programme rather than the module or course level (Price *et al.*, 2008). Hence, for some students, effective research skill development requires an iterative approach, based around the model of a spiral curriculum (Bruner, 1977). This approach requires scaffolding and by definition, must begin early in the degree programme.

From the perspective of curriculum design, it is vital that research skills are made explicit to students, not least so that students can identify what they have learnt and reflect on their learning. The mapping of research skills at the programme level is seen as being increasingly important in order to ensure a progression in skills development (DeVries & Downie, 1999). Online tools to audit research skills in the curriculum have been trialled in the biosciences (Fraser *et al.*, 2007), and such auditing is important to ensure adequate progression in those skills that are slowly developed and require practice.

By considering the way the skills are integrated with and build upon each other, it is possible to break down and model research skills early in an undergraduate degree programme. A learning approach that supports the acquisition of these skills should scaffold students as it is important for students to return to the same skills and practise them often as they become increasingly independent in their own enquiries. The more independence and autonomy the student has in making choices and interpretations, the closer to independent research the activity or course is perceived to be. In the following section, we provide some examples of how the development of several of the research skills we have discussed can be integrated into the undergraduate geography curriculum.

Case Studies

This section provides practitioners with several worked examples, and gives some insight into how to introduce and engage students with the research process through skills-based approaches. The sequence of the case studies starts from the lower level researcher skill development and progresses towards more complex higher-order researcher abilities.

Framing Research Questions

Research relies on the process of questioning. Framing questions was the second most frequent component skill mentioned by the INLT 2010 workshop participants as necessary for undergraduate students to become effective researchers. Usually, little time or attention is given to help students learn to frame questions. On the contrary, there are many forces at work at undergraduate level actively discouraging student questions (Brain, 1998).

Students in an introductory geography course at Texas A&M University, USA, are provided with a remotely sensed image of the USA–Mexico border towns of El Paso, Texas and Juarez, Mexico at night. In small groups, students were prompted to brainstorm questions related to the spatial pattern of light in the image, which displays obvious differences between the northeast and southwest sides of the border. This requires skills in identifying spatial patterns, making comparisons, being curious about the source of similarities and differences, and forming questions that may be feasibly explored. Tacit knowledge about what makes a 'good' research question is then shared through dialogue between the instructor and the students by considering 'what information is needed to answer the question?', 'is the information available or collectable?' and 'could alternative explanations be limited?'

In a comparative study by Berry and Chew (2008), undergraduate students who generated a minimum of three questions per week related to course content produced higher grades in examinations than the comparison group who generated no questions. They found a significant positive correlation between the number of questions generated and students' test scores. In a similar study by Harper *et al.* (2003), a significant positive correlation was found between the conceptual depth of the student-generated questions and improvements in learning.

Developing Constructive Criticism

Second and third-year geography students in a large sports geography course (about 100 students) at the University of New South Wales-ADFA in Australia were pushed to develop their abilities to both give and receive constructive criticism (i.e. formative feedback) while working in editorial review groups. The course is centred on a collective class project that creates an 'Atlas of Australian Sport'. In this project, each student compiles data on an aspect of Australian sport in which he/she is are interested and then designs a page for the atlas. The assessment of the student's contribution to the class project is broken into five parts: Topic proposal and data source identification, draft atlas page creation, work in editorial review groups to improve the page designs, reworking of draft pages to produce final page designs and writing an introduction to the review group's section of the atlas. The editorial review groups consist of four to six students, and are created by grouping students who have chosen similar topics to research. As groupings are based on atlas page topics, they typically include both second-and third-year students.

Editorial review groups were provided with a series of initial questions (based on graphicacy skills developed in other parts of the course) for evaluating the design of the draft pages, and were encouraged to develop and ask additional questions. Individual groups operationalize the process of providing critiques in different ways. On submitting the final page, students were required to supply a description of how they responded to the critiques of their pages. At the end of the assessment, each group member rated each other group member's ability to both give and receive constructive criticism. The instructor compiles each student's average rating and returns this feedback to the student along with their final mark. The assessment provides students with the opportunity to develop skills in constructive criticism while enabling students in a large class to receive formative feedback on their work. Most students reported benefits from working in the editorial review groups. For example, one student noted:

> Members were able to receive good honest constructive feedback, that wasn't judged or assessed. Members were also provided with a good opportunity to see their work from a different perspective, and therefore take the opportunity to make any necessary changes with which to improve it.

This student noted that the benefits from the editorial review groups may not be distributed evenly, another student saw at least some benefit from this, in that it encouraged deeper student engagement with the feedback received:

> The review groups provided a different kind of feedback to what you usually get from the lecturer. What the review groups did was to really allow discussion of each student's work, and, I think, less of a tendency to uncritically adopt suggested improvements. With the lecturer's feedback, you pretty much just take it on board—they're the ones marking it, but with the peer review, you think more about whether the suggestion is a good one or not. It also helps you know that you can get help from your peers as well as from higher up.

This student noted that his undergraduate thesis work benefited from the skills he practiced in the editorial review group:

> It helped me practice being critical of someone else's work – something I found valuable when reading research papers for my thesis, where critiquing the papers I read was an important part of the process.

Interestingly, this was an engineering student taking a geography course as an elective, indicating that the skills of giving and receiving constructive criticism were transferable to another discipline.

Spronken-Smith (2005) also reported using group work to provide students with formative feedback in a student-led research project. In her problem-based learning course, students provided feedback to group members on oral and poster presentations as part of a student research conference. This formative feedback could then be used to improve the quality of written group reports.

Critical Thinking

Critical thinking is developed when students examine a situation, ask questions, gather information, communicate their findings and derive a recommendation or explanation of the situation (Paul & Elder, 2008). Research-based learning is not the only means to develop critical thinking skills, but it does provide students with the opportunity to build and refine this skill. Seymour *et al.* (2004, p. 512) state that 43 per cent of their survey respondents gained critical thinking and problem-solving skills as a part of their undergraduate research experience. To develop this skill, students need practice, first starting with guided practice and moving to untethered enquiry. Students in the Economic Geography class at the University of North Alabama, USA, were guided into critical thinking through an assignment that requires them to develop a 'continuum of development' by choosing criteria to differentiate the level of development of selected countries. First, students were required to create a definition of development that is informed by relevant literature and to create a list of characteristics that define, explain or indicate some aspect of development. Next, students find numerical or categorical data that, when combined in a manner that the students themselves determine, allow them to place a set of countries they select into their personal 'continuum of development'. The students learn that there is no definitive answer but that their continuum must be justified by the data, literature and their own explanation and interpretation. One undergraduate student noted:

> The skills I learned have greatly helped me in other classes, as I now have been trained to see things from a new perspective. I can look at a set of data, determine which factors are most important, why they're important, and how they should be used and analysed. It is the difference in knowing how to do something and knowing *how it works*. This experience provided valuable skills that supported and strengthened my Capstone Project

Understanding the Research Process

Research skills development within the Geography programmes at the University of Chester, UK, is part of a continuum from first to final year, with a second-year module 'Enquiry and Research Design' providing undergraduates with a deeper understanding of research philosophies, methodologies, data collection, problem formation and design processes. This ultimately feeds forward into the student's final year optional dissertation module, where there is a concurrent support programme of researcher skills training. In the first term, with the assistance of the tutor, the students work in small self-selecting groups, for 8 weeks, on a project to develop research objectives and to collect primary data on a range of topics. This is supported through fieldwork and weekly methodological workshops or laboratory sessions. Students submit an individual 2000-word report, which demonstrates their understanding of the underpinning research philosophies and methodologies through the analysis of primary data.

In the second term, the research process is developed further, where students work individually with tutors on pre-defined or self-directed research topics. These are narrowed down to measurable research questions through the development of a literature review. Regular small group tutorials provide students with a mechanism to share and get feedback on their ideas and findings through annotated bibliographies and informal discussions. Ultimately, the students submit a 2000-word research proposal for a hypothetical dissertation topic, identifying specific research aims, activities, ethical considerations, health and safety issues, and a research question with associated literature.

A key to successful researcher skills development for students is an understanding of the research process, from thinking critically about research questions to taking a research proposal forward to a dissertation project. The Enquiry and Research Design module scaffolds this understanding for students so that they arrive at the final year dissertation as well prepared as possible. The students registered for the 'Enquiry and Research Design' module can elect to pursue a research-led dissertation or additional taught final-year modules. Over the last three years (see Table 3), the conversion rate for students selecting the geography dissertation option has increased from 83 to 100 per cent, suggesting the positive impact of engaging in the research design process. End of module feedback typically highlights this:

> I liked how the research projects involved new research techniques and helped me with my dissertation planning.

Table 3 The conversion rate of combined honours students opting for the geography dissertation at Chester University, UK

Academic year	Students enrolled on the module 'Enquiry and Research Design'	Dissertation conversion rate (%)
2007–2008	18	83
2008–2009	14	100
2009–2010	13	100

Note: The conversion rate is the percentage of students registered on the course 'Enquiry and Research Design' who subsequently enrolled for the optional final-year geography dissertation module.

Teaching Reflective Learning

To be effective lifelong learners, students need to have well-developed skills in self-reflection. Such skills are essential to give our graduates the capacity to 'learn by doing' to deal with the increasingly complex world around us (Bar-Yam, 2004). Such learning is much more than just learning from our mistakes, it is about keeping an open mind to how we can learn from our interactions with the world around us and in so doing have the skills to learn from every situation we find ourselves in.

The Viet Nam Field School at Australian National University takes 20–25 Australian students on an 18-day field course. A reflective learning journal is a key assessment item for the course. As a new form of assessment for the majority of the students, the instructors spend a lot of time modelling reflective practice in the course by summing up discussions in the form of hypothetical learning portfolio entries. Students are encouraged to read examples of learning portfolios from previous courses, which are available on the course website (http://fennerschool-people.anu.edu.au/richard_baker/ENVS2017/students work/learningportfolios.html)

An important means of establishing the overall ethos in the course on reflection is the detailed Internet resource, produced by former students, and updated each year by that year's class on how to carry out group research. A key requirement of the learning portfolio reflects on the interconnections between different parts of the course and on what they have learnt from their peers' research presentations at the end. There is a strong focus on developing research skills and thinking critically about what research is. Students at the beginning of the course are required to produce an individual introductory-learning statement where they give a personal definition of what research is, and then throughout the course are encouraged to reflect on their changing understandings of the nature of research. A key final task is to reflect on what they have learnt about doing research.

Self-Authorship

Providing students with an opportunity for self-authorship, as first discussed by Kegan (1994), is critical to their preparation for the future. Numerous universities in the USA sponsor a day to showcase student research, such as 'Scholar's Day' at SUNY Cortland and 'Research Day' at the University of North Alabama. Many Geography departments have dedicated annual events for their own students during Geography Awareness Week. Students are encouraged to display their research in poster format and interact with an audience composed of their peers, professors, deans and the public about their work. Some students present projects developed in the classroom, their independent-study projects, and some even create new work specifically for the event. As discussed in Walkington & Jenkins (2008), displaying work in a public domain is equivalent to publication in the broadest sense. Student researchers recognize the importance of these events and strive to present high-quality relevant work.

Instructors at SUNY Cortland have extended the poster session model for use with in-class research projects. Student work is displayed throughout the semester in the geography department hallway, a frequently visited area. This often puts additional pressure on the student because their friends and other professors will see their work and often comment on it. Students and professors can be seen looking at the maps in the hallway every day and sometimes enter the geography department to enquire about the projects they were viewing. Several students in every semester receive e-mails from professors enquiring about getting a copy of the maps they viewed or the corresponding data set for use in a different classroom. Other students have self-reported that they have enrolled in courses based on interest in student projects they saw in the hallway.

As a result of this public display, the quality of student work improved. For one assignment, students submitted maps and were then told that they would be displayed in the hallway. When given the option to create a second draft without any instructor feedback, all students in the course chose to further edit their maps before they were displayed. Both their first and second drafts of the assignment were graded based on established criteria. All students improved their grade, 84 per cent improved their grade by one letter grade or more, 8 per cent improved by two-thirds of a letter grade (B to A− for example) and the remaining 8 per cent improved by one-third of a latter grade (B to B + for example).

Interestingly, students asked whether their work could be displayed for future assignments and they often spent extra time working on those projects outside of class time because of the public display element. Student comments on course evaluations have also noted the significance and importance of the public display of their work and how that causes them to strive to submit higher quality work. An added benefit of this practice is that the research projects are often improved and extended based on feedback the original author receives or because a different student wants to continue with the topic. This provides evidence to support a more formalized feedback mechanism for campus events, such as Scholar's Day, as well as for the informal display of student work. This feedback could be in the form of verbal comments or written evaluations. Bringing in local experts to provide feedback for students could not only help students improve their work, but could also lead to community partnerships.

Scaffolded Research Skills Development

The commitment to thoroughly preparing students for independent research work in their final year can be built into programme level skill requirements in a holistic way. At Oxford Brookes University, UK, an institutional commitment has been developed to embed undergraduate research pathways through all degree programmes (Huggins *et al.*, 2007). As a result of the restructuring process, the geography programme now includes a carefully scaffolded progression of activities to develop research skills incrementally. Starting in year 1, students begin with desktop research to try to answer provided research questions before they engage in framing their own questions for team-based data collection in the field in year 2. Student teams present their preliminary findings to a panel

Table 4 A representative selection of the skills which students reported they had developed in 2008 as a result of writing a journal article for an undergraduate research journal in module U21180 at Oxford Brookes University

"Individual work: by taking this research module, it was up to me to go out there and make what I wanted of it. I think that by doing this it showed that I can work on my own, develop my own ideas and complete my own research to a high standard. Communication: constantly met with my tutor, discussing ideas for my paper. Researching: I think that because this was your own piece of work, it was therefore down to you to go to the library/WWW and find what was relevant. This wasn't easy and was time consuming."
"The fact I was writing for a publication did raise the standards I used when writing the article. I spent more time checking spelling, punctuation and making sure the research was sound. I think the skills I developed mostly revolved around self-checking and research skills."
"I have developed my written skills and having to write for a different audience."
"I have learnt to be critical not only of others' work, but also my own work. I learnt also to take my time researching, not only before I started writing, but I found myself doing ongoing research, and finding things not only relevant to the specific publication, but for other work that may link in."
"The ability to spend more time thinking about important decisions (although I'm still not able to make them)"
"Keeping to strict guidelines [for a journal article]. Writing that is more academic although that comes with practice."
"This paper has definitely *helped me understand the layout I should use, which is helping as I am currently writing my dissertation.* It was also a good refresher of statistical analysis and I think it helped develop my research skills further."
"Analytical skills have received the most benefit from this module. Critically evaluating every aspect of mine."
"Objectivity, looking at both sides of the argument and writing it up in a legible manner."
"Ability to express and expand new ideas/theories. Capability of drawing on other researchers' work and applying their findings to my own ideas (seeing where they agree/disagree!) Structuring a large content of research into the allocated word limit."

of faculty as a formative feedback exercise. This precedes the summative assessment several weeks later in a conference style event. There is sufficient time for critical reflection on feedback and the refinement of communication skills. Early in year 3, the individual authoring of journal articles based on the collected field data is then mentored by a tutor. Table 4 shows quotes highlighting the skills that students from one cohort perceived they had learnt as a result of writing for a real journal. The best work is published in 'GEOverse' an online journal (Walkington, 2008). All these activities take place before students complete their dissertations. An extra curricula departmental research conference to showcase student research findings is attended by students from all year groups, helping to create a sense of identity as geography researchers (McGuinness & Simm, 2003; Walkington & Rushton, 2008). Indeed, enquiry-based learning can help to draw students into a community of practice, both within the discipline, within the classroom itself and within the broader university, by providing a shared experience upon which students can draw (Garde-Hansen & Calvert, 2007).

Conclusions

An overwhelming amount of literature argues that embedding research skills at undergraduate level is beneficial for the student-learning experience. However, there are many different curriculum models capable of producing good graduates. This article clarifies the skills which can be achieved by embedding research opportunities much earlier in the curriculum than the traditional final year project or dissertation. It has also acknowledged the challenges that arise as institutions engage with transforming their curricula. It has reported data from 52 international geography faculty members who identified critical thinking and asking and framing questions as the research skills that are most difficult to teach because of the interrelated challenges posed by institutional structures such as large class sizes, passive modes of student engagement and assessment models based solely on examinations. We have gone on to provide examples of tried and tested cases where these skills are addressed in our own institutions and to demonstrate the progression of skills teaching that can provide a scaffolded approach to the curriculum for student researchers. The case studies reveal that a dialogic and collaborative approach seems to characterize teaching that embeds research skills. This collaboration is one way to deal with large classes, but also the dialogic approaches support students as they become researchers.

We have argued that all undergraduate students will benefit from being exposed to early attempts to develop skills in enquiry and research which are embedded into geography curricula. Many of the skills required for research are complex, and the evidence from our evaluations suggests that students feel better prepared for undertaking the independent research that is required as part of undergraduate theses, dissertations and capstone courses if given the opportunity to practice research skills in advance. By developing embedded opportunities for practising research skills, we hope that students will be able to take greater responsibility for their learning on the way to becoming independent researchers, valued employees and lifelong learners.

We have provided an example of the way in which some of the skills needed for undertaking independent research (as noted by Willison & O'Regan (2007) and by INLT participants in 2010) can be broken down into their component parts and mapped for curriculum planning purposes. Case studies of component research skills have demonstrated the potential for scaffolding student exposure to research, and include implementations that can be carried out in courses at different levels (first-year, large classes, at the departmental/school level or throughout an entire institution). Although the best results may come from curricula for which faculty have mapped out and carefully planned student opportunities to develop research skills, we would encourage faculty at all levels to adapt and evaluate the strategies we have suggested in their own courses.

By embedding undergraduate research and enquiry into the geography curriculum, the student-learning experience is enhanced in a number of ways. First, the curriculum takes on a sense of 'real world' research and a relevance to the student and their learning community. Second, it develops a critical approach to the literature and a much deeper understanding, through experiential learning, about the way in which data are generated, analysed and interpreted. In an information age, this is a vitally important graduate attribute. Third, where research results in dissemination of findings through making them public, whether to the study group, cohort or more widely available, the skills learnt in 'publishing' and representing research are central to employability in the workforce. Finally, the sense of self-esteem as a result of a student becoming a producer rather than simply a consumer of knowledge can engender a sense of lifelong learning and becoming a self-sustainable learner.

The benefits of undergraduate research and enquiry transcend the discipline of geography in many ways. However, identifying oneself as a geographer often results from an engagement with research because of the geographical perspective that is developed through engaging critically with spatial patterns and processes. By embedding sufficient support in the curriculum for discovery-based learning and stressing the value that students themselves bring to the curriculum, as producers of knowledge, a genuine learning community can be created. Geography has a crucial role to play in disseminating this learning community approach to other discipline areas.

References

Bar-Yam, Y. (2004) *Making Things Work: Solving Complex Problems in a Complex World* (Cambridge, MA: NESCI, Knowledge Press).

Berry, J. & Chew, S. (2008) Scholarship of teaching and learning exemplars: Improving learning through interventions of student-generated questions and concept maps, *Teaching of Psychology*, 35, pp. 305–312.

Boyer Commission (1998) *Reinventing Undergraduate Education: A Blueprint for America's Research Universities* (Princeton, NJ: Carnegie Foundation for the Advancement of Teaching). Available at http://naples.cc.sunysb.edu/Pres/boyer.nsf/673918d46fbf653e852565ec0056ff3e/d955b61ffddd590a852565ec005717ae/$FILE/boyer.pdf (accessed June 2010).

Brain, M. (1998) *Emphasis on Teaching: The Importance of Questions* (Raleigh, NC: BYG Publishing, Inc).

Brew, A. (2006) *Research and Teaching: Beyond the Divide* (London: Palgrave Macmillan).

Brown, J. & Isaacs, D. (2005) *The World Café: Shaping our Futures through Conversations that Matter* (San Francisco, CA: Berrett-Koehler).

Bruner, J. (1977) *The Process of Education* (Cambridge, MA: Harvard University Press).

DeHaan, R. (2009) Teaching creativity and inventive problem solving in science, *CBE—Life Sciences Education*, 8(3), pp. 172–181.

DeVries, P. & Downie, N. (1999) Ensuring skills' breadth across the curriculum: A case study and a method, *UltiBase Articles*. Available at http://ultibase.rmit.edu.au/Articles/dec99/deVries2.htm (accessed June 2010).

Elsen, M., Visser-Wijnveen, G. & van Driel, J. (2009) How to strengthen the connection between research and teaching in undergraduate university education, *Higher Education Quarterly*, 63(1), pp. 64–85.

Evans, R. & Witkosky, D. (2004) Who gives a damn what they think anyway? Involving students in mentored research, *National Social Science Journal*, 23(1), pp. 21–30.

Findley, K., Green, K., Brown, G. & Bahr, D. (2007) Re-evaluating the objectives for undergraduate research, *in: 37th ASEE/IEEE Frontiers in Education Conference*, Milwauke, October 2007, Paper F1D-18. Available at http://fie-conference.org/fie2007/papers/1189.pdf (accessed June 2010).

Fraser, G., Crook, A. & Park, J. (2007) A tool for mapping research skills in undergraduate curricula, *Bioscience Education*, 9, p. 9-1. Available at http://www.bioscience.heacademy.ac.uk/journal/vol9/beej-9-1.aspx (accessed June 2010).

Garde-Hansen, J. & Calvert, B. (2007) Developing a research culture in the undergraduate curriculum, *Active Learning in Higher Education*, 8(2), pp. 105–116.

Hackling, M. & Fairbrother, R. (1996) Helping students to do open investigations in science, *Australian Science Teachers Journal*, 42(4), pp. 26–34.

Harper, K., Etkina, E. & Lin, Y. (2003) Encouraging and analyzing student questions in a large physics course: Meaningful patterns for instructors, *Journal of Research in Science Teaching*, 40, pp. 776–791.

Harris, G. (2007) *Seeking sustainability: In an age of complexity* (Cambridge: Cambridge University Press).

Harrison, M. & Whalley, B. (2008) Undertaking a dissertation from start to finish: The process and product, *Journal of Geography in Higher Education*, 32(3), pp. 401–418.

Healey, M. (2005a) Linking research and teaching: Exploring disciplinary spaces and the role of enquiry-based learning, in: R. Barnett (Ed.) *Reshaping the University: New Relationships between Research, Scholarship, and Teaching*, pp. 30–42 (Maidenhead: McGraw-Hill/Open University Press).

Healey, M. (2005b) Linking research and teaching to benefit student learning, *Journal of Geography in Higher Education*, 29(2), pp. 183–201.

Healey, M. & Jenkins, A. (2006) Strengthening the teaching-research linkage in undergraduate courses and programmes, in: C. Kreber (Ed.) *Exploring Research-Based Teaching: New Directions for Teaching and Learning*, pp. 45–55 (San Francisco, CA: Jossey-Bass/Wiley).

Healey, M. & Jenkins, A. (2009) *Developing Undergraduate Research and Enquiry* (York: Higher Education Academy). Available at http://www.heacademy.ac.uk/assets/York/documents/resources/publications/DevelopingUndergraduate_Final.pdf (accessed June 2010).

Hodge, D., Haynes, C., LePore, P., Pasquesi, K. & Hirsh, M. (2008) From enquiry to discovery: Developing the student as scholar in a networked world, in: Learning Through Enquiry Alliance (LTEA) Conference 2008, Sheffield. Available at http://cilass-resources.group.shef.ac.uk/ltea2008/Hodge_CETL_2008_keynote.pdf (accessed June 2010).

Hodge, D., Pasquesi, K., Hirsh, M. & LePore, P. (2007) From convocation to capstone: Developing the student as scholar, in: Association of American Colleges and Universities Network for Academic Renewal Conference, April 2007, Long Beach, California. Available at http://www.miami.muohio.edu/president/reports_and_speeches/pdfs/From_Convocation_to_Capstone.pdf (accessed June 2010).

Huggins, R., Jenkins, A. & Scurry, D. (2007) Developing undergraduate research at Oxford Brookes University—Recommendations and models for future development, Internal paper available at http://www2.warwick.ac.uk/fac/soc/sociology/research/cetl/ugresearch/developing_ug_research_at_brookes.pdf (accessed June 2010).

Jenkins, A., Healey, M. & Zetter, R. (2007) *Linking Teaching and Research in Departments and Disciplines* (York: The Higher Education Academy).

Karukstis, K. & Elgren, T. (2007) How to Design, Implement, and Sustain a Research-Supportive Undergraduate Curriculum (Washington, DC: on Undergraduate Research). Available at http://www.cur.org/publications/compendium.html (accessed June 2010).

Kegan, R. (1994) *In Over Our Heads: The Mental Demands of Modern Life* (Cambridge, MA: Harvard University Press).

Kindon, S. & Elwood, S. (2009) Introduction: More than methods—Reflections on participatory action research in geographic teaching, learning and research, *Journal of Geography in Higher Education*, 33(1), pp. 19–32.

Knight, P. & Yorke, M. (2004) *Learning, Curriculum and Employability in Higher Education* (London: Routledge).

Kuhn, D. & Pease, M. (2008) What needs to develop in the development of enquiry skills? *Cognition and Instruction*, 26(4), pp. 512–559.

Lovitts, B. (2005) Being a good course-taker is not enough: A theoretical perspective on the transition to independent research, *Higher Education*, 30(2), pp. 137–154.

McGuinness, M. & Simm, D. (2003) Linking teaching and research through departmental research conferences for student project work, *Planet*, Special Edition 5, pp. 21–24.

Mitchell, M. (2009) *Complexity: A Guided Tour* (Oxford: Oxford University Press).

National Research Council (2006) *Learning to Think Spatially: GIS as a Support System in the K-12 Curriculum* (Washington, DC: National Academies Press).

Panelli, R. & Welch, R. (2005) Teaching research through field studies: A cumulative opportunity for teaching methodology to human geography undergraduates, *Journal of Geography in Higher Education*, 29(2), pp. 255–277.

Paul, R. & Elder, L. (2008) *The Miniature Guide to Critical Thinking: Concepts and Tools* (Dillon Beach, CA: The Foundation for Critical Thinking).

Price, M., O'Donovan, B., Rust, C. & Carroll, J. (2008) Assessment standards: A manifesto for change. Internal paper available at http://bejlt.brookes.ac.uk/article/assessment_standards_a_manifesto_for_change/ (accessed June 2010).

Scheyvens, R., Griffin, A., Jocoy, C., Liu, Y. & Bradford, M. (2008) Experimenting with active learning in geography: Dispelling the myths that perpetuate resistance, *Journal of Geography in Higher Education*, 32(1), pp. 51–69.

Scott, P. (2002) A lot to learn: We are all researchers now, *The Guardian*, 8 January, p. 13, Available at http://www.guardian.co.uk/education/2002/jan/08/highereducation.uk1 (accessed June 2010).

Seymour, E., Hunter, A., Laursen, S. & Deantoni, T. (2004) Establishing the benefits of research experiences for undergraduates in the sciences: First findings from a three-year study, *Science Education*, 88(4), pp. 493–534.

Spronken-Smith, R. (2005) Implementing a problem-based learning approach for teaching research methods in geography, *Journal of Geography in Higher Education*, 29(2), pp. 203–221.

Walkington, H. (2008) Geoverse: Piloting a National e-journal of undergraduate research in Geography, *Planet*, 20, pp. 41–46.

Walkington, H. & Jenkins, A. (2008) Embedding undergraduate research publication in the student learning experience: Ten suggested strategies, *Brookes E-journal of Learning and Teaching*, 2(3), http://bejlt.broo

kes.ac.uk/article/embedding_undergraduate_research_publication_in_the_student_learning_experi/ (accessed June 2010).

Walkington, H. & Rushton, E. (2008) Undergraduate research conference. A first for the department of Anthropology and Geography, *Teaching News*, 2(2), pp. 11–12.

Willison, J. & O'Regan, K. (2007) Commonly known, commonly not known, totally unknown: A framework for students becoming researchers, *Higher Education Research and Development*, 26(4), pp. 393–409.

Reviewing the 'Research Placement' as a Means of Enhancing Student Learning and Stimulating Research Activity

FIONA TWEED & ROB BOAST
Department of Geography, Staffordshire University, Staffordshire, UK

ABSTRACT *The links between research and teaching have been the subject of much debate and controversy over the need for both activities to take place in universities remains current. This paper reports on the development and evaluation of a research placement module, which aimed to provide opportunities for students to work as research assistants on small-scale projects. This paper outlines the development of the research placement and reviews staff and student experience of the module. The research placement illustrates the multiple benefits of linking teaching and research in terms of enhancing learning, employability and boosting the research culture.*

Background

The links between teaching, learning and research in higher education have been the subject of debate across many discipline areas. Much of this work refers to the meta-analysis of Hattie and Marsh (1996), who found no relationship between the research productivity and the effectiveness of teaching, leading to the recommendation that universities should try to increase the links between teaching and research. Recent literature has considered the relationship between research and different modes of learning in Geography, with 'inquiry-based learning', the 'research teaching nexus' and 'research-informed teaching' being the subjects of extensive discussion and debate (e.g. Jenkins, 1995; Jenkins, 2000; Perkins, 2002; Scott, 2002; Healey *et al.*, 2003a; Jenkins, 2003; Griffiths, 2004; Lee, 2004; Healey, 2005; Spronken-Smith, 2005; Le Heron *et al.*, 2006; Healey & Jenkins, 2009; Haslett, 2009). The connection between teaching and research is generally assumed to be positive, with excellence in research often claimed as a clear link to teaching excellence (e.g. Jenkins *et al.*, 2003). Scott (2002) argues that students need to be researchers and researchers need to be teachers, underscoring the view that research and teaching are inextricably linked (Brew, 2006). Some have said that to make an issue of such links implies that they are not inherent and emphasizes the possibility of the separation of research and teaching by those driving higher education policy, particularly in the UK (e.g. Lee, 2004). Whatever the outcome of such debates, it seems clear from most of the literature of the last few years that a shift has been advocated in the way in which curricula are designed; this shift puts inquiry-based learning at the core of the undergraduate experience and encourages the idea of students as active stakeholders or scholars in a research community, reflecting the practice of academic staff (e.g. Hodge *et al.*, 2008; Healey & Jenkins, 2009).

There is ample evidence that involvement in inquiry-based learning and research activity engenders student ownership of learning and enhances knowledge and research skills acquisition (e.g. Goodlad, 1998; Bauer & Bennett, 2003; Chang, 2005; Justice *et al.*, 2007; Willison & O'Regan, 2007; Healey & Jenkins, 2009). When attempting to develop curricula that encourage inquiry and research skills development, extend students to their limits, foster independent learning and cultivate employability, finding new ways for staff and students to work together seems to be regarded as a key challenge (Healey *et al.*, 2003b; Healey, 2005; Healey & Jenkins, 2009). Although inquiry-based learning can occur at any point in the undergraduate programme (e.g. Willison & O'Regan, 2007; Healey & Jenkins, 2009), the final year is often regarded as a place where it can flourish, building on previous learning experience. The dissertation, or independent research project, is seen as an opportunity to engage with the research process from start to finish and is widely considered to be the peak of undergraduate student study in geographical, earth and environmental sciences (Harrison & Whalley, 2008). Apprenticeships, internships or work placements are becoming increasingly common especially towards the latter part of undergraduate programmes and give students an opportunity to 'learn by doing', develop a range of transferable skills and gain experience relevant to a potential career (Morgan & Turner, 2000; Neill & Mulholland, 2003; Poulter & Smith, 2006). Students are sometimes directly involved in staff research when they are employed as research assistants, typically during vacation time. However, it is rare to find student research assistantships built into undergraduate curricula in the UK; exceptions include the University of Leeds, which has an optional research placement in Geography (Palmer, 2009) and the University of Gloucestershire, where students in geography and environmental sciences can choose to become involved in projects linked to external organizations and companies (McEwen *et al.*, 2004).

Aims

This paper reports on the development and implementation of a 'research placement' module in Geography at Staffordshire University. Elements of the piloting of this module were supported by Staffordshire University as part of a university-wide research-informed learning and teaching initiative, seeking to better understand the ways in which effective research-informed teaching can be deployed to enhance student learning. This paper aims to (i) share the practice of designing and delivering the research placement within a modular framework, (ii) review the experiences of students and staff involved in the research placement and (iii) discuss the potential transferability of this research placement model to other subject areas.

Institutional and Departmental Context

Staffordshire University was established in 1992 and is one of the 'new' universities in the UK. Undergraduate teaching is organized within a modular framework, each single module accounting for 150 hours of study time. Undergraduates typically take four modules per semester, making a total of eight per level of study.

Geography at Staffordshire University is nationally recognized for its teaching excellence. The Quality Assurance Agency for Higher Education in the UK (QAA) assessment in 2004 commended our focus on learning and teaching, praising our high-quality learning materials, strong skills and fieldwork delivery, innovative and diverse assessment, and clear development of student knowledge and skills. Several National Student Satisfaction Surveys (managed by the Higher Education Funding Council for England (HEFCE) and reported in the Times Higher Education Supplement) have ranked Geography highly. We regard inquiry-based and research-led learning and teaching as intrinsic to what we do (see Scott, 2002; Lee, 2004), and these approaches have been embedded in our curricula for many years. Recently, the university has awarded support for projects seeking to better understand the ways in which effective research-informed teaching can be employed to enhance student learning. A number of projects of this type have been funded in Geography (for example, see Harris & Tweed, 2010).

The tradition and output of research in Geography at Staffordshire University has been modest compared with that at some other institutions. In the UK Research Assessment Exercise (RAE) in 2000, we achieved a 3b. Although some key research-active staff members have since departed, diminishing the department's research capacity, there remain pockets of sustained and internationally recognized research endeavour, along with some intensive consultancy-based work, and most members of staff are engaged in scholarly activity. We are beginning to translate this into a more vibrant and sustainable research culture by implementing a variety of strategies to enhance our research position.

Two fundamental needs, therefore, inspired the development of the research placement: first, to maintain and enhance our reputation for high-quality learning and teaching by further developing our inquiry-based learning curricula (see Healey, 2005; Healey & Jenkins, 2009) and, second, to utilize research-informed teaching to help strengthen elements of the research culture in Geography.

Designing the Research Placement Module

The research placement was devised to give students with the aptitude for independent study an opportunity to gain and apply skills associated with the execution of a research project and effectively involves students acting as research assistants to members of staff. It was designed to enhance student skills, knowledge and employability and to help staff with research. The project with which a particular student assists is expected to be centred on a problem or issue that can be examined through fieldwork and/or by library/archival investigation or by the analysis and/or presentation of data that have already been collected. The placement could, therefore, involve adapting to a project or initiative that has already begun. Projects can involve staff and external clients; in circumstances where external clients are driving the research, staff member(s) act as supervisors. The module was conceived not only to enable an individual to increase understanding of a particular research topic but also to offer students some wider insights into the process of academic research and consultancy. The research placement was, therefore, seen as particularly suitable for those wanting to go on to do some form of inquiry-based work or a research degree after graduation and was predicted to boost employability. Comments at Geography's recent subject review also contributed to the rationale and justification for the research placement.

External assessors recognized that we invest both time and resources in 'lower achievers'; to match this, they wished to see us providing more means by which we might 'stretch' our more able students, especially in their final year as undergraduates.

A range of issues had to be addressed to enable the research placement to function within the modular framework, to dovetail with existing and planned curricula and to assist in meeting the overall learning outcomes for the level of study. These issues were considered alongside the need to retain the necessary flexibility for the research placement to service the requirements of students and staff and act as viable vehicle for research assistantship.

The research placement was designed to be offered as a Level 3 (third or final year) undergraduate option module in Geography and to sit alongside more 'traditional' thematic modules from which students select their final-year options. The dissertation constitutes 30 Credit Accumulation and Transfer Scheme (CATS) points of the 120 CATS points required for Level 3 study. All students must do a dissertation and currently take the Advanced Field Class module worth 15 CATS points. This leaves them five option modules to choose to fill the 75 CATS points remaining; the Geography and Environment Research Placement is one such option. The research placement was developed to enable it to function as a single-semester module (12 weeks) or as an activity that could run over two semesters, providing that the work contained within it totalled 150 hours. To correspond to assignment weightings on other modules at Level 3 in Geography, assessment for the research placement is by means of a 30-minute oral presentation to discuss key findings of the project work and a reflective report, including a work diary. The presentation constitutes 30 per cent of the marks for the module; the reflective report is worth 70 per cent of the module marks and has a 1750-word limit. On completion of the module, the learning outcomes for the module are that students should have

1. gained an advanced understanding of the intellectual and methodological basis of a particular research question as practised by researchers within higher education;
2. assisted in undertaking a substantive piece of research work, guided by a suitable supervisor;
3. gained and applied skills associated with the execution of a research project and
4. gained and applied skills associated with the presentation of research to an appropriate audience.

Students choose their modules on a year-by-year timescale, and uptake onto the research placement is necessarily defined by projects available in any one year. The undergraduate award team was keen to try to simulate real-world competitive application processes in order to further develop employability skills. Projects were advertised and potential student applicants were invited to tender an appropriate résumé (*curriculum vitae)* and a covering letter. Checking applicants' skills and aptitudes against the criteria for each placement was a key part of the selection process, as it would be for employment. Interviews for the individual placements could be scheduled if decisions could not be made on the basis of the written application. Students were counselled about this process when making module choices, as were any students who were unsuccessful in their applications. Staffordshire University has an equality policy to which the selection process for the research placement is subject.

The module was designed so that support was given by the student research supervisor in the form of one-to-one sessions for specific project guidance and preparation for assessment. These sessions also provided the student with regular formative feedback and allowed supervisors to evaluate progress, which was recorded via a meeting log that both supervisor and student signed as representing an accurate record of what took place at the meeting. This is similar to the process used in Geography to record student progress on the undergraduate dissertation. As students were working under the guidance and supervision of a member of Geography staff, this staff member advised on issues of safety and insurance and assessed the need for the completion and submission of a risk assessment or ethical statement for any work in which the student was engaged as part of the research placement. These assessments were processed in accordance with established procedure for such activities on other modules in Geography.

Running the Research Placement: The Trial, the Pilot and the Option Module

The research placement module was initially offered to final-year undergraduate Geography students in 2007–2008 for two specific projects with which staff required assistance. The research placement module had, at this point, just been validated, and research opportunities arose which justified using the framework provided by the module to appoint two undergraduates to carry out research work. One project involved working with a local village heritage trust to gauge village inhabitants' views and attitudes on the preservation and sustainability of its community. The research placement student helped to identify the aims of the survey, formulated appropriate survey questions, advised on the sampling methodology, entered the survey data into SPSS and conducted data analysis (descriptive statistics and cross-tabulations) and interpretation. The second project entailed acting as a researcher to help gather information and materials to support a new undergraduate module, 'Sustainable Communities', designed to deliver key research inquiry skills, including interviewing and questionnaire design and survey, within the context of acting as an environmental consultancy for the university. This new module was itself a receiver of funding from Staffordshire University's research-informed learning and teaching initiative. The research placement student conducted a desk-based study of campus sustainability experiences and identified feasible themes for the first delivery of the module scheduled for 2008–2009. Both of these research placements ran in the second semester. In this paper, we refer to the above two research placements as the 'trial' phase.

The module was then more formally piloted in 2008–2009 as part of a funded research-informed teaching venture, using the opportunities presented by three different research projects. We ran a further placement in association with the 'Sustainable Communities' module; this time, work involved managing information (literature, websites and data) that students gathered as part of consultancy on the module and reporting on a student conference. The placement student helped compile a report of the module's findings for dissemination within the university. This project ran over both semesters. A placement entitled 'Environment Chamber' involved the testing of clay–plasterboard blocks under a variety of temperature and humidity conditions simulated in an environmental chamber. The blocks were designed and produced by a local company for use in construction, and a report on their performance was submitted to the external clients on completion of the placement. The placement student also produced an accessible illustrated guide to using the environment chamber. This project ran in the second semester. The 'Scottish Highland

Deglaciation' placement involved analysis of aerial photographs, literature and maps and a period of field research in the Scottish Highlands to extend the understanding of the dynamic glacial history of a valley through investigation of landforms and sediments. Part of this research involved testing the ideas of previous workers. The work will contribute towards a field guide and a research paper as part of a wider project being undertaken in the area. This project ran in the first semester.

These three projects were selected as they broadly represent consultancy, applied research and pure research, respectively, and would, therefore, test the research placement's ability to deliver to the student learning and research agendas across different research types. The timing of the research element varied between the different projects; on the one hand, it involved work in concentrated episodes (e.g. fieldwork in Scotland) and, on the other hand, work conducted over a more extended period such as one day per week for ten weeks (e.g. work in the Environment Chamber). The work for the Sustainable Communities project was episodic in nature and over a longer timescale.

In 2009–2010, following the trial and the pilot project, we ran the research placement as standard option module recruiting five students to placements, details of which can be found in Table 1.

Research Placement Evaluation

The experience of running the research placement was assessed in a number of ways. Two important tests of success were student performance on the module and their feedback on the experience. Furthermore, significant measures of the effectiveness of the research placement were derived from the staff who were involved in supervision of placement students and from the clients who were the recipients of research findings.

Student Performance

Clearly, the research placement is slightly different from other option modules; it attracts students who enjoy working independently and it is designed to have a small student

Table 1. Details of projects to which students were recruited in 2009–2010, the first year of the Research Placement as a standard option module

1. *Sustainable communities.* A further placement student was engaged to collate information and to report on the end of module student conference, in a similar way to the previous two Sustainable Communities placements.
2. *Ladyside wood.* The student recruited to this placement examined the biostratigraphy and sedimentology of Lower Carboniferous Milldale Limestones, prepared a detailed graphic sedimentary log, carried out specimen sampling and made provisional interpretations of data.
3. *Climate change adaptation.* A placement student assisted a local authority with the collection and analysis of climatic information, as a step towards developing climate change adaptation and mitigation strategies.
4. *Virtual Catalunya.* The student appointed to this placement helped staff by creating web pages and data visualizations, geo-tagging information to new layers in Google Earth and manipulating photographic images to assist in building a field course orientation package for future students.
5. *Green group.* This project engaged a student in consultancy-style scoping studies for the formulation of a team to promote a culture of sustainability within the Faculty of Sciences at Staffordshire University. The student determined relevant links and resources that the Faculty could utilize, identified and initiated baseline audits of 'green' activities and developed a web presence and terms of reference for the group.

uptake. All students performed very well, producing useful research outputs and achieving good marks (upper second and first class) for the module. Their presentations were effective, drawing on relevant illustrative material and succinctly summarizing the findings of their projects. The students also submitted excellent reports, putting their work into context and showing evidence of reflection and evaluation.

Six of 10 students achieved upper second class marks for the research placement, for the dissertation and for their degree overall. One student attained a first class mark for the research placement, dissertation and overall degree; another achieved a first class mark on all but the research placement, for which they got an upper second. One student secured a first class mark for the research placement, but an upper second for the dissertation and the degree; the last student achieved an upper second for the research placement and the degree, but a lower second for dissertation. We intend to monitor the research placement in forthcoming years to enable more critical evaluation of student performance.

Student Feedback

As part of the assessment for the module, all students produced a reflective report on their research placement work. This report had specific guidelines (see Table 2), recommending that the piece contained, among other sections, 'a statement of personal motivations and expectations' and 'a statement of personal evaluation and reflection'; both of these elements also enabled some evaluation of student experience. The two research placement students involved in the trial phase provided verbal feedback on their experiences through discussion with staff. The eight students who were part of the pilot project and the standard option module had the opportunity to evaluate the quality and effectiveness of the research placement through participating in a specifically designed feedback questionnaire. This questionnaire contained 18 questions, 8 of which were closed, consisting of a series of statements to which responses were obtained on a five-point scale, generally from 'strongly disagree' to 'strongly agree'. The other 10 questions were open response, allowing for comments in reaction to broader questions.

Given the nature of the module and the consequential small number of students, the closed question sections of the questionnaire do not merit statistical analysis; the responses are, therefore, summarized in Table 3. All students agreed that the module had been delivered in a different way to other Level 3 modules and that it had encouraged them to achieve a deeper level of knowledge generation and greater specialist knowledge when compared with other Level 3 modules. The majority of students thought that learning on the research placement module was harder than their learning elsewhere at Level 3 and that the bulk of that learning took place in independent study time. When asked to estimate the amount of time that they had spent on the module in terms of hours, six students selected 150 hours and two felt that it was closer to 300.

The structured responses are reinforced through feedback obtained via the open response questions on the questionnaire. All students further emphasized the fact that they had been encouraged to achieve a greater depth of specialist knowledge than that acquired on other modules and several said that they would have liked to have had more opportunities to pursue similar forms of study. Students felt that they had considerably improved their research, presentation, organization and time management skills. All students said that they appreciated the skills that their respective projects had enhanced, that they had enjoyed the opportunity to work as research assistants and that they believed

Table 2. Geography and Environment Research Placement reflective report guidelines

We might reasonably expect your report to contain the following:

Executive summary
You should provide 150-word summary of the key aspects of your research placement. This element is included in the word count of the report.

Statement of personal motivation and expectations
Describe your reasons for undertaking the research placement and outline what you expected from the module at the outset of the project.

Remit and client's requirements
In this section, you should outline the project aims and background (context) to the project that you undertook. If you were part of a larger project, briefly describe the project and make it clear what your role entailed.

Methods of inquiry
Give a short description of the data collection methods that you used, the nature of data you collected and the data analysis methods you undertook. Evaluate the appropriateness of methods used, consider what alternative methods were available and explain how the use of these alternative methods could have enhanced or improved the project.

Outcomes and achievements
In this section, you should briefly summarize the key findings of the project and outline the principal outcomes (sometimes these two elements may be the same). Critically evaluate the relationship between the applied study in which you have been involved and the theoretical aspects of the topic, with reference to relevant literature and/or other projects or case studies.

Further research
This section involves putting your research contribution into context. Briefly describe any further research that needs to be done as part of the project and identify any challenges that remain. How will this work be taken forward?

Personal evaluation and reflection
Evaluate your experience of the research placement. Did your experience differ from your expectations and if so, in what ways? In particular, you may wish to focus on the degree to which the research placement used your existing skills, highlight any new skills acquired, the potential for you to use knowledge or skills in the context of future employment, etc.

Work diary
This should literally take the form of a research diary containing a log of the work that you have done for the project. Specify dates, agreed goals, a summary of time spent doing each bit of research, brief details of work completed and any issues or problems. This element is not included in the word count of the report.

References
Include a full list of references cited in the report (again, the reference list is not included in the word count of the report).

that the module had enhanced their employability; they envisaged using the expertise and experience as part of further study and/or employment. Several students also remarked that doing the research placement had boosted their confidence, that they had enjoyed practicing professionalism and that they welcomed the sense of responsibility that the experience had given them.

Students also identified challenges involved in the research placements that they undertook. One student commented that getting started on the project had been difficult due to its wide-ranging nature and that, at times, it was difficult to know what was expected of them. The need for clear expected outcomes and regular progress reviews was

Table 3. Summary of closed question responses from the student feedback questionnaire administered in 2008–2009 and in 2009–2010

This module has been delivered in a different way to other level 3 modules:				
Strongly disagree	Disagree	Neutral	Agree	Strongly agree
0	0	0	4	4
This module has encouraged me to achieve a deeper level of knowledge generation compared with other level 3 modules:				
Strongly disagree	Disagree	Neutral	Agree	Strongly agree
0	0	0	5	3
In comparison with other level 3 modules, this module has enabled me to gain greater specialist knowledge:				
Strongly disagree	Disagree	Neutral	Agree	Strongly agree
0	0	0	3	5
In comparison with other level 3 modules, learning on this module was:				
Much easier	Easier	The same	Harder	Much harder
0	1	1	6	
I would have welcomed the opportunity to do more modules like this elsewhere on my course:				
Strongly disagree	Disagree	Neutral	Agree	Strongly agree
0	0	1	3	4
Estimate how much time you spent on this module in terms of credits:				
	Under 15	15	30	Over 30
	0	6	2	0
Where do you think that the bulk of your learning was achieved on this module:				
		Consultation with client/supervisor	Conducting the research project	Independent study
		1	2	5
I was broadly satisfied with the module:				
Strongly disagree	Disagree	Neutral	Agree	Strongly agree
0	0	0	3	5

Feedback elicited through additional open response questions is discussed more extensively in the text.

raised by one participant. Another student highlighted that work could be sporadic, underscoring the need for careful time management. In giving advice to future research placement students, one individual felt that they should be aware of the need to 'serve two agendas'; doing the research work and fulfilling the criteria for success in assessment. Students also advised their successors that roles can be varied and, therefore, adaptation and organization were key skills.

The questionnaire also included a section in which the students were invited to make additional comments about their experience of the module (see Table 4). Comments reinforce several of the points already made regarding experience of the 'real world' and research processes, taking ownership and responsibility and enhancing knowledge and understanding.

Staff Feedback

Consultation with staff involved in supervising the research placements highlighted several key issues. Staff remarked that the research placement was rewarding and gave them a chance to refresh and reinvigorate their research, providing a basis for further development. In an environment where funding for small experimental research projects is scarce, the research placement is enabling valuable pump-priming activity to take place. Some placements have enabled us to 'plug a gap' in resources (for example, where there was a small amount of capital available for equipment, but insufficient funding to support a research assistant as part of a project). Engagement with the research placement has given staff a springboard from which to develop grant applications and proposals for funded

Table 4. Student comments from the open response section of the Geography and Environment Research Placement questionnaire

'The module was an overall good experience, which I enjoyed. I would recommend the module to any students considering it for next year.'
'It didn't always feel like a module, but a 12-week project that you could become fully involved with and hopefully make contributions towards that project.'
'It provided me with a greater knowledge and new skills, it was 'different' to conventional modules and it gave more freedom and independence to make the work my own.'
'I felt that the module gave me a chance to thrive in the academic environment it placed me. . . . and has granted me an insight into research lifestyles.'
'It was a good experience to take part in research that was not student led, but your efforts as a student could contribute to the understanding of that area of research.'
'I have developed my research study skills a fair bit and I also learnt to question some of the material that is presented to you as a student. The module also developed my analytical skills.'
'I feel that students should have more opportunities to do modules like this as they are more like the 'real world' dealing with real companies and working people, rather than just classroom learning.'
'I really enjoyed the different learning experience gained through the research placement module. The structure of the placement was good and the assessment methods were perfect for the work carried out within the module.'
'The module has allowed me to grow by being allowed to manage a project largely on my own. The module has real word expectations and research tasks helping me to experience something closer to a workplace environment.'
'The freedom was beneficial, at the same time knowing that I was working towards my degree.'

post-graduate research students; for example, on the basis of collaboration on the Environment Chamber research placement, the staff concerned have submitted two grant bids and an application for a PhD research studentship. In some cases, the relationships forged with outside clients have led to further collaborative work between staff, external clients and future research placement students; for example, on the basis of a successful research placement in 2009–2010, we have recruited a further student to continue work on climate change adaptation in collaboration with the local council. Four staff felt that the opportunity to engage in research in conjunction with the students, i.e. the 'co-learning experience' (Le Heron *et al.*, 2006), was particularly enjoyable. The results of some research placements are being written up for publication; where this is the case, the students involved will either be co-authors or acknowledged for their assistance, depending on the nature of the work that they did for the project.

Staff remarked that the research placement needs careful monitoring and planning to ensure that students have clearly defined tasks and outputs, that these are associated with specific deadlines and that the work corresponds to the 150 learning hours expected from the module. It was also commented that under circumstances where the client and the supervisor are the same individual, it may be desirable to have a third party to act as intermediary between the student and the person for whom the work is being carried out; this would simulate consultancy and help to ensure fair practice. One member of staff, in praise of the placement student with whom they had been working, commented that the success of the project work rested heavily on selecting the 'right' student. Staff also remarked that the experience of research placement supervision was different from that of supervising a dissertation student in that there was a sense of co-learning and a greater need to specifically direct some elements of the work.

Client Feedback

Six research placements involved outside clients, who were also encouraged to give feedback. Work as part of the Environment Chamber research placement resulted in a summary report compiled by the student, which the company found useful and was very pleased with. This work has led to the consolidation of working relationships between staff in Geography, the Engineering subject area and the external company; further collaborative research is already underway. The local village heritage group was pleased with the results of the collaboration. The use of a research placement student allowed the project to be delivered at an affordable cost for the client. The project provided valuable experience for both the student and the staff involved in dealing with external clients and working to an external brief.

Research placement students have been extremely valuable for the development and ongoing delivery of the Sustainable Communities module. Placement students have developed materials to be used in the module and then compiled work from the module for university-wide dissemination. This work has been instrumental in developing momentum for the university to deliver on its sustainability agenda with student and subject area assistance. The local authorities involved in the Climate Change Adaptation placement have provided us with comprehensive feedback on their experience. Their comments centred on the reciprocal nature of the work, the usefulness of being able to 'tap into student knowledge' and the fact that the placements 'help the client to conduct research in an environment where resources are scarce'. Feedback also underscored the need for effective time management on behalf of the client as well as the student and 'the need for the client to be sensitive to the student's array of work for other modules'.

Further Perspectives

The Undergraduate Experience

Student feedback indicates that research placement participants felt that they had acquired greater subject knowledge, and at a deeper level as compared with other Level 3 modules, confirming the essence of research undertaken by other workers (e.g. Bauer & Bennett, 2003; Chang, 2005; Justice *et al.*, 2007; Healey & Jenkins, 2009). In particular, students remarked on the acquisition of a greater depth of specialist knowledge, the academic rewards of research, improved transferable skills, enhanced employability, a sense of responsibility and the opportunity to practice professionalism as positive aspects of the research placement. Challenges were also identified, students specifically noting the need for time management and adaptability. It is clear that the research placement is not suited to all students; it is self-selecting and it will attract a particular type of individual who relishes the opportunity to work independently and develop specific sets of skills.

It should be emphasized that the Geography and Environment Research Placement does not seek to exclusively recruit those who are regarded as academically 'high-achieving'. The research placements to date have each required different skills and aptitudes. Some placements have called for, for example, good oral communication skills or particular types of IT experience over written ability or examination performance. We have recruited students to the research placement module who have, at the point of appointment, averaged a lower second class mark; however, their aptitudes in particular areas secured them the placement. The research placement advertising process focuses on the skills that a successful applicant will possess, encouraging students to reflect on their own suitability when formulating their written application.

Students who have undertaken research placements have taken their skills in a variety of directions. For example, one of the students from 2007 to 2008 has gone on to do a fully funded PhD and two of the three students from 2008 to 2009 left intending to pursue masters level study. Of the five students who did placements in 2009–2010, one is embarking on a career in teaching, one has just started a Master's degree and another has been doing short-term environmental consultancy work following the 'Green Group' research placement.

Staff Reflection on Student Experience

Given the nature of the module and, consequently, the nature of the students that it attracts, it is perhaps unsurprising that students produced useful research outputs and attained good marks on both pieces of assessment. Continued monitoring of the research placement in forthcoming years will enable us to derive a larger data set from which we may be able to draw some more robust conclusions on student's performance.

The nature of the research work varies between placements; sometimes clients or supervisors require a written report as the main research product and sometimes the output from the placement can be in another format (for example, a series of web pages, collated materials or a set of schematic diagrams). Both the oral presentation and the reflective report draw on the work that was done on the placement, standardizing the means by which students are assessed. When work is completed for external clients, the client attends the oral presentation and completes an assessment marking proforma for the presentation. The client's views are, therefore, reflected in the oral presentation mark. The reflective report is assessed by the supervisor and double marked by another member of staff in line with departmental assessment policy. This system was perceived to work effectively by staff and students.

Research placement supervisors were encouraged by student feedback on the placement experience, especially given the drivers underpinning its development. It is imperative that there is a good fit between the skills and aptitudes of the student selected for the placement and the nature of the research activity. In managing the research placement, it is important to develop clearly defined tasks, to monitor student workload and to have clear oversight of the timing of activities. There is a particular need to be sensitive to workload issues, given the demands being placed on students by the other modules that they are taking alongside the research placement. Part of the rationale behind the student keeping a work diary is so that the supervisor can monitor the amount of time that certain elements of the research placement are taking. This is similar to filling out a time sheet in a working environment, and staff commented that regular review of the work diary helped both supervisor and student to manage the placement.

There is also a subtle difference between acting as a research assistant to a member of staff and doing research for an undergraduate dissertation; it is, therefore, imperative that realistic expectations on behalf of both staff and students are built into the process. On the basis of our experience to date, students sometimes work on research placements that are thematically related to their dissertation (for example, the student who worked on the Scottish Highland Deglaciation placement did a dissertation on glacial trimlines in the Lake District and a student whose dissertation was concerned with media attitudes to wind farms pursued a placement concerned with green issues within our Faculty); however, we have had students working on placements that are thematically different from their dissertations. It is necessary to exercise judgement concerning the degree of overlap between dissertation topics and research placement activities as this could result in unfair advantage in assessment.

Research and Teaching: Cementing the Links

Jenkins (2002) comments that students are frequently unaware of staff research or think that it is something staff do 'surreptitiously late at night and at weekends' (p. 3). The experience of running the research placement suggests that it could be one of a range of student learning activities that could help to demystify research. It could also assist in changing the perception of learning and teaching from being regarded as a transmission of 'what is known' into a more complex process involving shared exploration. The research process involves learning, so actively researching staff members are engaged in the same activity as their students (Brew & Boud, 1995; Brew, 2003); in addition, if students are involved in research activities, research can inform practice in facilitating learning (Brew, 1999). Research placement activity provides an innovative way in which students and staff can work together (Healey, 2005; Healey & Jenkins, 2009), enabling students to become stakeholders in research. Such activity engenders some student ownership of learning and gives them first-hand experience of research helping to redress student perceptions of themselves as 'primarily recipients of research, rather than actors in its production' (Healey, 2005, p. 194). The research placement has also helped to reinvigorate elements of the research culture in Geography by encouraging staff to develop and extend their own research efforts (see Evans & Witkosky, 2004) and to capitalize on collaborative links outside the department.

Research Placement Sustainability and Transferability

The research placement module is relatively straightforward to run as it essentially constitutes a framework that enables research assistantship to fit into a modular undergraduate degree structure. There is a need to safeguard teaching quality by developing clear criteria for the operation of the module and for the assessments; for example, there are tactics that can be employed to prevent accusations of bias or 'cosiness'. Good practice suggests that a staff member external to a particular placement project assists in marking the assessments for that project and that a similarly 'independent' staff member should also be available to liaise with the student in order to advise them, if necessary, on aspects of managing the project. The module is designed to sit alongside other Level 3 option modules in Geography, offering an additional opportunity for students and enabling them to get formal credit for being a research assistant. We see the module as building on our strong framework of inquiry-based learning, which starts at Level 1 and runs throughout our Geography undergraduate programme.

Geography at Staffordshire University typically has 30–40 students in the final year; we would normally envisage four to six students engaged in placements in any one year, contingent upon our current staff base and the availability of appropriate projects. So far, we have not had to 'search' for research placement projects, neither have we struggled to fill those positions for which we have advertised. Staff in Geography have welcomed the module and generated a suitable range of placements in any one year; staff now actively consider research placement possibilities when considering potential research and scholarship directions.

Geography is in an interesting position at the intersection between the physical sciences, the social sciences and the arts (Healey, 2005); this interdisciplinarity presents fertile ground for a range of projects and different ways of working together, which should help to sustain the research placement module and to provide different experiences for students. Staffordshire University is currently exploring the transferability of the research placement model to other subject areas; it has already been adopted in three Biological Sciences Masters Programmes.

The research placement module was developed to add to a curriculum that already had a strong research-informed learning and teaching approach and has a reputation for stimulating students to become effective self-learners. It is thought that the research placement has the potential to be adopted elsewhere within the higher education sector, with some adaptations to suit the requirements of different curricula and administrative structures. Identifying new ways in which staff and students can work together is particularly timely in a UK context, given the drivers to further develop undergraduate research and inquiry (e.g. Healey *et al.*, 2003b; Healey, 2005; Healey & Jenkins, 2009); the research placement could form part of an array of inquiry-based opportunities in the undergraduate learning portfolio.

Conclusions

The experience of running the research placement outlined in this paper demonstrates some of the mutual benefits of incorporating research into learning and teaching. There is evidence that students and staff alike feel that research placement activity enhances student learning and employability through knowledge and research skills acquisition, building experience and confidence for research after graduation. Involvement of students as research assistants encourages student ownership of learning and develops an active stakeholder culture. The research placement also helps to engender a more vibrant research environment.

Several elements are critical to the effective running of research placement activity. We feel that it is most appropriate for adoption in the final year of undergraduate study, as this permits the development of knowledge and skills underpinning the kinds of work with which a research placement student might become engaged. There is a need to introduce elements of research-informed or inquiry-based learning early in the curriculum so that students have experience of this type of independent work when they encounter the option to do a research placement. Delivered as discussed in this paper, the research placement needs careful monitoring and planning to ensure that students have clearly defined tasks and outputs and that the work corresponds to the learning hours expected from the module. The key to a successful research placement relies strongly on the selection of the most appropriate student for the task; as in postgraduate research and in the workplace, the relationship between the researcher and supervisor is important in the success of the work undertaken. The research placement model discussed in this paper has potential transferability to other subject areas; in essence, there is scope for its adoption wherever members of staff have elements of research with which they need assistance.

Acknowledgements

We would like to thank Pat Cossey, Tim Harris, Jon Fairburn, Dave Moreman, Derek Pratts, Janet Wright and Mike Anderson for their assistance in research supervision. Samuel Doyle, Shani Gallagher, Samantha Jones, Piers McDonald, Neil Wootton, Ian Brooke, Simon Hargreaves, Christine Kinsella, Neil Lee and Tom Ramsay gave us valuable feedback on their experiences as research placement students. Level 2 Geography undergraduates and a range of Staffordshire University employees gave their time to Sustainable Communities. Akristos Ltd. were the interested recipients of the report on their clay–plasterboard blocks, and Tim Deville and Doveridge Heritage Trust worked with us as part of the village community survey. Landowner Tommy Girvan gave permission to use some of the mountain tracks in Glen D'ho for the Scottish Highlands Deglaciation project. We thank Alan Carr from Staffordshire County Council for his comments on the experience of research placement supervision. The piloting of the research placement module was supported by funding from the HEFCE, which was deployed by the Academic Development Institute at Staffordshire University as part of an initiative to enhance links between learning, teaching and research. We would also like to thank three anonymous reviewers whose constructive comments helped us to improve this manuscript.

References

Bauer, K. & Bennett, J. (2003) Alumni perceptions used to assess undergraduate research experience, *The Journal of Higher Education*, 74, pp. 210–230.

Brew, A. (1999) Research and teaching: Changing relations in a changing context, *Studies in Higher Education*, 24, pp. 293–301.

Brew, A. (2003) Teaching and research: New relationships and their implications for inquiry-based teaching and learning in higher education, *Higher Education Research and Development*, 22, pp. 3–18.

Brew, A. (2006) *Research and Teaching: Beyond the Divide* (Hampshire: Palgrave Macmillan).

Brew, A. & Boud, D. (1995) Teaching and research: Establishing the vital link with learning, *Higher Education*, 29, pp. 261–273.

Chang, H. (2005) Turning an undergraduate class into a professional research community, *Teaching in Higher Education*, 10, pp. 387–394.

Evans, R. & Witkosky, D. (2004) Who gives a damn what they think anyway?: Involving students in mentored research, *National Social Science Journal*, 23, pp. 21–30.

Goodlad, S. (1998) Research opportunities for undergraduates, *Studies in Higher Education*, 23, pp. 349–356.

Griffiths, R. (2004) Knowledge production and the research-teaching nexus: The case of the built environment disciplines, *Studies in Higher Education*, 29, pp. 709–726.

Harris, T. & Tweed, F. (2010) A research-led, inquiry-based learning experiment: Classic landforms of deglaciation, Glen Etive, Scottish Highlands, *Journal of Geography in Higher Education*, 34, pp. 511–528.

Harrison, M. E. & Whalley, W. B. (2008) Undertaking a dissertation from start to finish: The process and product, *Journal of Geography in Higher Education*, 32, pp. 401–418.

Hattie, J. & Marsh, H. W. (1996) The relationship between research and teaching: A meta-analysis, *Review of Educational Research*, 66, pp. 507–542.

Healey, M., Jordan, F., Pell, B. & Short, C. (2003a) The student experience of research and consultancy, in: *SEDA-Society for Research into Higher Education Joint Conference on The Scholarship of Academic and Staff Development: Research, Evaluation and Changing Practice*, Bristol, April 2003, pp. 9–11.

Healey, M., Blumhof, J. & Thomas, N. (2003b) The research-teaching nexus in Geography, Earth and Environmental Sciences (GEES), *Planet*, Special Edition 5, pp. 5–13.

Healey, M. (2005) Linking research and teaching to benefit student learning, *Journal of Geography in Higher Education*, 29, pp. 183–201.

Healey, M. & Jenkins, A. (2009) *Developing Undergraduate Research and Inquiry* (York: The Higher Education Academy).

Haslett, S. K. (2009) Unpicking the links between research and teaching in Higher Education, *Newport CELT Journal*, 2, pp. 1–4.

Hodge, D., Haynes, C., LePore, P., Pasquesi, K. & Hirsh, M. (2008) From inquiry to discovery: Developing the student as scholar in a networked world, Learning Through Enquiry Alliance (LTEA) conference, *Inquiry in a Networked World*, Sheffield, June, pp. 1–23.

Jenkins, A. (1995) The impact of research assessment exercises on teaching in selected geography departments in England and Wales, *Geography*, 80, pp. 367–374.

Jenkins, A. (2000) The relationship between teaching and research: Where does geography stand and deliver? *Journal of Geography in Higher Education*, 24, pp. 325–351.

Jenkins, A. (2002) Developing the research/teaching nexus: Suggestions for course teams, *Teaching News*, February, pp. 3–5.

Jenkins, A. (2003) Designing a curriculum that values a research-based approach to student learning in Geography, Earth and Environmental Sciences (GEES), *Planet Special Edition*, 5, pp. 2–5.

Jenkins, A., Breen, R., Lindsay, R. & Brew, A. (2003) *Re-shaping Higher Education: Linking Teaching and Research* (London: Routledge/SEDA).

Justice, C., Rice, J., Warry, W., Inglis, S., Miller, S. & Sammon, S. (2007) Inquiry in higher education: Reflections and directions on course design and teaching methods, *Innovative Higher Education*, 31, pp. 201–214.

Lee, R. (2004) Research and teaching: Making or breaking the links? *Planet*, 12, pp. 9–10.

Le Heron, R., Baker, R. & McEwen, L. (2006) Co-learning: Re-linking research and teaching in geography, *Journal of Geography in Higher Education*, 30, pp. 77–87.

McEwen, L., McAllister, B., Booth, D. & Hunt, J. (2004) Promoting sustainable development of built and human-altered environments, *CEBE News Update*, 9, pp. 19–21.

Morgan, A. & Turner, D. (2000) Adding value to the work placement: Working towards a professional qualification in an undergraduate degree programme, *Education and Training*, 42, pp. 453–460.

Neill, N. T. & Mulholland, G. E. (2003) Student placement – structure, skills and e-support, *Education and Training*, 45, pp. 89–99.

Palmer, S. (2009) GEOG3820 Research Placement. Available at: http://webprod1.leeds.ac.uk/banner/dynmodules.asp?Y=200910&F=P&M=GEOG-3820 (accessed October 2009).

Perkins, J. (2002) Linking research and teaching and re-designing the undergraduate modular programme, *Teaching News*, February, pp. 1–2.

Poulter, E. & Smith, B. (2006) A work placement review from an undergraduate's perspective, *Planet*, 16, pp. 43–45.

Scott, P. (2002) High wire, *Guardian Education*, 8 January, p. 13.

Spronken-Smith, R. (2005) Implementing a problem-based learning approach for teaching research methods in Geography, *Journal of Geography in Higher Education*, 29, pp. 203–221.

Willison, J. & O'Regan, K. (2007) Commonly known, commonly not known, totally unknown: A framework for students becoming researchers, *Higher Education Research and Development*, 26, pp. 393–409.

Index

Note: Page numbers in **bold** type refer to figures
Page numbers in *italic* type refer to tables
Page numbers followed by 'n' refer to notes

For Product Safety Concerns and Information please contact our EU
representative GPSR@taylorandfrancis.com
Taylor & Francis Verlag GmbH, Kaufingerstraße 24, 80331 München, Germany

www.ingramcontent.com/pod-product-compliance
Ingram Content Group UK Ltd.
Pitfield, Milton Keynes, MK11 3LW, UK
UKHW050053200926
475702UK00019B/2821